Nanofabrication

PRINCIPLES TO LABORATORY PRACTICE

Optical Sciences and Applications of Light

Series Editor
James C. Wyant
University of Arizona

Nanofabrication: Principles to Laboratory Practice, *Andrew Sarangan*

Blackbody Radiation: A History of Thermal Radiation Computational Aids and Numerical Methods, *Sean M. Stewart and R. Barry Johnson*

High-Speed 3D Imaging with Digital Fringe Projection Techniques, *Song Zhang*

Introduction to Optical Metrology, *Rajpal S. Sirohi*

Charged Particle Optics Theory: An Introduction, *Timothy R. Groves*

Nonlinear Optics: Principles and Applications, *Karsten Rottwitt and Peter Tidemand-Lichtenberg*

Photonics Modelling and Design, *Slawomir Sujecki*

Numerical Methods in Photonics, *Andrei V. Lavrinenko, Jesper Lægsgaard, Niels Gregersen, Frank Schmidt, and Thomas Søndergaard*

Please visit our website ***www.crcpress.com*** *for a full list of titles*

Nanofabrication

PRINCIPLES TO LABORATORY PRACTICE

Andrew Sarangan
University of Dayton, OH, USA

CRC Press
Taylor & Francis Group
Boca Raton London New York

CRC Press is an imprint of the
Taylor & Francis Group, an **informa** business

CRC Press
Taylor & Francis Group
6000 Broken Sound Parkway NW, Suite 300
Boca Raton, FL 33487-2742

First issued in paperback 2019

ISBN-13: 978-0-4987-2557-6 (hbk)
ISBN-13: 978-0-367-87319-6 (pbk)

Library of Congress Cataloging-in-Publication Data

Names: Sarangan, Andrew, author.
Title: Nanofabrication : principles to laboratory practice / Andrew Sarangan.
Description: Boca Raton : CRC Press, Taylor & Francis Group, 2017. | Series: Optical sciences and applications of light | Includes bibliographical references and index.
Identifiers: LCCN 2016008970 | ISBN 9781498725576
Subjects: LCSH: Nanolithography. | Nanostructured materials--Design and construction. | Semiconductors--Etching. | Nanoelectronics--Experiments.
Classification: LCC TK7874.843 .S27 2017 | DDC 621.3815/31--dc23
LC record available at https://lccn.loc.gov/2016008970

Visit the Taylor & Francis Web site at
http://www.taylorandfrancis.com

and the CRC Press Web site at
http://www.crcpress.com

Contents

Series Preface

Optics and photonics are enabling technologies in many fields of science and engineering. The purpose of the Optical Sciences and Applications of Light series is to present the state of the art of the basic science of optics, applied optics, and optical engineering, as well as the applications of optics and photonics in a variety of fields, including health care and life sciences, lighting, energy, manufacturing, information technology, telecommunications, sensors, metrology, defense, and education. This new and exciting material will be presented at a level that makes it useful to the practicing scientist and engineer working in a variety of fields.

The books in this series cover topics that are a part of the rapid expansion of optics and photonics in various fields all over the world. The technologies discussed impact numerous real-world applications, including new displays in smartphones, computers, and televisions; new imaging systems in cameras; biomedical imaging for disease diagnosis and treatment; and adaptive optics for space systems for defense and scientific exploration. Other applications include optical technology for providing clean and renewable energy, optical sensors for more accurate weather prediction, solutions for more cost-effective manufacturing, and ultra-high-capacity optical fiber communication technologies that will enable the future growth of the Internet. The complete list of areas where optics and photonics are involved in is very long and keeps expanding.

Preface

This book grew out of the nanofabrication laboratory that I created at the University of Dayton, in addition to a number of academic courses, theses, and dissertations on the subject. When I started this journey about 20 years ago, I had very little cleanroom experience, having spent most of my academic career on modeling and simulation. During the intervening years, I had to follow a large number of textbooks, journal articles, and websites and frankly it was a struggle to put all the pieces of information together. The biggest problem I've had is that while journal articles dive into one topic in great depth, the reviews and overviews often go the other extreme and gloss over major critical factors. It has been difficult to find sources of information in the middle for those who want to understand the big picture, but also want to dive into the details and work with their hands on a project by themselves. Honest perspectives have also been hard to find, since most articles tend to emphasize the beneficial aspects of the author's own specialties.

At most educational institutions, cleanroom work is done almost exclusively by graduate students. Recipes and procedures are developed and handed down from one generation of students to the next. Faculty involvement in this area is rare, and their proficiency in the cleanroom is even rarer. As a result, much of the practical must-know information seldom make their way into the research literature, and can only be found by talking to people directly.

My philosophy has been the opposite. I endeavor to spend as much time in the cleanroom as possible, even more than my graduate students. This experience has been rewarding, because this gives me the opportunity to assemble vastly different levels of information, from details such as how to dismantle and repair a vacuum pump to understanding major technology trends in the business. As I see it, fun is all in the action, and I'd rather be in the lab and be part of the exciting work than find out about it from a desk.

While there are a number of great books already written on nanofabrication, this book was developed in the hope that it fills that niche gap between academic research and laboratory practice. The book is also written from the perspective of electrical engineering and photonics, due to my own background in these areas. Many of the results shown in this book, including SEM images and numerical calculations are from work performed in my own lab, so hopefully they will come across with some personal insights and experience. My hope is that this book will be a companion for those just starting to work in a cleanroom. This is the first edition of this book. Undoubtedly it will not be perfect but I hope it still serves as a useful reference for those starting their journey into this exciting field.

Andrew Sarangan, PhD, PE
University of Dayton
Dayton, Ohio

Author

Andrew Sarangan is a professor and associate director of the Electro-Optics Graduate Program at the University of Dayton, Dayton, Ohio. He received his BASc, MASc and PhD in electrical engineering from the University of Waterloo, Waterloo, Ontario, Canada in 1991, 1994 and 1997, respectively.

Dr. Sarangan's current research interests include semiconductor optoelectronics, nanofabrication, and computational electromagnetics. Starting from a small research laboratory at the University of Dayton he gradually developed the laboratory's capabilities, with expertise ranging from optical thin films and plasma processes to photolithography and laser interference lithography. In addition to maintaining a funded research program, he personally maintains everything in the laboratory, including equipment repairs and building new tools, and enjoys spending most of his time working with his hands and honing his skills.

Dr. Sarangan primarily teaches graduate students. He has developed graduate courses in nanofabrication, nanophotonics, optical thin films, and integrated optics. He is a senior member of the IEEE and the SPIE and a registered professional engineer in the state of Ohio. He is also an accomplished pilot and a certificated flight instructor.

1 Introduction to Micro- and Nanofabrication

1.1 INTRODUCTION TO MICRO- AND NANOFABRICATION

There is currently neither a widely accepted definition of nanofabrication nor a definition of where the boundaries between nano- and microfabrication lie. Although one would be tempted to define nanofabrication as the process of making things that are in the single-digit nanometers, in practice a much wider range of dimensions are accepted as nano. The term "nano" started to emerge in the popular and scientific media as the size of electronic devices started to drop below 1 μm. Component sizes that used to be several micrometers gave way to single-digit micrometers, then hundreds of nanometers, and finally down to a few tens of nanometers where they are today. What used to be called "microfabrication" was rebranded as nanofabrication, although many of the basic principles and techniques have remained essentially the same. Although the main driver of this technology has been the manufacture of integrated circuits, there have been tremendous fallout benefits to other areas, including photonics and micro-electro-mechanical systems (MEMS).

This book is intended for scientists and engineers who are new to the field of micro- and nanofabrication, as well as beginning graduate students or senior undergraduate students who will be designing and building devices in a cleanroom as part of their academic research. It is based on over 20 years of experience in research and teaching, as well as building, maintaining, and operating a fully functional nanofabrication cleanroom.

The science and technology of device fabrication can be overwhelming to a new student because it involves a large number of techniques and terminology that may not be covered in standard academic curricula. Device fabrication draws from material science, electromagnetics, electronics, and chemistry in a multidisciplinary manner, and one needs to become comfortable connecting all these disciplines together to perform a useful function without having to engage in an extended specialization in any one of these areas.

Nanofabrication can be loosely divided into four major areas: thin films, lithography, etching, and metrology. These are vast subject areas. The goal of this book is to cover the essential principles and laboratory practices, to outline the pros and cons of commonly used techniques, and to offer some practical tips for someone new to these areas. I hope it provides a broad foundation on which further exploration or training can be built. Although this book is primarily intended for those who will be working directly inside a cleanroom, even those who design and model electronic or photonic devices may find this book useful to make informed choices in their designs.

1.1.1 Importance of Understanding the Techniques

The importance of understanding the tools and techniques and their limitations can be illustrated through an ordinary example of designing a part for machining. The designer might draw a 90° corner in a milled slot because it is easier to draw on paper, but he or she might be unaware that a curved slot is actually a lot easier to make due to the cylindrical shape of the cutting tools. The machinist, however, might take the design literally and end up spending unnecessary effort and expense to achieve that 90° corner. Similar things are true in device fabrication as well. Not everything that can be drawn or conceptualized can be easily made in the laboratory, while something that may appear complicated at first may in fact be quite easy to make. Standard techniques in device fabrication include lithography, etching, doping, deposition, etc. A good understanding of what is easy and what is difficult will allow the scientist and engineer to estimate the time and costs required to complete a project more accurately.

1.1.2 Creative Problem Solving

Engineering is about creative problem solving. This requires experience and insights, which cannot be acquired by reading and listening alone. Knowledge is acquired by "reading and listening" which when followed by "seeing and doing" results in insights. Insight is when a light bulb goes off in your head. We cannot be creative without insights. Creativity cannot be taught, but it can be inspired by insights, which in turn are inspired by "seeing and doing." Obviously, a book cannot provide this experience, but hopefully you will gain some appreciation of the importance of "seeing and doing" from this book. Unfortunately, in many engineering curricula today the importance of "doing" has become significantly diminished, especially at the graduate level.

Creative problem solving also requires one to use all five senses. Far too often we consider creativity as a purely cerebral activity. This may be true with philosophy, history, or law, but engineering is the art of creating machines that can do useful things, and it is difficult to achieve mastery of this art while sitting at a desk far from the laboratory. Being able to organize and delegate tasks are great qualities, but they alone do not define an engineer. The ability to sense sounds, texture, smells, and visual appearance should not be left to the shop floor technicians alone. The engineer and scientist should also be proficient in interpreting the signals from all our human sensors and applying them to solve problem that need solving.

1.1.3 What Has Been Done by Others versus What You Can Do

There is a big difference between what someone else has done versus what we can do here and now. This fact is often under appreciated by scientists. A good example is the complementary metal oxide semiconductor (CMOS) technology, which is in everyone's cell phones, laptops, and electronics. A student who learns about the latest CMOS technology could arrive at the false impression that it is a mature and accessible technology because it is so highly mass-produced and cheap. Nothing could be further from the truth. The 14 nm CMOS technology is nothing short of a grand achievement through collective development efforts and billions of dollars of investment. Although it is a mature technology that is being mass-produced, it is extremely difficult for a laboratory or even a medium-sized company to reach that level of technology. When reading the literature, we have to be cautious in making assumptions about what we can realistically do.

1.1.4 Experiment versus Project

The term "experiment" is often used interchangeably with "project" especially in the science disciplines. In engineering, the distinction is clearer. An experiment is a task undertaken for studying a certain phenomenon or to answer a specific question. For example, one performs an experiment to study how the carrier lifetimes vary in a semiconductor as a function of doping density. A project usually requires producing a working unit, such as a machine or a computer program. In other words, we perform "scientific experiments," or "engineering projects," and not "scientific projects" or "engineering experiments." However, a project may require performing a number of experiments, and an experiment may also require projects (such as designing and building a test bed in order to carry out the experiment). Even though the distinction may seem moot, it is useful to have a clear idea of what we are hoping to produce. Projects are what ultimately benefit society, and these are what engineers are expected to produce.

1.1.5 Nano and the Media

The media is full of hyped-up announcements of inventions and discoveries, most of which we will never hear about after the initial news release. The vast majority of these inventions and discoveries eventually fade away because they are just flashy experiments, not viable projects. That does not necessarily mean we should entirely disregard these announcements, just that we need to use our experience and judgment instead of taking them at face value. Many investors and start-up companies have fallen prey to hyped-up claims, investing money in technology that had little hope from the beginning. On the other hand, some technologies that arrive without any announcement end up becoming major disruptive innovations. The ability to distinguish between hype and reality is a valuable skill that can only be acquired through experience.

1.1.6 Carbon versus Silicon and Self-Assembly versus Micromachining

There are two siblings of nanotechnology that are often not clearly articulated in the literature. One is carbon nanotechnology. Here, carbon is processed in different ways to make new materials with exotic properties, such as carbon nanofibers and nanotubes, which are then used to make advanced composite structures. This is often a "self-assembly" process because the material is synthesized by flowing carbon-containing gases and liquids to allow the carbon atoms to "settle on their own" in the desired arrangement. This is known as the bottom-up approach. The other approach is to start with a block of carbon and machine it down to the desired shape. This is also known as the "top-down" approach. Carbon nanostructures are almost never made using top-down methods because the applications do not require the structures to have exact sizes, locations, or orientations.

Silicon nanotechnology is the other sibling. Here, the top-down approach is the norm. We start with a block of silicon and then machine it down to the desired shape. When making MEMS, electronic or optical devices, we need to have precise control over the location, orientation, and size of the parts. As of today, this can only be achieved with a top-down approach.

This, however, does not mean top-down methods cannot be used with carbon, or bottom-up methods with silicon. Bottom-up methods are cheaper, faster, and scalable to higher volumes compared to top-down approaches. Research is ongoing to develop these methods where silicon structures can be grown by self-assembly with precise control over their shape, location, and size. One example of this is called directed self-assembly (DSA), which is a lithography process for making ultrasmall structures using a block copolymer. Graphene, which is an allotrope of carbon, has opened up a new area of electronics where carbon is used instead of silicon using top-down methods.

The terms "top-down" and "bottom-up" can sometimes lead to inaccurate descriptions. A better descriptor is self-assembly versus micromachining. The focus of this book will primarily be on top-down micromachining approaches.

Some technologies are a hybrid of both approaches. An example is epitaxial crystal films. These are better described as self-assembly processes. However, these films are eventually patterned and etched in a top-down fashion to make functional devices.

1.1.7 Nanotechnology Is Old

Manmade nanotechnology has been around for centuries. The different colors in stained glass windows were created by impregnating the glass with different-sized gold nanoparticles. So why is nanotechnology considered a new subject just now?

The answer is in the tools. Although our ancestors made gold nanoparticles, they were not able to observe them under a microscope, manipulate them, or have much control over them. Developments in the last two decades have produced advanced microscopy where particles down to a single atom can be imaged. We have developed tools that can define structures down to a few nanometers in size. Despite how it is portrayed in the media, nanotechnology is not a new creation that began at a specific time. What used to be microtechnology slowly transitioned to become nanotechnology, just like what used to be macromachining evolved into microfabrication in the 1950s and 1960s. In fact, many laboratories simply changed the name of their microfabrication laboratories to nanofabrication laboratories with essentially no other fundamental change. Although this change in name was partly driven by politics and economics, microfabrication had advanced and had slowly morphed into the nanoscale, so it was time to change its name as well.

1.1.8 Moore's Prediction and Driving Forces

Gordon Moore was a cofounder of Intel who predicted that the number of components in an electronic integrated circuit chip will double every two years. Another one of Moore's prediction (Rock's law) is that the cost of setting up a semiconductor chip fabrication plant will double every four years. Nevertheless, the cost of transistors in a chip has been steadily declining over the years, which is why a single cell phone today contains more than the entire computing power that was used by the Appollo mission. Electronic chip manufacturing arguably has had the most impact on human civilization in recent history. It has been the driving force behind much of our development in telecommunication, information storage, Internet, and entertainment. Just like in the U.S. space program, many of the tools and techniques that were developed to serve the electronics industry have had beneficial effects on a number of other industries. For example, the way we make MEMS and integrated photonics heavily rely on borrowed techniques originally developed to serve the electronic chip making industry. Without the massive volume in consumer electronics, none of these techniques that we take for granted in the laboratory would be as affordable as they are today.

1.1.9 Why Components Have to Be Small

Current CMOS technology has gate widths smaller than 20 nm. For those unfamiliar with CMOS devices, a gate can be thought of literally as a mechanical gate that stops or opens the flow of electrons between the source and destination (drain). The smaller the gate, the more quickly and with less energy it can be opened/closed, which is the reason we want to go toward smaller and smaller gate sizes.

Besides speed and power considerations, there are other factors that have been driving the component sizes smaller. One of these is the demand for greater and greater functionality from the same chip. A mobile phone is not just a telephone. It is a computer, camera, media player, and storage device all in one. The old microprocessors could only perform simple arithmetic operations, but the chips inside today's portable electronics can do a lot more. The only way to achieve more functionality is by increasing the number of transistors in the chip. One would expect that this means the chips would have to grow in size. But that is not a desirable option. In order to keep manufacturing yield high, the chip size has to remain the same, or get even smaller. This doubles the pressure on device sizes—shrink the device size to fit more devices on a single chip, and shrink the devices even further to make the overall chip size smaller. The yield is defined as the number of good chips from a wafer. For example, if there are 1500 chips in a wafer, and if there are 50 defects per wafer, that means we get approximately 1450 working chips, or 96.7% yield. On the other hand, if the chip size is larger, and we can only fit 500 chips on the same wafer, the same 50 defects will produce 450 good chips, or a 90% yield. With all other things being equal, to get higher yield, we need smaller chip sizes and a lower defect density. The wafer sizes have also been getting larger due to their increased production efficiencies. All of these driving forces have led us to where we are today—300 mm wafers and 14 nm gates. It is important to note that these are leading-edge technologies done at large manufacturing facilities. They are generally beyond the capability of nearly all small laboratories.

Another surprising fact is that all of these nanoscale devices are made using 193 nm UV radiation, far below the diffraction limit of light. For example, the 14 nm devices are smaller than a tenth of the 193 nm wavelength. How this is done, not one time in the laboratory, but continuously in a production environment is something we will discuss in a later chapter.

1.1.10 Nanofabrication Is a Multidisciplinary Science

Nanofabrication is difficult to summarize and deliver in a single course because it actually contains several large disciplines. Lithography, thin films, pattern transfer and metrology can be loosely defined as the major categories of nanofabrication. "Lithography" is the science of printing fine traces on the wafer. "Thin films" is the process of depositing metal or dielectric films. "Pattern transfer" is the process of transferring the printed image from lithography to the substrate or thin film.

Inspection and characterization is another major area known as "metrology." The components that we fabricate are often too small to accurately examine with an optical microscope. Different tools are used, such as electron microscopy and atomic force microscopy. Other properties such as film thickness, stress, and refractive index may also have to be measured and monitored. All of these require specific tools and techniques.

1.1.11 Units of Measure

SI unit has become the accepted standard at many scientific journals and institutions. However, other units of measures are still widely used in laboratories and industry due to their legacy, convenience, and existing tools. For example, "mil" (one thousandth of an inch) is used in many areas of machining to specify material thicknesses and is used even in lithography processes in device fabrication. Thin films are widely measured in Angstroms rather than nanometers. Compressed gas pressures are most commonly measured in pounds per square inch (psi). Gas flows are measured in standard cubic centimeters per second (sccm) or liters per minute (lpm). Vacuum is measured in Torrs in the United States and Pascals or millibars in other parts of the world. Resistivities are measured in Ω·cm. Semiconductor doping is measured in cm^{-3}. Most of these are not SI units. Time is measured on a base 60 system, even by the most ardent proponents of the decimal system. It is the author's personal view that we should appreciate and respect the standards that have evolved in different areas. Which units are most appropriate should really be left for those doing the actual work in their respective fields. Every engineer and scientist should be comfortable converting between different units. The purpose is to clearly communicate a quantifiable unit of measure, not to make the conversion arithmetic simpler. Most importantly, the minor inconvenience of converting between units should not be the sole reason to force conformity to a single standard. In that spirit, we will use the most commonly used units of measure for each of the areas covered in this book.

1.2 CLEANROOMS FOR DEVICE FABRICATION: BASIC CONCEPTS

The properties of a cleanroom differ depending on the industry, such as pharmaceutical, semiconductor device fabrication, biotechnology, space technology, or a myriad of other industries [1, 2]. In device fabrication, we are primarily interested in reducing airborne particles. Of secondary concerns are chemical and radiation contaminations.

In a typical device fabrication cleanroom, airborne particles are reduced by filtering the air that enters the cleanroom. This is done with high-efficiency particulate air (HEPA) filters with an integral fan, known as fan filter units (FFUs). In a typical cleanroom, these units are mounted on the ceiling and the clean air blows downward. Most cleanrooms are sealed environments, although it is possible to also operate it as an open environment. In a sealed environment, the return air can be recirculated back into the FFU that will significantly reduce the workload on the filter units. In an open environment, the filters have to work continuously to clean the outside air and that can quickly saturate the filter units. One drawback of a closed environment is that any chemical vapors will continuously recirculate until they slowly leak out of the laboratory. A secondary chemical filtration can be used, but it is rarely done. Typically, chemical contamination is controlled by limiting the use of chemical vapors to an exhausted fume hood.

The FFUs create a higher pressure inside the cleanroom compared to the ducts or plenum above the cleanroom. The return air from the cleanroom is fed to an air handling unit (AHU), which regulates the temperature and humidity and pumps it back into the ceiling plenum. The AHU also draws air from the outside to make up for air leaks and maintains a higher air pressure in the plenum than the outside ambient air pressure. This positive pressurization ensures that air only flows out of the laboratory and never into the laboratory, which prevents outside contaminants from diffusing into the cleanroom. Even in a sealed environment, there are always inevitable openings such as pass-throughs, doors, gas pipings, etc. The AHU also provides make-up air from the outside to compensate for air leaking out of the cleanroom. A prefilter is used before the outside air is sent to the AHU. It is also common practice to design an intentional air leak in the cleanroom, either via exhaust ducting or with openings. This will reduce the latency of chemical vapors, although it comes at the expense of increased loading on the HEPA filters.

The airflow in a cleanroom can be a laminar flow configuration or a semilaminar flow configuration. A laminar flow configuration, shown in Figure 1.1, requires a raised perforated floor to allow the airflow to enter a subfloor plenum before being returned to the air handler. The parallel flow will help to maintain a uniform particle count throughout the laboratory environment. In a semilaminar flow configuration, shown in Figure 1.2, the return air exists through ventilation ducts at specific locations in the walls and is fed through the wall cavity back to the air handler. This configuration is less expensive than a laminar configuration because it does not require a raised perforated floor, and hence requires a smaller vertical clearance. However, since the airflow is not uniform, there will be turbulence, which can create areas of low flow and even stagnation. In these areas, the particle count will be significantly higher.

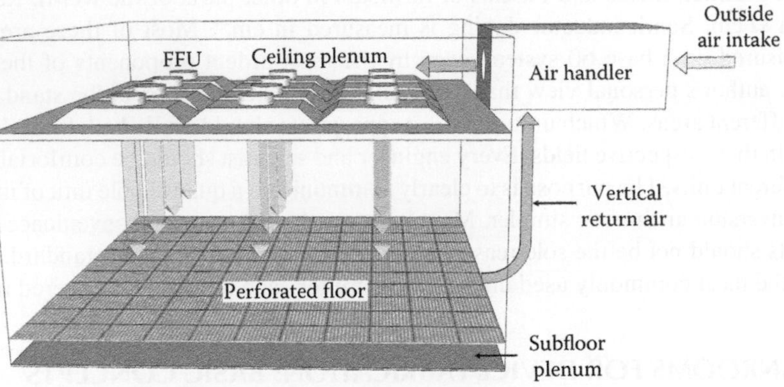

FIGURE 1.1 Laminar flow cleanroom configuration with a raised floor.

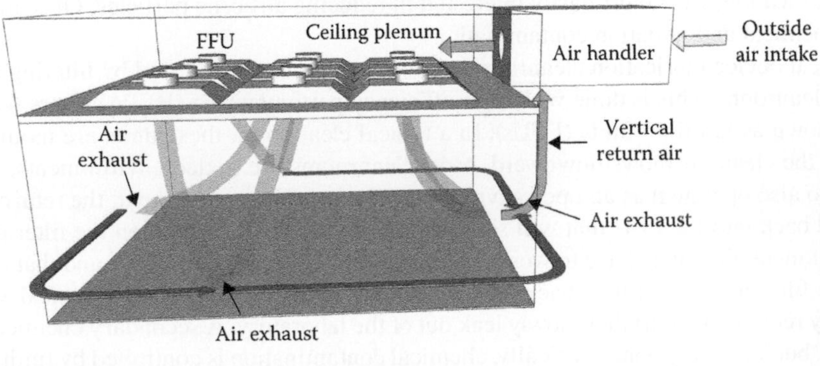

FIGURE 1.2 Semilaminar flow cleanroom configuration.

1.2.1 Cleanroom Classification and Airflow Rates

The grade of a cleanroom is defined in terms of the size distribution of particles within a fixed volume of air at atmospheric pressure. The classification standard was originally based on the U.S. Federal Standard 209E [3], but has since been superseded by an equivalent ISO standard [4]. Nevertheless, the terminology used in the original standard (class-100, class-1000, etc.) still continues to be used.

A class-100 cleanroom contains fewer than 100 airborne particles larger than 0.5 μm in size, or fewer than 3500 airborne particles larger than 0.1 μm in size within a cubic foot of air. A class-1000 cleanroom contains fewer than 1,000 airborne particles larger than 0.5 μm in size, or fewer than 35,000 airborne particles larger than 0.1 μm in size within a cubic foot of air. Class-10 is one of the lowest levels one is likely to encounter in research laboratories. In integrated circuit manufacturing environments, it is not uncommon to have class-1 or even class-0.1. There is also an equivalent metric standard and an ISO standard for cleanroom classification. Table 1.1 lists the classification criteria using both standards. Most cleanrooms have multiple areas with different classifications, depending on the applications. For comparison, a typical office will be about class-1 million.

This relationship is expressed as:

$$C_n = \left(\frac{0.1}{D}\right)^{2.08} 10^N \tag{1.1}$$

where

C_n is the number of particles in a cubic meter
D is the particle size in micrometers
N is the ISO classification number

We can verify that with $N = 5$ we can get 3516.7 particles/m³, which is equivalent to 100 particles/ft³ (class 100).

Humans are the largest source of particles in a cleanroom. Regardless of personal hygiene, we always emit skin cells, hair, oil, and fibers from our clothing, which is the main reason for donning cleanroom suits. These suits are for protecting the laboratory from people; not for protecting the people from laboratory hazards. Even with a full cleanroom suit, humans are still the largest source of contamination. When we move, we create 10 times more particles than when we are stationary. In addition, some materials like books, paper, and cardboard can emit a copious amount of particles. Most cleanrooms strictly prohibit the use of these items. Cleanroom notebooks are made of a special paper that is coated on both sides to reduce particle emission. For this to work, the sheets should not be cut or torn.

TABLE 1.1
Cleanroom Classifications Based on Particle Counts

	Number of Particles/ft³		
Class	**>0.1 μm**	**>0.5 μm**	**ISO Standard**
10	350	10	ISO 4
100	3,500	100	ISO 5
1000	35,000	1000	ISO 6

Source: Adapted from Whyte, W., *Cleanroom Technology: Fundamentals of Design, Testing and Operation*, John Wiley & Sons, New York, 2010.

The steady-state particle count in the laboratory is a dynamic balance between the emission rate of particles from contamination sources and the elimination rate of the particles by air filtration. The elimination rate is a function of the volumetric airflow rate and the efficiency of the filtration units. Figure 1.3 shows a depiction with airflow into and out of the cleanroom as F (in ft^3/min) and the particle generation rate as P (number of particles/min). The particle concentration in the cleanroom is C_n (number of particles/ft^3), the volume of the cleanroom is V (in ft^3), and the efficiency of the filtration is η. This can be represented as

$$V\frac{dC_n}{dt} = P + FC_n(1-\eta) - FC_n \tag{1.2}$$

which simplifies to

$$\frac{dC_n}{dt} = \frac{P}{V} - \frac{\eta F}{V}C_n. \tag{1.3}$$

The time-dependent solution of this equation is:

$$C_n = \frac{P}{\eta F} - \frac{P - \eta F C_{n-1}}{\eta F}e^{-t/\tau} \tag{1.4}$$

where C_{n1} is the initial particle concentration and the lifetime of the particles is

$$\tau = \frac{V}{\eta F}. \tag{1.5}$$

The steady state solution becomes:

$$C_{n0} = \frac{P}{F\eta} \tag{1.6}$$

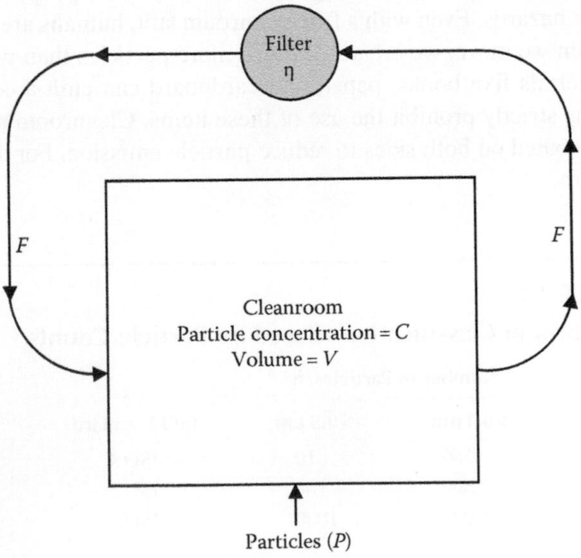

FIGURE 1.3 Particulate balance in a cleanroom.

or, equivalently

$$C_{n0} = \left(\frac{P}{V}\right)\left(\frac{V}{F}\right)\frac{1}{\eta} \tag{1.7}$$

where

P/V is the particle emission rate per unit volume,
F/V is the number of air changes per minute.

From Equation 1.6, we can see that to achieve a low particle count, one has to (1) reduce the emission rate, *P*, of particle contaminants, (2) increase the airflow rate *F*, and (3) increase the efficiency of filtration. One can compute the required flow rate, and hence the number of FFUs needed to achieve that flow, based on the expected particle emission rate. The latter mostly depends on the number of people expected to work in the cleanroom. An average person walking around will generate roughly 5×10^6 particles per min [5]. With proper cleanroom attire, the same movement may produce an order of magnitude smaller number of particles, or 5×10^5 particles per min. To achieve a class-100 condition, the filtration airflow rate required will be 5×10^3 ft³/min/person, assuming a filtration efficiency of 100%. If a single HEPA filter has a flow rate of 1000 ft³/min, this would translate to 5 filters/person. If the volume of the cleanroom is 12 ft × 12 ft × 7 ft, the flow would have to produce 5 air changes per min per person, or 300 air changes per hour per person. However, this model assumes a laminar flow. Any obstructions can create turbulence and areas of low flow. Common obstructions include equipment and tables. For this reason, cleanroom tables are often made of perforated tops to allow the air to flow through instead of around. Nevertheless, some obstructions are unavoidable, so the particle count will never be uniform across the entire cleanroom volume.

Even in a well-maintained cleanroom, the particle count will be higher near the entry door and progressively decrease toward the interior. This is taken into consideration when deciding where to place equipment. The equipment or process that requires the greatest level of cleanliness is typically placed farthest from the entry door.

Cleanroom furniture is almost never made from wood or fiberglass due to particle emission considerations. Stainless steel is the most common material. Cleanroom walls are made of aluminum or ordinary concrete or drywall with an epoxy surface sealant. In the latter case, it is important to not puncture or drill into the wall. Similarly, solid floors are also sealed with an epoxy.

1.2.2 Particle Count Measurement

A particle counter is an important piece of equipment in any cleanroom to monitor the integrity of the air filters. The number of airborne particles in a volume of air is typically measured with a laser particle counter. Air is pumped through a small chamber with a laser and the number of off-axis light scattering events is used to count the number of particles [6–8]. A HeNe laser (632 nm) can be used to measure particles down to 0.5 μm in size, and UV lasers can be used to measure particles down to 100 nm in size.

1.2.3 Service Access

Cleanroom processes require a number of service tools, such as vacuum pumps, chillers, and gas tanks. For ease of access, repair, and contamination control, these items are best kept outside the cleanroom environment. The most common practice is to place these items in a service area immediately adjacent to the cleanroom and connect them by making penetrations in the separating wall. This allows services to be performed without personnel having to enter the cleanroom.

Nitrogen gas is used in copious amounts in any cleanroom for drying, cleaning, and cooling. This is not the same as compressed air. Compressed air is generated by oil-sealed compressors and almost always contains water and oil, which can be disastrous for device fabrication. However, compressed air can be used for pneumatic actuation in various pieces of equipment.

1.2.4 Humidity, Temperature, and Lighting

In addition to pressure and flow rate, the AHU typically also controls the temperature and humidity. Some units only dehumidify, and will not humidify. Some will do both. The temperature inside the cleanroom is typically maintained slightly cooler than an ordinary office building to account for the cleanroom suits one has to wear. In addition, the FFUs and various pieces of equipment also produce heat. It is not unusual for the class-100 area, where there will be more FFUs, to be 1° or 2°C warmer than a class-1000 area.

Photolithography processes are sensitive to ambient temperature and humidity. The photo-acid generation requires a certain level of ambient moisture. What works in the summer when the humidity is 70% may not work in the winter when the humidity is 10%. On the other hand, cross-linking epoxy resists such as SU-8 can fail if the humidity is too high. In addition, too low a humidity can create static charges that can be destructive to certain electronic components. For this reason, the laboratory temperature and humidity should be recorded during all photolithography processes so that the process outcomes can be compared against the environmental conditions.

Photolithography is done in a dark room or "amber" room because photoresists are sensitive to ultraviolet radiation. Since ordinary fluorescent lights produce ultraviolet radiation, they have to be used with a filter, which is what makes the lighting appear amber in color. Photoresists are sold and stored in dark bottles. Furthermore, photoresists degrade over time, but their shelf life can be significantly extended by storing them in a refrigerator. Most laboratories store large bottles of photoresists in a refrigerator and dispense them into smaller bottles for regular use.

1.2.5 Safety

A fully operational cleanroom will have nearly all hazards imaginable—chemical, mechanical, electrical, and fire. To make matters worse, due to the expense of the cleanroom structure itself, equipment is often placed more densely than in an ordinary laboratory, and this can create obstructions to normal personnel movement. This requires one to be extra vigilant while working in a cleanroom.

Extremely flammable solvents, such as acetone and alcohols, are routinely used for photoresist stripping and for general cleaning. Hot plates, ovens, and furnaces are often left to run continuously. Extremely corrosive and toxic acids, such as sulfuric and hydrofluoric acids, are routinely used in cleaning. High-power UV lamps and lasers are used during photolithography. Electron beam accelerators that produce 10,000 V DC are routinely used in thin film processes. RF radiation of several thousand watts is used in plasma processes. A safety briefing is almost always required, and new personnel are rarely allowed to work alone without supervision and training.

PROBLEMS

1.1 Consider a cleanroom sized 15 ft × 15 ft × 7 ft. Assuming HEPA filters with an efficiency of 99% and three people in the laboratory generating 5×10^5 particles per min each (using 0.5 μm particle size), calculate the total airflow and the number of air changes per hour required to achieve a class-100 level. If the particle count rises to 1×10^7 particles/ft^3 due to some temporary maintenance work inside the cleanroom, use Equation 1.4 to calculate how long it would take before the particle count would fall below the class-100 level with one person in the cleanroom with a filter efficiency of 80%.

1.2 Based on the contamination level of the outside air that is used for making up air losses, the filtration is assumed to lose 10% flow rate per year. The size of each filter is 2 ft × 4 ft, and the linear velocity of the air exiting the filter (when brand new) is 100 ft/min. Calculate how many filters should be installed so that it would meet the class-100 specifications for at least 5 years.

LABORATORY EXERCISE

1.1 Take measurements of particle counts in your cleanroom at various locations starting from the entry doorway to the interior, and evaluate the reasons for the different readings.

REFERENCES

1. W. Whyte (ed.), *Cleanroom Design*, 2nd edn., Wiley, New York, 1999.
2. W. Whyte, *Cleanroom Technology: Fundamentals of Design, Testing and Operation*, John Wiley & Sons, New York, 2010.
3. Federal Standard: Airborne particulate cleanliness classes in cleanrooms and clean zones, FTD-STD-209E.
4. International Organization for Standardization, Classification of air cleanliness by particle concentration, ISO 14644–1:2015. http://www.iso.org.
5. D. W. Cooper, Particulate contamination and microelectronics manufacturing: An introduction, *Aerosol Sci. Technol.*, 5 (1986) 287–299.
6. Particle Measuring Systems, *Beginner Guide to Particle Technology, Particle Measuring Systems*, Boulder, CO, pmeasuring.com.
7. D. L. Black, M. Q. McQuay, and M. P. Bonin, Laser-based techniques for particle-size measurement: A review of sizing methods and their industrial applications, *Prog. Energy Combust. Sci.*, 22(3) (1996) 267–306.
8. W. W. Szymanski and Benjamin Y. H. Liu, On the sizing accuracy of laser optical particle counters, *Part. Part. Syst. Charact.*, 3(1) (1986) 1–7.

2 Fundamentals of Vacuum and Plasma Technology

2.1 FUNDAMENTALS OF VACUUM

Though at first it may seem unrelated to nanofabrication, vacuum is an integral part of many device fabrication processes, such as thin film deposition and etching. Vacuum is required for achieving high-purity conditions during a process and also for reducing molecular scattering. Therefore, an understanding of vacuum chambers, vacuum pumps, and their principles becomes imperative for anyone working with device fabrication and process development.

Vacuum is the absence of air. The standard atmospheric composition is 78% N_2, 21% O_2, and 1% Ar with some amount of H_2O that can vary significantly. Atmospheric air pressure is measured using a number of different units—14.6 psi, 760 Torr, 29.9 in. Hg, 1013 mbar, or 101 kPa—depending on the area of application and also the geographic region. In vacuum science, Torr is widely used in the United States. In some applications, atmospheric pressure is used as the zero reference with positive numbers representing above-atmospheric pressures and negative numbers representing below-atmospheric pressures. In vacuum measurements, an absolute unit is used, with zero pressure representing a perfect vacuum with positive numbers representing the presence of gas molecules.

The purpose of creating a vacuum is to reduce the number of air molecules that may interfere with the fabrication process. A perfect vacuum can never be created in practice. This will require every single molecule of air to be evacuated. However, the density of air molecules can be reduced to the extent where it starts to matter less and less. At 760 Torr air contains 2.7×10^{19} molecules/cm^3 (which can be obtained from the relation that 1 mole of an ideal gas occupies 22.4 liters). At 0.1 μTorr, the same air will contain 3.5×10^{9} molecules/cm^3. A few billion molecules in a cubic centimeter may still seem like a very large number, but it is actually low enough to be considered ultra-high vacuum.

To put things in perspective, at an altitude of 30,000 ft where airlines fly, the pressure is 225 Torr. At 38,000 ft, it drops to 160 Torr. The partial pressure of oxygen at sea level is $760 \times 21\% = 160$ Torr. Therefore, at 38,000 ft, one would require 100% supplemental oxygen to be equivalent to sea level conditions. Despite these dramatic physiological effects, in vacuum science both these pressures are considered to be nearly indistinguishable from atmospheric pressure. At 80,000 ft, the pressure drops to 20 Torr. This is the same as the vapor pressure of water at room temperature. In other words, water (and bodily fluids) will boil at 80,000 ft if held at room temperature. This is the reason why astronauts and U2 pilots have to wear a pressure suit. At an altitude of 50 miles above the earth, which is defined as the beginning of space, the pressure drops to 1 mT. At 230 miles above the earth, where the international space station orbits, the pressure is around 1 nT. At 22,000 miles, near the geosynchronous orbit, the pressure drops to 10 pT. The pressure on the lunar surface is also about 10 pT. In the laboratory, we can reasonably achieve pressures of around 0.1 μT. Pressures lower than 10^{-8} Torr become significantly more difficult to create and maintain, but it is nevertheless required in some applications.

In thin film deposition, material is removed from the source one atom at a time and deposited on the substrate. If air is present in the environment, the depositing atoms can react with the air molecules and form oxides or nitrides. The probability of this reaction depends on a number of factors. The first factor is the probability that the atom will collide with an O_2, N_2, or H_2O molecule while in transit to the substrate or after arriving at the substrate and form a compound. This is also dictated by the probability that the atom will form a compound when it comes in contact with an air molecule. Some atoms like titanium are very reactive and form compounds immediately, while

TABLE 2.1
Mean Free Path × Pressure and Molecular Weight of Common Gases

Gas	$\lambda \cdot P$ (cm × Torr)	g/mol
H_2	9.0×10^{-3}	2.0
He	13.5×10^{-3}	4.0
Ne	9.2×10^{-3}	20.1
Ar	4.8×10^{-3}	40.0
O_2	4.9×10^{-3}	32.0
N_2	4.6×10^{-3}	28.0
CO_2	3.0×10^{-3}	44.0
H_2O	3.0×10^{-3}	18.0
Cl_2	2.3×10^{-3}	70.9
Air	5.0×10^{-3}	28.9

Source: Oerlikon Leybold Vacuum, *Fundamentals of Vacuum Technology*, 2007, Table III.

others like gold do not react. The speed (energy) of the atom can also affect the probability of compound formation. A useful quantity here is mean free path. This is the average distance a molecule travels before colliding with another molecule. When creating a vacuum, we are making the mean free path very large such that the probability of collision with an air molecule becomes small. If the mean free path becomes larger than the distance between the source and substrate, one can reasonably assume that no collisions will take place during transit. The mean free path of several gases is shown in Table 2.1 expressed as a product $\lambda \cdot p$, where λ is the mean free path and p is the pressure. For example, at a water vapor pressure of 10 μT the mean free path will be 300 cm. This is larger than most vacuum chambers, so we can assume the transport to be collision-free at this pressure.

The second factor is the bombardment of the gas atoms on the substrate. Most solids have an atomic density of around 10^{23} atoms/cm^3. From this, we can infer that the thickness of one monolayer will be about 2 Å, and the surface density of atoms will be about 2×10^{15} atoms/cm^2. If a thin film is being deposited at 1 Å/s, which is a fairly typical rate, it corresponds to about ½ monolayer per second, or an incident flux density of 1×10^{15} atoms/cm^2/s. This assumes that 100% of the incident atoms stick to the substrate and do not rebound. There is an equivalent relationship between the impingement rate Z_A of the atoms on a surface and the background gas pressure P, expressed as [1]

$$Z_A = \frac{N_A P}{\sqrt{2\pi MkT}} \tag{2.1}$$

where
- N_A is the Avogadro constant (number of constituent particles in 1 mol: 6.02×10^{23}/mol)
- k is the Boltzmann constant
- M is the molecular weight of the gas species
- T is the absolute temperature

Using more convenient units, this works out to

$$Z_A = 3.5 \times 10^{22} \frac{P}{\sqrt{MT}} \tag{2.2}$$

where
Z_A is the rate in atoms/cm²/s
P is the pressure in Torr
T is in Kelvin
M is in g/mol

At a pressure of 10 μT, assuming the background gas is oxygen, the gas molecules will impinge all surfaces at a rate of about 3.5×10^{15} molecules/cm²/s. Therefore, if the residual oxygen partial pressure is 10 μT, and if titanium is being deposited at 1 Å/s, oxygen molecules will be arriving on the substrate at three times the rate of the titanium atoms. This will result in virtually every titanium atom being oxidized. Therefore, one has to consider both the mean free path and the surface impingement rate when determining the lowest pressure required in a process. At very low pressures, the surface impingement rate will dominate the interaction, whereas at higher pressures the mean free path will dominate.

In some processes, the vacuum chamber is filled with a gas, either to sustain a plasma or to produce a chemical reaction. The purity of the gas also plays a critical role. Before filling a chamber with a high-purity gas, it must first be evacuated to a sufficiently low vacuum. The lowest pressure reached before filling with the process gas is known as the base pressure. A low base pressure ensures a higher purity environment during subsequent processing. For example, consider the sputter deposition of titanium using 10 mT of argon pressure, at a rate of 1 Å/s. Based on the calculation described earlier, in order to prevent oxidation, the gas must contain less than 0.1 μT of oxygen (which would be equivalent to 33 titanium atoms arriving on the substrate for each oxygen molecule). That means, the chamber must first be evacuated to a base pressure lower than 0.1 μT and filled with argon that contains less than 1 part oxygen in 10^5 parts argon, or 10 ppm of oxygen. This corresponds to a purity of 99.999%, which is abbreviated as 5N purity (for five nines). Furthermore, if the process involves the dissipation of heat into the chamber, one also needs to be careful that oxygen (or water vapor) is not released from various parts of the chamber during the process.

A vacuum system consists of a vacuum chamber, vacuum pumps, connecting hoses, vacuum measuring devices, and other accessories. In the following sections, we will examine the principles of vacuum creation and the factors that influence the lowest pressure that can be achieved in a practical system.

2.1.1 Conductance

Gases have to flow through connecting hoses. An important concept in vacuum hoses and connectors is the conductance, C. This is equivalent to the concept of conductance (or resistance) in electrical circuits. Just like Ohm's law, the following relationship is used:

$$Q = C\Delta p \tag{2.3}$$

where
Q is the gas flow measured is pressure * volume/time
Δp is pressure difference across the component

The unit for conductance is volume/time, typically L/s. Unlike in electrical circuits, the conductance C is not a constant, but it is a function of pressure. Its value generally falls as pressure decreases. The decrease in conductance can be viewed as arising due to the number of molecules colliding with the walls compared to the number of molecules colliding with each other. At higher pressures, molecules collide with each other more than with the walls, and the "friction" due to the walls will be small and conductance will be high. As pressure drops, a larger fraction of molecules will collide with the walls and the conductance will decrease. As pressure continues to fall, conductance reaches a minimum value and eventually becomes independent of pressure. The region where conductance is linearly dependent on pressure is known as the "viscous" region, and the region where it is independent of pressure is known as the "molecular" region. The viscous region can be thought of as a "sticky" fluid and the molecular region as consisting of non-interacting molecules. These effects can be accurately modeled using fluid equations, but for this purpose, we can empirically represent the equation that describes conductance as a function of the diameter and length of a tube as [1]

$$C = 2942\frac{d^4 p}{l} + 78\frac{d^3}{l}\left(\frac{1+648dp}{1+800dp}\right) \tag{2.4}$$

where

- p is in Torr
- d is the diameter in inches
- l is the length in inches
- C is the conductance in L/s

For example, Figure 2.1 shows the plot of conductance vs. pressure of a 3 in. diameter × 24 in. long vacuum hose.

2.1.2 Pumping

As described in Equation 2.3, the flow through a passive tube can be written as $Q = C(p_i - p_o)$, where p_i is the intake pressure and p_o is the outlet (exhaust) pressure. This expression can be expanded as $Q = Cp_i - Cp_o = Q_i - Q_o$. The flow quantities Q_i and Q_o can be viewed as two opposing flows with $Q_i > Q_o$ to create a net flow in the direction of the negative pressure gradient.

A vacuum pump pump can be viewed as an active tube where the pumping process produces $Q_o > Q_i$ to create a flow in the same direction as the pressure gradient (i.e., opposite the normal direction of flow). This can be represented as $Q = C_i p_i - C_o p_o = Q_i - Q_o$ with asymmetric conductances such that $C_i \gg C_o$. This allows Q to be positive ($Q_i \gg Q_o$) even when $p_i \ll p_o$. We can consider $Q_i = C_i p_i$ as being due to the pumping action and $Q_o = C_o p_o$ as being due to the backward flow

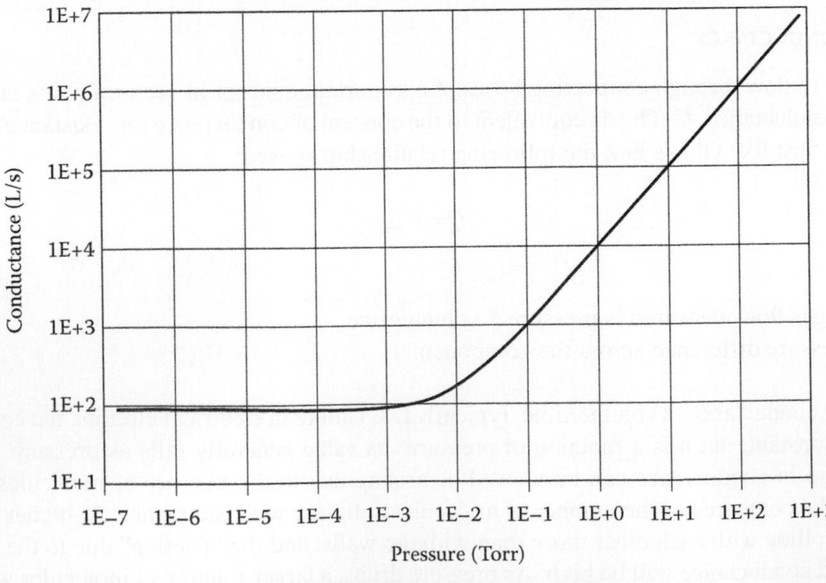

FIGURE 2.1 Conductance of a 3 in. diameter × 24 in. long vacuum hose as a function of pressure.

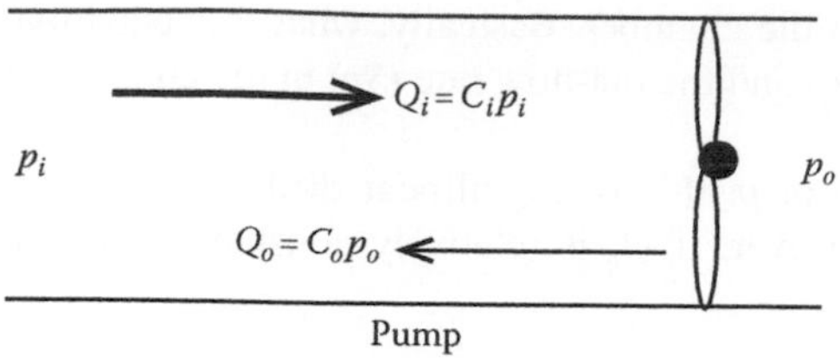

FIGURE 2.2 Schematic representation of a vacuum pump in terms of conductances.

through the pump. In this case, the net flow is from the lower pressure p_i toward the higher pressure p_o. This is illustrated in Figure 2.2.

The net flow divided by the intake pressure is defined as the pumping speed, which can be written as

$$S = \frac{Q_i - Q_o}{p_i} = \frac{C_i p_i - C_o p_o}{p_i}. \tag{2.5}$$

The value of S will be a combination of pump geometry, rotation speed, and pressure. Furthermore, C_i and C_o will not be constants, but they will depend on pressure. In viscous pumps, conductances fall as pressure drops. $C_i p_i$ will fall faster than $C_o p_o$ because the exhaust pressure p_o will be held relatively constant, whereas the inlet pressure will decline due to the pumping action. Therefore, at some minimum pressure, $C_i p_{i,\min}$ will become equal to $C_o p_o$, and the net flow of gas Q through the pump will cease. The pumping speed will then become zero. This corresponds to the lowest achievable pressure at the pump's inlet, assuming there are no intentional gas flows or leaks. The ratio $p_o/p_{i,\min}$ when $S = 0$ is known as the maximum compression ratio. This ratio is

$$K = \frac{p_o}{p_{i,\min}} = \frac{C_i}{C_o}. \tag{2.6}$$

The pumping rate equation can be derived from the ideal gas law $pV = \nu N_A kT$. In the presence of an external gas flow Q_g into the chamber, the net flow $\Delta Q = Q_g - Q_i + Q_o$ will result in a change in the mole number ν as described by the following equation:

$$\Delta Q = Q_g - Q_i + Q_o = \frac{d\nu}{dt} N_A kT. \tag{2.7}$$

From the ideal gas law,

$$\frac{d\nu}{dt} N_A kT = \frac{dp}{dt} V + \frac{dV}{dt} p. \tag{2.8}$$

V can be considered to be a constant because it is the volume of the vacuum chamber. Therefore, assuming, for now, that the chamber pressure p is equal to the inlet pressure at the pump p_i

$$\frac{dp}{dt} V = Q_g - Q_i + Q_o = -Sp + Q_g \tag{2.9}$$

$$\frac{dp}{dt} V = -Sp + Q_g. \tag{2.10}$$

This is the pumping rate equation. S is the pumping speed in volume/time, V is the volume of the system being pumped, p is the pressure (assumed to be the same as the pump's inlet pressure), and

Q_g is the gas flow rate into the chamber. Basically, what this equation says is that the difference between the in-flow rate (Q_g) and the out-flow rate (Sp) in the chamber gives rise to a change in the chamber pressure.

Because S is a function of p, this is a nonlinear differential equation. Therefore, it can only be solved numerically. However, if Q_g is relatively constant in time compared to dp/dt, that is, $\frac{d^2p}{dt^2}V \ll \frac{dQ_g}{dt}$, we can get

$$p = \frac{Q_g}{S}. \tag{2.11}$$

Since S is a function of p, this equation still cannot be readily solved. Nevertheless, we can qualitatively see that it is necessary to have a high pumping speed S and a low gas flow rate Q_g in order to achieve a low pressure. In the absence of any intentional gas flow ($Q_g \to 0$) the solution converges to the same $p_{i,\min}$ when S converges toward zero, as discussed earlier.

2.1.3 Effect of a Vacuum Hose

In the earlier mentioned pumping equation, we have assumed that the pressure at the pump's inlet is the same as in the vacuum chamber. We have ignored the effects of any connecting hoses between the vacuum pump and the chamber. Now consider a hose that has a conductance C_h. The pumping equation becomes

$$\frac{dp_c}{dt}V = -S_c p_c + Q_g \tag{2.12}$$

where

- p_c is the chamber pressure
- S_c is the effective pumping rate in the chamber (which is different from the pumping rate S at the pump's inlet)

The chamber pressure p_c will be higher than the pump's inlet pressure by Q_g/C_h. That is,

$$p_c = p_i + \frac{Q_g}{C_h} \tag{2.13}$$

and the pressure in the hose can be written as an average of the chamber pressure and the pump inlet pressure:

$$p_h = \frac{p_c + p_i}{2}. \tag{2.14}$$

Furthermore, we will assume that the volume of the hose is small compared to the volume of the vacuum chamber, so that the overall volume V remains unchanged. Since the flow rate leaving the chamber will be equal to the flow rate entering the pump, this becomes

$$S_c p_c = S p_i. \tag{2.15}$$

Combining all of these equations, the pumping rate equation can be expressed as

$$\frac{dp_c}{dt}V = -Sp_c + Q_g\left[1 + \frac{S}{C_h}\right]. \tag{2.16}$$

Compared to Equation 2.10 earlier, Equation 2.16 behaves as if the gas load has increased by a factor of $\left[1+\frac{S}{C_h}\right]$. This has the effect of slowing down the effective pumping rate. The steady-state pressure in the chamber will also increase by the same factor for the same gas flow Q_g. By setting $dp_c/dt = 0$, we can show that

$$p_c = \frac{Q_g}{S}\left(1+\frac{S}{C_h}\right) = Q_g\left[\frac{1}{S}+\frac{1}{C_h}\right]. \tag{2.17}$$

Compared to Equation 2.11, we can see that the chamber pressure will be higher by a value of Q_g/C_h, which should not be a surprising result.

The factor $\left[\frac{1}{S}+\frac{1}{C_h}\right]$ is often referred to as the *effective pumping rate* in the chamber due to the hose conductance, such as

$$\left[\frac{1}{S}+\frac{1}{C_h}\right] = \frac{1}{S_c} \tag{2.18}$$

where S_c is the effective pumping rate in the chamber. Again, we can verify that in the absence of any gas flow Q_g, the solution converges to the same $p_{i,\min}$ as before when S becomes zero. The application of these equations is illustrated through the examples in the next section.

2.1.4 Rough Vacuum

Rough vacuum is defined as ranging from atmospheric pressure down to about 1 mT. The flow characteristics in this range can be treated as viscous.

Oil-sealed positive displacement vacuum pumps (rotating and piston types) can evacuate from atmospheric pressure down to about 1 mT. These are the types of pumps most commonly found in general industrial applications for generating low levels of vacuum. The operating principle of these pumps is based on mechanically expanding a cavity volume to create a low-pressure region, filling that region with gas molecules from the chamber, sealing off that region, compressing that volume to increase the pressure to a value higher than atmospheric pressure and finally expelling it through the exhaust port.

Oil-sealed rotary vane pumps, like the one shown in Figure 2.3, consists of a rotor on an off-axis center inside a cavity. The vanes are located in radial slots in the rotor and ride along the sidewalls of the cavity sealed by oil. They can be spring loaded to push outward or they can extend outward simply by centrifugal force. The expanding cavity sucks air from the intake, and the shrinking cavity pushes the compressed air out through the exhaust. In a two-stage pump, the exhaust of the first stage is connected to the intake of the second stage, allowing a greater overall compression to be achieved. Most oil-sealed rotary pumps have pumping speeds S at near atmospheric pressures of the order of 1–10 L/s and a maximum compression ratio of about 10^6. All roughing pumps operate in the viscous regime. As a result, the conductances of the internal parts of the pump will decline as the pressure drops. This has the effect of reducing the pumping speed S as pressure drops. This is one of the reasons why roughing pumps cannot get below about 1 mT.

In addition to declining conductances, there are other factors that limit the ultimate lowest pressure that can be achieved with roughing pumps. Leaks inside the pump can introduce an additional gas load. The vapor pressure of the vacuum oil can also introduce a gas load, especially at higher operating temperatures. Vapor pressure of some oils can be as high as 10 mT at 100°C. Moisture desorption from the pump chambers as well as the oil can also limit the ultimate pressure. A pump that has been sitting idle for some time will most likely contain moisture in the oil and will take

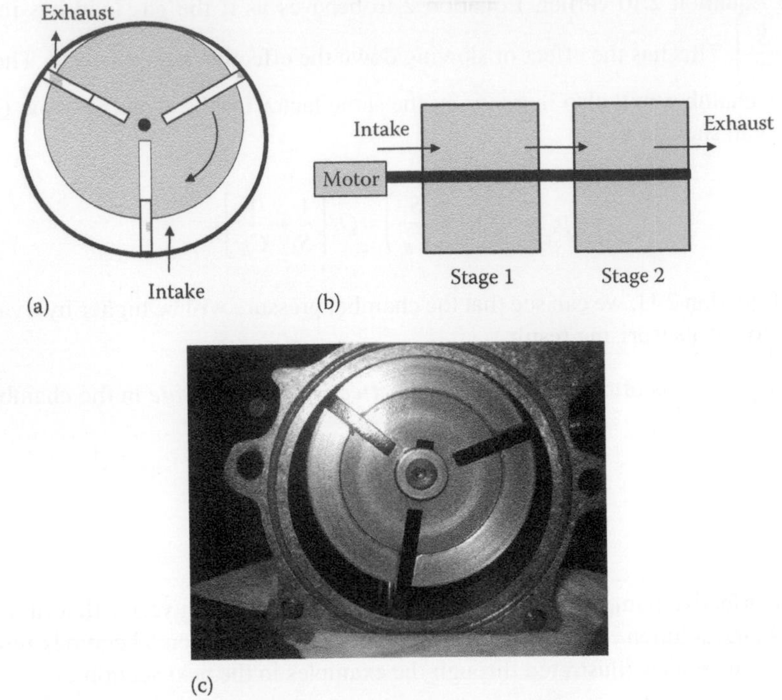

FIGURE 2.3 Oil-sealed rotary vane pump. (a) Cross section of the rotor with vanes, (b) two-stage rotary pump, and (c) photograph of an actual rotor.

some time to fully desorb. Running the pump continuously has the advantage of keeping the oil hot to prevent moisture from accumulating.

The reverse conductance C_o also has an important effect on the ultimate performance of the pump. The rotating vanes have to be sealed well against the chamber walls in order to achieve a low reverse conductance. This is one of the main functions of the pump oil. It enables a tight seal to be achieved while maintaining the surfaces to slide with low friction. As the pump ages, or as the oil accumulates contaminants, the effectiveness of this seal will deteriorate and the reverse conductance will increase, resulting in a reduction in the pumping speed S.

An example of pumping speed vs. pressure (S vs. p) characteristic of a typical rotary vane pump is shown in Figure 2.4. The pumping rate at high pressures is 3.3 L/s, declining to 1 L/s at 10 mT, and then rapidly falling to insignificantly small values below 10 mT. The lowest pressure from the chart is 0.5 mT. Since the exhaust pressure is normally held at atmosphere (760 Torr), the ultimate compression ratio can be calculated as $760/(0.5 \times 10^{-3}) = 1.5 \times 10^6$.

Now consider this pump connected to a 100 L vacuum chamber with a 1 in. diameter × 24 in. long hose. If there were no intentional or unintentional gas flows, the chamber pressure should reach the same 0.5 mT ultimate pressure of the pump's intake port, though that may take a long time. However, in practice, there will always be some unintentional gas flows. This could be due to outgassing from various surfaces of the vacuum system or small leaks. If these unintentional gas flow rates in the chamber amounts to 1 mT · L/s, the lowest achievable pressure in the chamber can be calculated from Equation 2.17, which was $p_c = Q_g\left[\frac{1}{S}+\frac{1}{C_h}\right]$. However, this is not a straightforward substitution because S, P_c, and C_n are all functions of the pump pressure p_i. So it would require some type of iterative solution. Alternatively, we can also numerically solve the pumping rate equation (Equation 2.16) using finite difference discretization by stepping through time until the pressure

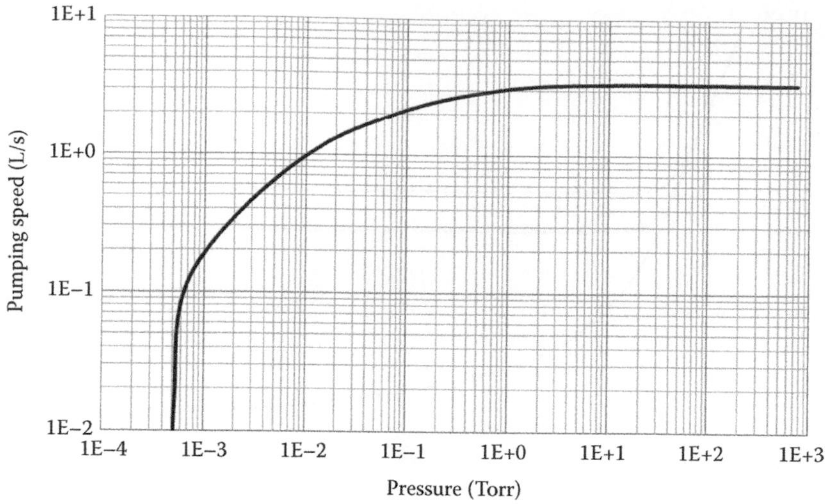

FIGURE 2.4 Pumping speed vs. pressure for a typical rotary vane pump.

reaches a steady state. This is actually a lot easier than attempting an analytical solution, and the method is briefly outlined next.

We can discretize the pumping rate equation (Equation 2.16) into the following:

$$\frac{p_c(t+\Delta t)-p_c(t)}{\Delta t}V=-S(p_i)p_c(t)+Q_g(t)\left[1+\frac{S(p_c)}{C_h(p_c)}\right] \tag{2.19}$$

where

$p_c(t)$ is the chamber pressure at time step t

$p_c(t+\Delta t)$ is the chamber pressure at the next time step $t+\Delta t$

$S(p_i)$ is the pumping rate when the inlet pressure is p_i (which depends on the chamber pressure through Equation 2.15

$C_h(p_c)$ is the hose conductance that depends on the chamber pressure (or inlet pressure)

$Q_g(t)$ is the gas flow rate at time t

This can be rearranged such that

$$p_c(t+\Delta t)=\left\{\frac{Q_g(t)}{V}\left[1+\frac{S(p_c)}{C_h(p_c)}\right]-\frac{S(p_i)p_c(t)}{V}\right\}\Delta t+p_c(t). \tag{2.20}$$

This equation relates the pressure at the time step $t+\Delta t$ to the conditions at the previous time step t. The time increment Δt must be small enough such that the change in pressure between time steps is not too large to prevent errors and instabilities. Using this equation, we can start from the initial atmospheric pressure and step through time to find how the pressure progresses as a function of time, taking into account that S and C_h depend on pressure as well.

Figure 2.5 shows the effect of gas flow on the pressure vs. time curve. With no gas flow ($Q_g=0$), we can see that the chamber pressure reaches the ultimate value of 0.5 mT in approximately 40 min. If a gas flow is introduced, the lowest achievable pressure also increases. With $Q_g=1\,\text{T}\cdot\text{L/s}$ (or equivalently $79\,\text{atm}\cdot\text{cm}^3/\text{min}$, which is typically the unit used for gas flow, and is designated as standard cubic centimeters per second, or sccm), the ultimate pressure rises to 395 mT. If the flow increases to 395 sccm, the ultimate pressure also rises to 1.6 T.

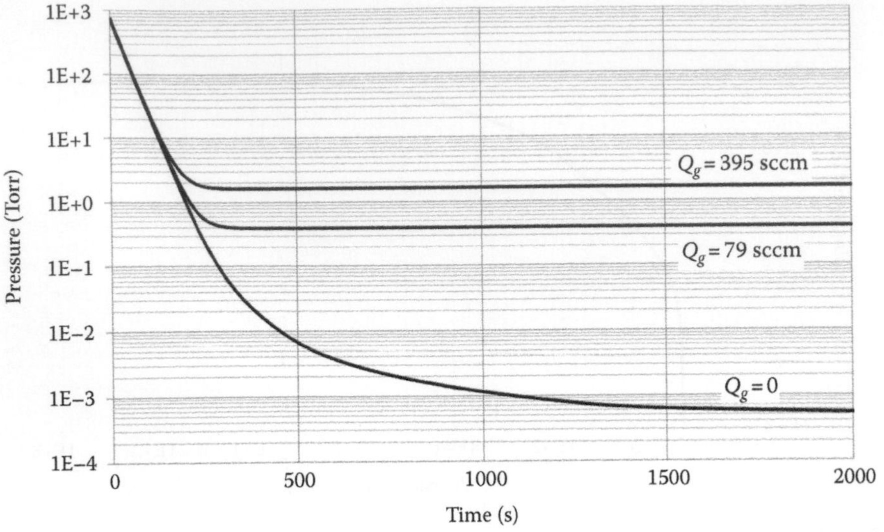

FIGURE 2.5 Chamber pressure vs. time for different gas flow rates.

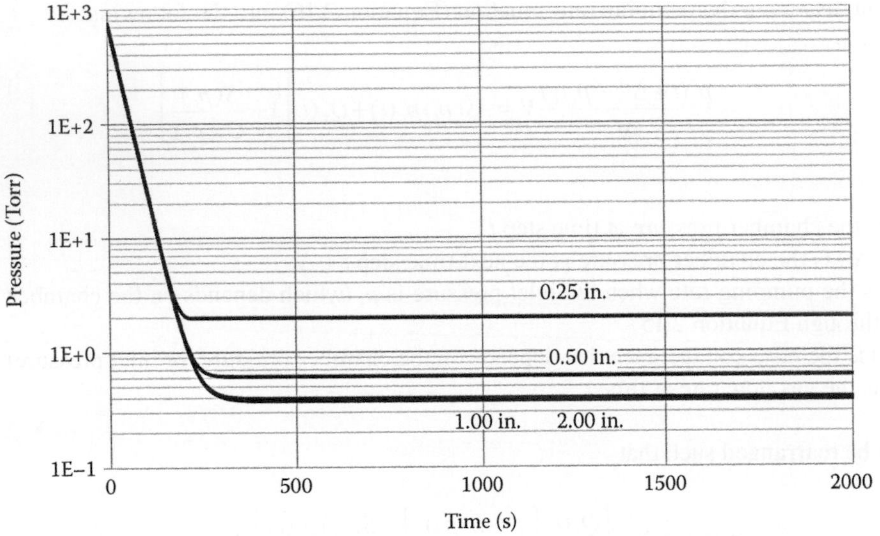

FIGURE 2.6 Chamber pressure vs. time for 79 sccm flow with different hose diameters. The 1 in. and 2 in. diameter pressure curves are nearly on top of each other.

Figure 2.6 shows the effect of the hose diameter for a gas flow rate of 79 sccm. The 1 in. diameter hose is able to reach a much lower pressure than the 0.25 in. diameter hose. We can also see that 2 in. diameter hose does not show a noticeable improvement over the 1 in. diameter hose. As vacuum hoses are typically made of aluminum or steel and tend to be bulky and inflexible, it is useful to perform this calculation to have an idea of what length and diameter hose is necessary for a specific application.

Even in the best case, the ultimate pressure that can be achieved with the roughing pump is only of the order of a few mT. In order to reach pressures below mT levels, a high-vacuum pump has to be used.

2.1.5 High-Vacuum Pumps

Roughing pumps operate in the viscous regime where we can think of the gas as a "sticky" fluid. When gas is removed at one end of the chamber, the remaining gases rush in to fill that low-pressure region. This is due to the frequent collisions between the molecules that make the fluid behave as a connected system. In the molecular region, each gas molecule behaves independent of the other molecules. Removing gases at one end will not make the other gas molecules to follow. Instead, we have to wait for gas molecules to randomly arrive at the pump's inlet port, and redirect or trap those molecules once they arrive. This requires a completely different pumping mechanism than the roughing pumps. Turbo molecular pumps and cryo pumps are two of the most common pumps used in high-vacuum applications. Of course, there are many other types of pumps as well, and an interested user is directed to the references at the end of this chapter.

2.1.5.1 Turbo Molecular Pumps

The turbo molecular pump is an example of a high-vacuum pump. Unlike the roughing pump, the pumping speed here is fairly constant through most of its operating pressure range. A very high inlet conductance C_i is achieved by using a large inlet diameter and high spin speeds. To minimize the effects of conductance, the pump is connected directly to the chamber without any hoses. Typically, no oil is used in these pumps, eliminating any vapor pressure issues. Although the pumping surfaces are built with tight tolerances, they are not tightly sealed with vanes and oil as in roughing pumps. This allows the blades to spin at extremely high velocities without mechanical friction. Despite the poorly sealed surfaces and wide gaps, the reverse conductance C_o in these pumps is very low due to the low operating pressures.

The turbo pump basically looks like a fan that spins at a high speed, in the range of 500–1000 Hz. Compression is achieved by imparting a velocity component to the incident molecules and redirecting them toward the pump exhaust. Hence, these are also known as kinetic pumps. For example, a 6 in. diameter pump rotating at 500 Hz (30,000 RPM) will have a maximum blade velocity of 240 m/s. It will alter the momentum of a 500 m/s gas molecule by about 50% and redirect its velocity along the axis of the pump toward the exhaust port. Multiple layers of vanes are used, with stationary vanes tilted in the opposite direction between the rotating layers of vanes. In order for the momentum transfer to occur, the distance between the layers has to be smaller than the mean free path. Otherwise, the molecules will be deflected in random directions due to collisions before reaching the next set of vanes. Furthermore, since pressure will gradually increase along the length of the pump toward the exhaust port, the distance between the adjacent layers of vanes is also progressively reduced. A typical cross section of a turbo pump is shown in Figure 2.7. The root-mean-square molecular speeds for various gas species at room temperature is shown in Table 2.2.

Pumping speeds are in the range of 100–1000 L/s with compression ratios of about 10^9. Compared to rotary vane roughing pumps, this is several orders of magnitude higher. However, as we will see in the numerical examples following this section, at the low pressure ranges used for these pumps the pump down pressure curve is primarily dictated by the desorption rates of residual gases and moisture, so the high pumping speed does not necessarily translate into a shorter pump down time. But the high compression ratio is important as it is what allows a lower ultimate pressure to be reached.

An important aspect of molecular flow is that each molecule will flow independent of the others, and it moves in accordance with its own molecular mass and temperature. Therefore, the conductances will also be different for each type of gas. Hydrogen, due to its low mass, has a higher conductance than nitrogen. As a result, the effective pumping speed will be slightly higher for hydrogen compared to nitrogen. Additionally, the velocity component imparted by the pump to hydrogen molecules (which travel at 1761 m/s at room temperature) will be significantly less than for nitrogen molecules (which travel at 471 m/s). In other words, after entering the pump, nitrogen molecules will have a greater component of their velocity along the pump's axis than hydrogen. This can also be viewed as hydrogen having a greater reverse conductance C_o compared to nitrogen. As a result,

FIGURE 2.7 Interior view of a turbomolecular pump. (Reproduced from Wikimedia Commons, CC BY-SA 3.0. User: Liquidat, November 2005.)

TABLE 2.2
Mean Thermal Speeds of Various Gas Species at Room Temperature

Gas	RMS Speed (m/s) at 300 K
H_2	1761
He	1245
H_2O	587
N_2	471
O_2	440
Ar	394

nitrogen will have a greater compression ratio than hydrogen. The compression ratio for nitrogen can be as much as 1000 times greater than hydrogen. Therefore, pumping parameters such as speed and compression have to be specified for each gas type. The pumping speed quoted in the manufacturers' specifications is typically for nitrogen.

The layout of a typical vacuum system is shown in Figure 2.8. Initially, roughing pumps are used to reach from atmosphere down to about 10 mT, and then a high-vacuum pump is initiated to reach lower pressures while the roughing pump continues to pump the exhaust port (foreline) of the high-vacuum pump. A high-vacuum pump cannot be used to pump down directly from atmosphere. Since turbo molecular pumps are constructed from lightweight aluminum to allow

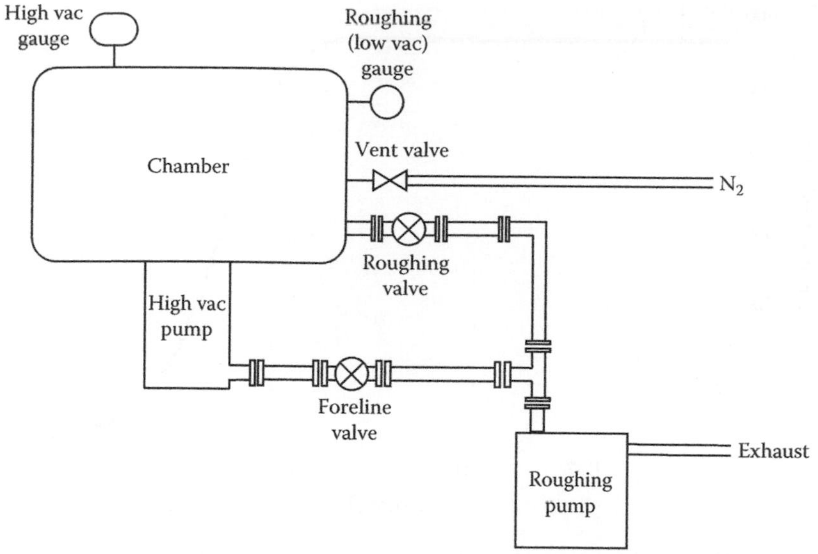

FIGURE 2.8 Layout of a typical vacuum system.

TABLE 2.3
Typical Performance Specifications of a Turbo Molecular Pump

	N_2	He	H_2
Pumping speed	550 L/s	600 L/s	510 L/s
Compression ratio	10^9	10^7	10^6

high rotation speeds, they are not strong enough to withstand the forces that will occur at higher pressures. In fact, sudden venting of the pump that is running at full speed can cause catastrophic damage to the vanes. On the other hand, if a pump is started at rest from atmosphere, it will never reach full rotation speed due to the excessive drag forces. Additionally, due to the open construction of the pump, at higher pressures and short mean free paths, the compression ratio will also be extremely poor.

Table 2.3 shows the typical performance specifications of a turbo molecular pump, and Figure 2.9 shows the pumping speed vs. pressure of the same pump, which indicates a declining pumping speed above 1 mT. This is due to a combination of several factors, such as excessive drag forces on the pump and the higher reverse conductance that will reduce the net forward flow through the pump. At lower pressures, the pumping speed levels off, which in this example is 550 L/s.

As the pressure drops further toward the maximum compression ratio, the pumping speed will decline again toward zero (not shown in Figure 2.9). The pumping speed vs. pressure for this low-pressure regime can be derived from $S = \frac{C_i p_i - C_e p_e}{p_i} = C_i - C_e \frac{p_e}{p_i}$. Since S falls to zero at the maximum compression ratio K, it can be shown that $C_e = C_i / K$ where K is the maximum compression ratio. Therefore, the pumping speed in the molecular flow regime can be written as $S = C_i \left(1 - \frac{1}{K} \frac{p_e}{p_i} \right)$.

For the pump characteristics shown in Figure 2.9, we can calculate the pumping speed vs. pressure for different gases, assuming a constant exhaust pressure at atmospheric composition. Standard

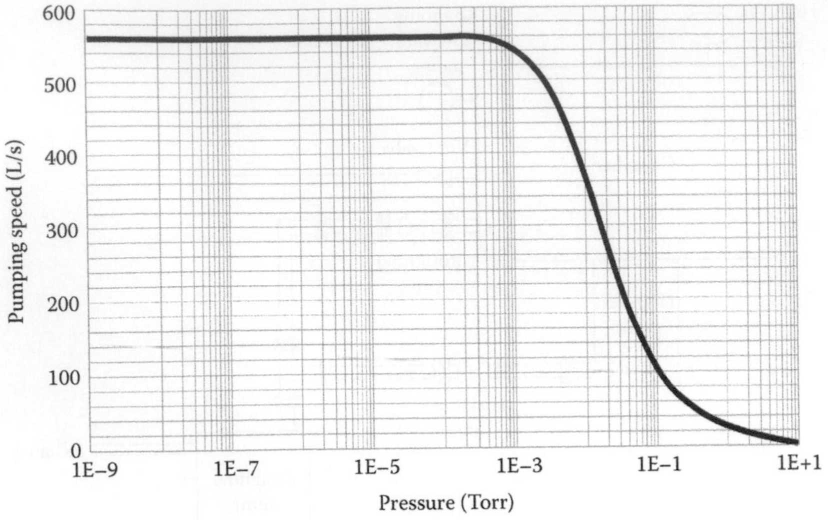

FIGURE 2.9 Pumping speed vs. pressure for a turbo molecular pump for N_2.

atmosphere contains 78% nitrogen and 5.24×10^{-4}% helium If one can reach the maximum compression ratio (which would require a zero gas flow, and would be nearly impossible in almost all practical situations), and assuming a 10 mT pressure at the foreline, the inlet pressure will be $10\ \text{mT} \times 0.78 \times 10^{-9} = 10\ \text{pT}$ for nitrogen, and $10\ \text{mT} \times 5.24 \times 10^{-6} = 10^{-7} = 5\ \text{fT}$ for helium. The nitrogen/helium ratio is 1560. In standard atmosphere, nitrogen/helium ratio is 150,000. Therefore, the helium content has increased by a factor of 100 in the chamber. This is the expected result in the high-vacuum range—the gas composition will change and the molar fraction of the light gases will increase.

2.1.5.2 Cryo Pumps

Cryo pumps are high-vacuum pumps with very few moving parts. They consist basically of a set of ultracold surface panels that condense and trap any incident molecule on them. Hence, they are also known as entrapment pumps. The mechanisms are cryocondensation, cryosorption, and cryotrapping [4].

The coldest part of the cryo pump is at a temperature of about 10–15 K, which will condense all of the atmospheric gases except hydrogen and helium. The system consists of a helium compressor, connecting hoses and the pump module that is attached to the vacuum chamber. Helium is compressed and flowed to the pump through hoses. The cooling action is performed by expanding the compressed helium inside a motor driven piston in the pump's head. The returning piston stroke will push the expanded helium out through the hoses back to the compressor. In that sense, it is very similar to the household refrigeration cycle, except the refrigerant used here is helium instead of hydrofluorocarbons.

The cross-sectional cutout of a typical cryo pump is shown in Figure 2.10. The saturation vapor pressure of atmospheric gases and their freezing temperatures are shown in Figure 2.11 and Table 2.4, respectively. The top cryopanel, which is held around 80 K, will condense water, carbon dioxide, and some argon. The remaining gases will make their way down toward the lower cryopanel that is held at a much lower temperature of around 15 K. However, some fraction of these gases may also reflect and escape back into the vacuum chamber. The lower cryopanel will condense nitrogen. oxygen, and argon. The light gases, such as helium and hydrogen cannot be condensed even at 15 K. These gases are eventually trapped by adsorbants coated on the inside panels and the radiation shields by cryo-adsorption. The most common adsorbants are activated charcoal.

The pumping speed of a cryo pump is related to the available surface area of the cooling panels for each type of gas (i.e., upper panel area for water and carbon dioxide and lower panel area for

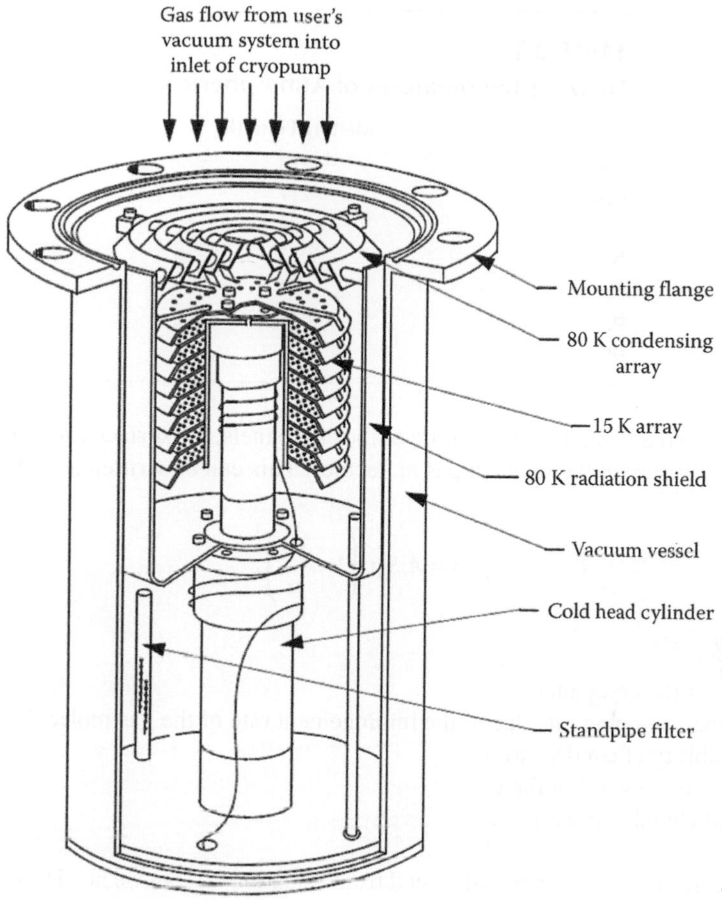

FIGURE 2.10 Cutout of a typical cryo pump. (Reproduced with permission from *Cryo-Torr Pump Installation, Operation and Maintenance Manual*, Brooks Automation Inc., Chelmsford, MA, 2013.)

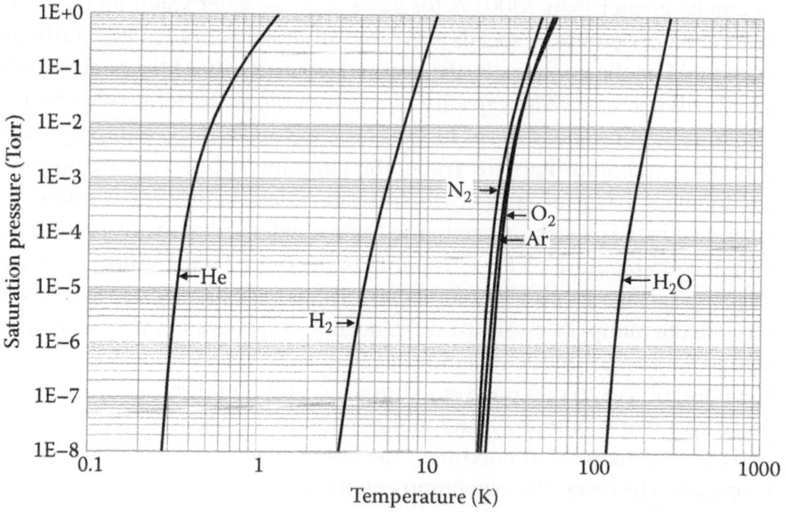

FIGURE 2.11 Saturation vapor pressure of common gases at low temperatures.

TABLE 2.4
Freezing Temperatures of Atmospheric Gases

Gas	Melting Point (K)
H_2O	273
CO_2	217
Ar	84
N_2	63
O_2	54
H_2	14
He	1

nitrogen, oxygen, and argon), the temperature of the cryopanels, the surface impingement rate of the gas, and the temperature of the incident gas molecules. This can be written as follows [1]:

$$S = A_c S_A \alpha \left(1 - \frac{p_u}{p}\right) \tag{2.21}$$

where

A_c is the area of the cryopanel
S_A is the surface pumping rate due to the impingement rate of the gas molecules
α is the probability of condensation
p_u is the ultimate pressure of the gas
p is the actual chamber pressure

The surface impingement rate can be calculated from Equation 2.1, but we need to know the temperature of the incident gas molecules, which can vary for each gas type because they are condensed by different cryopanels in the pump. Water is the first to be condensed, so its incident temperature can be assumed to be 300 K. At this temperature, the value of S_A is 14.6 L/s/cm^2. Nitrogen, on the other hand, is likely to be at a lower temperature by the time it arrives at the second cryopanel. Assuming 80 K for the incident temperature of nitrogen, the value of S_A is 6.1 L/s/cm^2. The overall pumping speed of typical cryo pumps can be greater than 2000 L/s for gases such as water vapor and nitrogen. Therefore, cryo pumps can have higher pumping speeds and lower ultimate pressures than turbo pumps.

As condensates build on the cryopanels, the thermal conductivity of the panels will decrease and the pumping speed will decrease. The capacity of a cryopump is the total volume of a gas that can be trapped before the pumping speed drops by 50%. For example, if the pumping speed is 2000 L/s with a capacity of 1000 standard-liters for argon, the pumping speed will decline to 1000 L/s after it has condensed to about 1.8 kg of argon (which is equivalent to 1000 L of argon at atmospheric pressure). Regeneration is the process of removing these condensed gases from the cryopanels. This is done by raising the temperature of the cryopump up to room temperature and purging it with a heated dry nitrogen gas. Regeneration has to be performed periodically depending on the gas load being introduced into the pump. In most cases, the need for regeneration is indicated by a rise in the base pressure due to the decline in the pumping speed.

The operational use of a cryo pump is very similar to that of a turbo pump. The cryo pump has a foreline (exhaust) port that is pumped by a roughing pump, and the inlet has a large diameter. It is connected directly to the chamber to maximize the conductance and probability of gas molecules entering the cryopanels. However, the cryopump cannot be exposed to atmosphere when the chamber is being vented unless the pump is also being regenerated at the same time. An isolation valve has to be installed at the pump's inlet and closed off during the venting process. A turbo molecular pump, on the other hand, can be exposed to atmosphere as long as its rotation speed has slowed down sufficiently. The turbo pump can also be stopped and started frequently, but a cryopump's compressor and cold head have to be left running continuously. Furthermore, a turbo pump does not have a gas capacity limit and can be run indefinitely with gas flows without requiring periodic regeneration. As a result, for applications requiring high gas flow rates, such as sputtering and reactive processes, a turbo molecular pump may be more appropriate. However, for a comparable size and inlet diameter, a cryopump has a much higher pumping speed and a lower ultimate pressure, so it can reach lower pressures faster than turbo pumps.

2.1.6 Leaks

Just like intentional gas flows, leaks can be considered as unintentional gas loads and represented separately as Q_L. There are two kinds of leaks: real leaks and virtual leaks. Real leaks are due to atmospheric gases flowing into the chamber through poor seals and manufacturing defects. By using better materials and seals, it should be possible to eliminate real leaks almost entirely.

Virtual leaks are caused by internal sources. They are due to the slow release of gases trapped in various components inside the vacuum chamber. For example, gases could be trapped in screw holes, in rubber seals and behind adhesive tapes. The conductance from these locations to the pump's inlet could be very low, and it may therefore take a long time to pump out these gases. For this reason, it is common practice to use vented screws (which have drilled axial holes) for fastening all interior components. The use of adhesive tapes should also be minimized, and all interior parts should have a wide and clear path towards the pump's inlet.

One distinguishing characteristic between real and virtual leaks is that virtual leaks will eventually stop, whereas a real leak will not stop. However, it is not always easy to diagnose and distinguish leaks this way. A helium leak detector is an essential tool for detecting real leaks. The instrument is essentially a mass spectrometer that has been tuned to detect trace amounts of helium. The instrument is installed on a port to detect the gases inside the vacuum chamber, and helium is sprayed from the outside in the areas suspected of having leaks. Alternatively, helium gas can be flowed inside the chamber and the detector can be used as a sniffer from the outside to detect the presence of leaks. Figure 2.12 shows an example of a portable helium leak detector that includes a mass spectrometer with a self-contained turbo molecular pump and a roughing scroll pump.

Regardless of how well a vacuum chamber is manufactured, one always need openings in the chamber to send electrical power, receive measurement signals, send fluids such as water or nitrogen for cooling, and transmit mechanical motion to substrate rotation and shutters. The vacuum pump itself requires an opening. All of these are done through what is generally referred to as ports. Various fixtures can be attached to these ports, such as pumps, vacuum gauges and gas flow controllers as well as electrical, mechanical, fluid, or optical feedthroughs. A view port is also necessary for visual inspection of the chamber's interior. These ports can also become sources of leaks, especially those employing mechanical feedthroughs. Proper sealing is essential to prevent real and virtual leaks from appearing at these sites. The flanges and attachment mechanism used in these ports normally follow industry-standard protocols for sizes, configurations, and seals. The most common configurations are CF (Conflat), QF/KF (Quick Flange or Klein Flange), and ISO flange. The mating flange surfaces are sealed with metal or rubber gaskets. Metal gaskets are used one time only, but they tend to work better than rubber gaskets, especially if they will be exposed to elevated temperatures.

2.1.7 Adsorption and Desorption

Adsorption and desorption of gases are important factors that affect the time it takes to pump down a vacuum chamber. Gas molecules are continuously attaching and detaching themselves from the various surfaces inside the vacuum chamber. Adsorption is the attachment process and desorption is the detachment process.

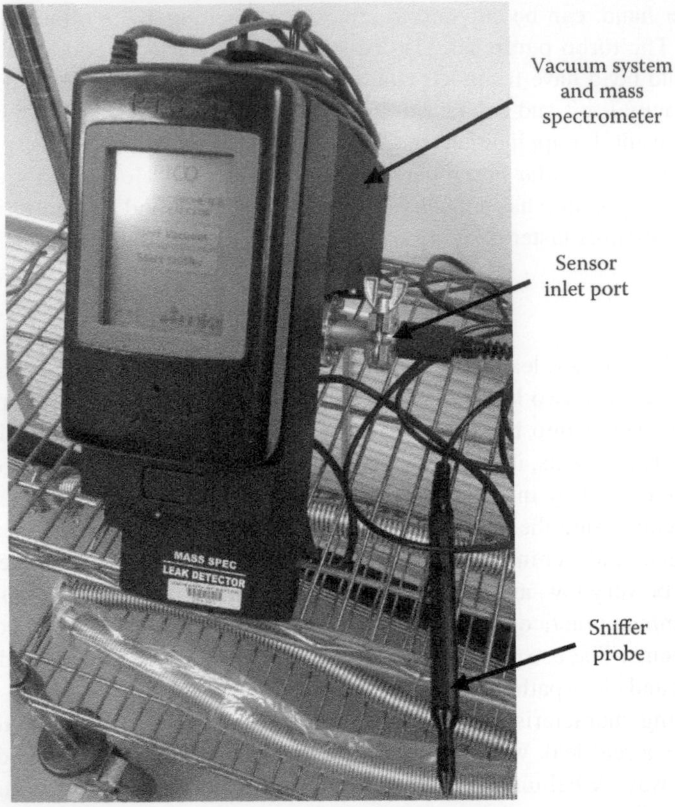

FIGURE 2.12 A portable helium leak detector.

Under atmospheric conditions, gas molecules will adsorb to the chamber surfaces due to physisorption or chemisorption processes. During the pumping process, these adsorbed molecules will desorb from the surfaces and will be released to the pump as a gas load, which can be represented as Q_d. The rate of desorption widely varies depending on the chemical and physical properties of the gas. Water molecules are particularly problematic in vacuum systems because they get strongly attached to the inside walls of the vacuum chamber (usually stainless steel), and desorb at much slower rates compared to other atmospheric gases such as oxygen and nitrogen. Therefore, the water vapor gas loading will be a small but persistent factor that will take a long time to desorb completely [5]. The actual elimination rate of these gases also depends on the conductance between the various desorbing areas and the inlet of the pump. Desorption from a surface that is hidden behind shields, screws, or other vacuum components will be eliminated significantly more slowly than the exposed sidewall surfaces closer to the pump.

In general, the desorption rate can be represented empirically as

$$Q_d = Q_{d0}\left(\frac{t - t_0 + \tau}{\tau}\right)^{-\alpha} \tag{2.22}$$

where

τ is a characteristic time constant
t_0 is the pumping start time
Q_{d0} is the initial desorption rate

For brief air exposures $\alpha = 1$ is a good approximation, with slightly larger values for prolonged exposures [6]. Considering the pumping rate equation (Equation 2.16), assuming $\frac{dQ_d}{dt} \ll \frac{d^2 p_c}{dt^2}$, its solution with the slowly varying gas flow rate will be

$$p = Q_{d0}\left(\frac{t - t_0 + \tau}{\tau}\right)^{-\alpha}\left[\frac{1}{S} + \frac{1}{C_n}\right]. \tag{2.23}$$

Therefore, to reduce pressure in the presence of desorption one has to increase the pumping rate S, or wait long enough for the desorption to subside. In nearly all cases, the pump down curve (pressure vs. time curve) of a typical vacuum system will be largely determined by Q_d more than any other single factor. For this reason, vacuum chambers' exposure to moisture should be minimized as much as possible. Venting should be done with dry nitrogen rather than air. The chamber walls can also be heated during pump down to accelerate the desorption rate, which effectively increases Q_{d0} and decreases τ. Known as the bake out process, this will result in a faster removal of the adsorbed gases and a shorter time to reach the ultimate pressure.

Consider a turbo molecular pump with the same characteristics as shown in Table 2.3 and Figure 2.9, with a 100L chamber, a desorption gas flow model as shown in Equation 2.22, and a fixed exhaust pressure of 5mT. Figure 2.13 shows the effect of different values of starting desorption rate Q_{d0} and α. It can be seen that both factors significantly affect the rate at which the pressure declines as a function of time. Nevertheless, Q_d has a finite duration, so it will eventually decline to zero. Therefore, after a very long pumping time, all the curves should converge toward the line with no gas flow.

Figure 2.14 shows the plot of actual data taken during the pump down of a vacuum chamber after prolonged exposure to ambient moisture in the air. The rough vacuum range and the switchover point to the high-vacuum pump are also shown.

2.1.8 Types of Pumps

Though we only discussed the most common type of pumps found in general cleanroom laboratories, there are also a large number of other pumps that we have not discussed in this chapter.

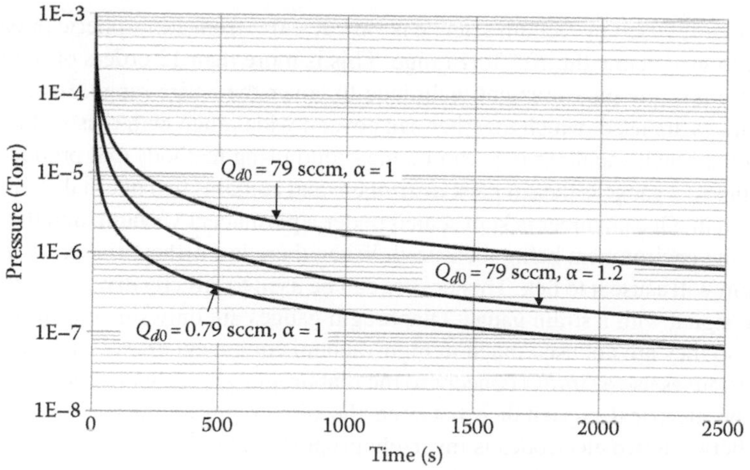

FIGURE 2.13 Plot of chamber pressure vs. time for different desorption rate models. The value of τ was equal to 1 s in all cases.

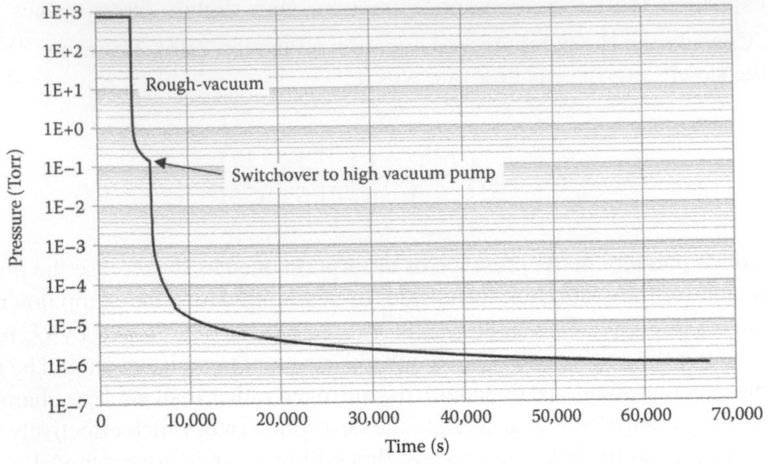

FIGURE 2.14 Actual measurement data of chamber pressure vs. time after prolonged exposure to ambient air.

TABLE 2.5
Major Categories of Vacuum Pumps

Positive Displacement Pumps	Kinetic Pumps	Entrapment Pumps
Oil-sealed rotary vane pump	Turbo molecular pump	Cryo pump
Oil-sealed rotary piston pump	Oil diffusion pump	Getter ion pump
Dry roots pump	Ejector pump	Sputter ion pump

Table 2.5 lists some of these pumps. For further information about these pumps and their principles, the reader is directed to the references at the end of this chapter [3,7–14].

2.2 PRESSURE AND FLOW MEASUREMENTS

2.2.1 Pressure (or Vacuum) Measurement

The range of pressures used in vacuum science is vast. It starts from atmospheric pressure (760 Torr) and goes down to well below the 10^{-9} Torr range. This is more than 12 orders of magnitude, so it is impossible to use a single measurement technique to cover the entire range. Instead, a number of different techniques are used, which together can cover a wide range with reasonable accuracy [15].

Pressures in the rough vacuum range can be measured using methods that produce a mechanical action. The diaphragm gauge is the simplest instrument and contains two internal chambers separated by a thin diaphragm. One chamber is kept sealed under a permanent vacuum, and the other chamber is connected to the environment being measured. When there is a differential pressure, the net force on the diaphragm will cause it to flex. This flexure can be used to directly drive a pointer or converted to an electronic signal with a strain gauge. Diaphragm gauges are only suitable from 760 Torr down to about 1 Torr, so they are not used much in high-vacuum equipment except as a coarse indicator.

Capacitance gauges use a similar concept to the diaphragm gauge, but the deflection is measured indirectly by the electrical capacitance between the diaphragm and a stationary electrode. The capacitance C between two electrodes is inversely proportional to the distance d between them:

$$C = \frac{\varepsilon A}{d} \tag{2.24}$$

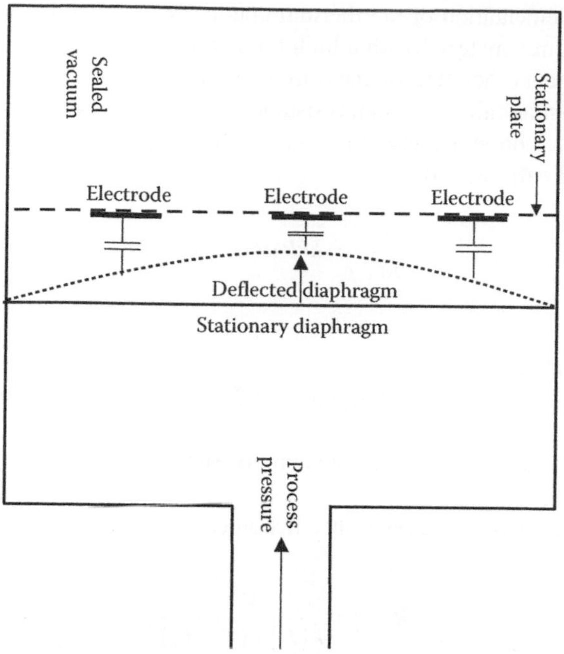

FIGURE 2.15 Capacitance gauge is based on measuring the capacitance between the deflecting diaphragm and stationary electrodes.

where

A is the area

ε is the dielectric constant (which in this case is assumed to be that of vacuum)

By measuring C, the distance d can be inferred. There are many different configurations used in the capacitance gauge, and one of them is shown in Figure 2.15. Several electrodes are placed on a fixed plate on the sealed vacuum side of the diaphragm, and the capacitance between the moving diaphragm and the fixed electrodes is measured to determine the position of the diaphragm and hence the differential pressure across the diaphragm. The sealed vacuum side is maintained at very low pressures (typically a few μT). Gettering elements such as titanium are also used in this chamber to eliminate small amounts of atmospheric gases that may leak into the chamber. These gettering materials are intended to react with gases and produce a solid byproduct, thereby removing these gases. Capacitance gauges are sensitive from atmospheric pressure down to about 1 mT and are ideally matched to the pressure range of roughing pumps. Hence, they are widely used for measuring the foreline pressures of high-vacuum systems.

It is also possible to construct a capacitance gauge to take advantage of the inverse relationship between C and d to boost the sensitivity at the low-pressure range. This would require the diaphragm to be closest to the electrodes when the differential pressure is zero and deflect away from the electrode at higher pressures.

Another popular vacuum gauge is the thermal conductivity gauge. The pressure in this gauge is derived by measuring the thermal conductivity of a gas. Thermal conductivity is directly related to pressure—a higher pressure will produce a greater thermal conductivity because there will be more molecules that collide with surfaces to transfer the thermal energy. The simplest implementation of this gauge is the thermocouple gauge. It uses a thermocouple directly attached to a heated wire filament to measure its temperature. The relationship between the heat produced in the filament (which is simply voltage × current) and the temperature of the filament will be related to the thermal conductivity of the gas, from which the pressure can be deduced.

A more elegant implementation of the thermal conductivity gauge is the Pirani gauge. In this, the filament is made from a material with a high temperature coefficient of resistance (TCR). The temperature of the filament is not directly measured, but indirectly deduced by adjusting the current through that filament to maintain a constant resistance (because the resistance depends on temperature). At steady state, the ohmic heating in the wire will be equal to the thermal energy lost due to conduction. We can write this as follows:

$$\frac{V^2}{R(T)} = k(P)\times(T - T_a) \tag{2.25}$$

where
- T is the filament temperature
- $R(T)$ is the filament resistance (which depends on T)
- V is the voltage
- $k(P)$ is the heat conductance (which depends on pressure)

Rearranging Equation 2.25, the wire resistance becomes

$$R(T) = \frac{V^2}{k(P)\times(T - T_a)}. \tag{2.26}$$

To maintain a constant resistance value, T has to be be kept constant. Therefore, the only variables in this equation are $k(P)$ and V. As pressure increases, $k(P)$ will increase, which will require a higher V to maintain the same resistance. Therefore, by measuring V, we can determine pressure. In practice, this is done by utilizing a Wheatstone bridge circuit with the heated filament as one element of the bridge, and adjusting the voltage to balance the bridge. The balanced condition will be indicated by a zero current through an ammeter across the bridge, indicated by A in Figure 2.16. The gauge

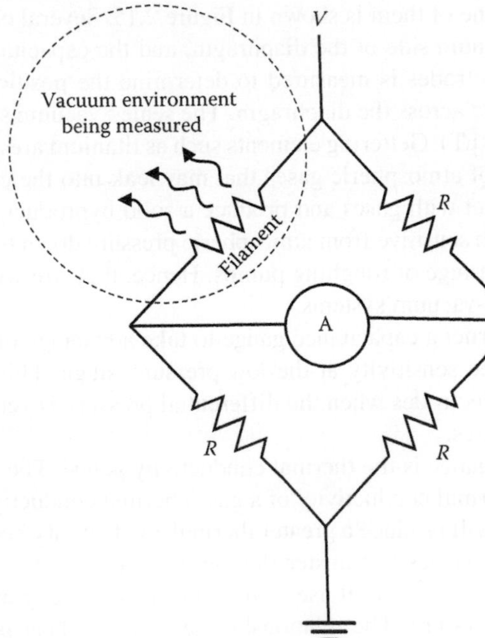

FIGURE 2.16 Balanced Wheatstone bridge used in a Pirani gauge.

also needs to be calibrated for each gas type since the thermal conductivity and heat capacity will be different for each gas. Pirani gauges can measure from atmospheric pressures down to about 0.1 mT and are used just as widely as capacitance gauges.

The advances in micro-electromechanical systems (MEMS) has made it possible to realize some of these vacuum sensors on a more compact and inexpensive platform. The diaphragm, capacitive, and the Pirani gauges have all been implemented using MEMS. The microphones in virtually all portable electronics use the same concept as the diaphragm pressure sensor. Piezoresistive strain gauges are fabricated on top of a suspended membrane to measure the deflection. Capacitive sensing on a MEMS platform allows fine spacing of the electrodes, low masses, and greater deflection sensitivities. The Pirani gauge has also been implemented on MEMS and is commercially available as a vacuum sensor. The MEMS Pirani gauge makes it possible to run the filament (which is actually a thin film resistor on a silicon surface) at a much lower temperature (50°C instead of 150°C). This results in faster response times and longer filament life.

For pressures below 0.1 mT, bulk gas effects like mechanical deflection and thermal conductivity will be far too small to be a useful measurement quantity. The most commonly used method is to ionize the gas with a high electric field and measure the ion current to determine the density of gas molecules.

The hot filament ion gauge shown in Figure 2.17, also known as the Bayard–Alpert gauge, is an improved version of the triode gauge originally used for vacuum measurements. In this gauge, a wire is heated to produce a cloud of free electrons by thermionic emission. These electrons are then accelerated toward the cylindrical wire grid maintained at a positive potential with respect to the heated filament. While some of the electrons will strike the grid wires and be lost, most of them will continue through the grid. When these electrons strike neutral gas molecules with sufficient energy, ions and additional electrons will be produced. The ions that are produced inside the cylindrical grid will be drawn toward the thin wire (ion collector) at the center which is held at a negative potential with respect to the grid, and the electrons will be drawn toward the grid wires. When the ions reach the ion collector, they will be neutralized by electrons and a current will flow between the grid and the ion collector. This current will be related to the density of the gas molecules in the

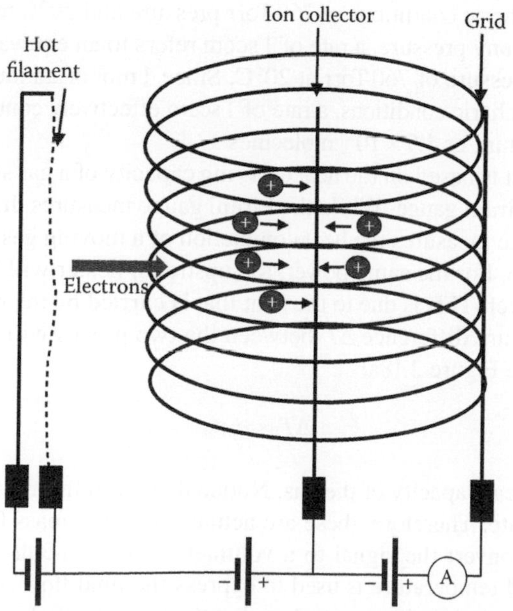

FIGURE 2.17 Hot filament ion gauge (also known as the Bayard–Alpert gauge).

environment because a fixed percentage of molecules will be ionized by this process. Nevertheless, the indication is not very linear and will depend on the gas pressure. At high pressures, the relatively short mean free paths will cause some of the ions to be lost to recombination before they reach the ion collector. The ionization cross section also depends on the gas type, so the gauge needs to be calibrated for each gas being used. Nitrogen is usually the standard reference.

Ion gauges can read from about 1 mT down to 10^{-10} Torr range. One of the disadvantages of the hot filament ion gauge is the hot filament temperature. This can cause the filament to evaporate or chemically react with the gas molecules and eventually burn out, like an incandescent light bulb. The hot filament can also produce outgassing that can reduce the sensitivity of the gauge. Excessive gas pressures can also accelerate the deterioration of the filament. For this reason, the ion gauge is designed to automatically turn off when the pressure exceeds a certain value. A separate gauge such as a Pirani gauge is used to monitor the pressure and activate or deactivate the ion gauge. Nevertheless, the hot filament is the weakest part of this ion gauge. A sudden increase in gas pressure or fast venting can cause the filament to fail.

Some of the problems with the hot filament ion gauges can be addressed with cold cathode ion gauges. Instead of emitting electrons from a hot filament by thermionic emission, electrons in cold cathode gauges are emitted by high electric fields. This is also known as field emission. However, this also results in a smaller number of free electrons being emitted and can lead to a smaller ion current. To increase the ionization rate, strategically placed magnets are introduced to make the electrons travel in a tight spiraling path. This greatly increases the collision rate of electrons within a volume, but the ultimate sensitivity of the gauge is still considered to be somewhat inferior to hot filament gauges. The primary advantage of cold cathode gauges is that they do not have filaments. Sudden changes in pressure are not a serious concern as they are with hot filament gauges. They also have less gas desorption issues due to the absence of a hot filament.

2.2.2 Gas Flow Rate Measurement

In the pumping rate equations discussed earlier in this chapter, the gas flow rates were represented by Q_g (with the dimensions volume × pressure/time). The most prevalent unit used in semiconductor processes is standard cubic centimeters per second (abbreviated as sccm), where the "standard" refers to standard atmospheric conditions of 760 Torr pressure and 20°C temperature. Although gas flows can be measured at any pressure, a rate of 1 sccm refers to an equivalent amount of gas that is contained in 1 cm³ at a pressure of 760 Torr at 20°C. Since 1 mol of an ideal gas occupies 22.4 L of volume at standard atmospheric conditions, a rate of 1 sccm effectively contains 4.46×10^{-5} mol/min, or 2.69×10^{19} molecules/min, or 4.5×10^{17} molecules/s.

Flow rate measurement is based on the heat carrying capacity of a gas stream. In that sense, it has some similarities to the Pirani gauge. While the Pirani gauge measures the heat conduction of a stationary gas, the flow gauge measures the heat conduction of a moving gas. Two resistive heaters are used to heat the flow tube. For the same power, the upstream heater will be at a lower temperature than the downstream heater. This is due to the heat that is carried by the gas from the first heater to the second. The temperature difference ΔT between the two probes will therefore be a function of the mass flow rate $\emptyset_m$ (see Figure 2.18a)

$$\Delta T \propto c_p \emptyset_m \tag{2.27}$$

where c_p is the specific heat capacity of the gas. Notice that ΔT relies on the mass flow rate and not on the volumetric flow rate. Therefore, these are actually thermal mass flow meters. If the density is known, it is easy to convert the signal to a volumetric flow. The density of the gas at normal atmospheric pressure and temperature is used to express the final flow rate in sccm. Most of these measurement units also include a flow control so that the user can dial in a desired flow rate. These are known as thermal mass flow controllers (MFCs).

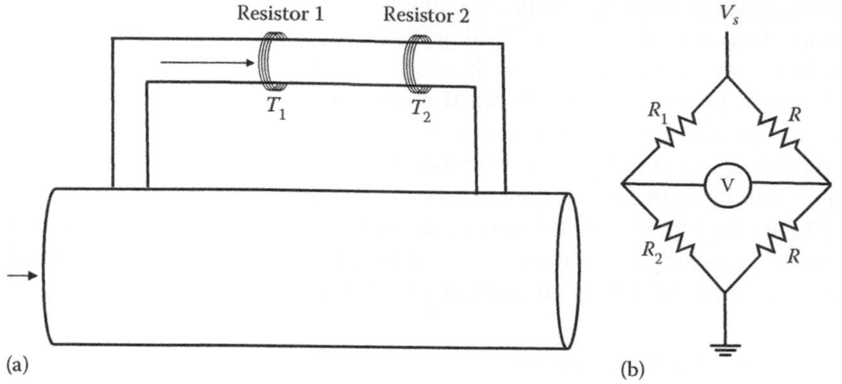

FIGURE 2.18 (a) Thermal mass flow controller with a bypass tube and (b) bridge sensing circuit.

In most thermal MFCs, the gas flow is split and the sensing action is performed in a smaller bypass tube that runs parallel to the main tube. The bypass tube is much smaller in diameter and is designed to have laminar flow characteristics. The smaller diameter also gives it a greater surface to volume ratio for more effective heat transfer between the heating elements and the gas. The heating resistors also perform the function of temperature sensing because they are designed using materials with a high temperature coefficient of resistance (TCR). The sensing is performed on a Wheatstone bridge circuit where one arm contains the two heating resistors R_1 and R_2, and the second arm contains two reference resistors. Any differences in R_1 and R_2 will be reflected as a voltage between the two arms, which can then be calibrated to read mass flow rates.

Since ΔT depends on the specific heat capacity of the gas, this technique will obviously be a function of gas type, just like the Pirani gauge. The correction factor for volumetric flow will be

$$F = \frac{\rho_1 c_{p1}}{\rho_0 c_{p0}} \tag{2.28}$$

where

ρ_0 and c_{p0} are the density and specific heat capacity of the original reference gas, respectively
ρ_1 and c_{p1} are for the gas being corrected for

2.3 FUNDAMENTALS OF PLASMAS FOR DEVICE FABRICATION

In most semiconductor processes, vaccum and plasmas go together because a vacuum system is necessary before creating a plasma. Therefore, having discussed vacuum systems, we will next discuss the basics of laboratory plasmas.

Plasma is an ionized gas and is often described as the fourth state of matter. For example, consider argon gas. Ordinarily each atom will be a neutral argon atom. When the gas is ionized, some of the atoms will lose an electron and become Ar^+. This is a higher energy state of the atom, and this ion will have a tendency to attract an electron and become neutralized. When it recombines with an electron, it releases energy, some of which will be in the form of visible photons. This is the characteristic glow that we see in plasmas. Fluorescent lights are common examples of everyday plasmas. Other examples include gas lasers (argon laser, helium-neon laser, etc.), and plasma displays. To sustain the plasma, power has to be continuously supplied to ionize atoms to make up for the loss of ions due to ion–electron recombination. This power source can be in the form of heat, DC, AC, microwaves, magnetic fields, etc. Most plasmas in the laboratory are ionized only to a very small extent. Typically, about 1 in 10^4–10^6 atoms will be ionized. In certain high-density plasmas, this ratio can approach 1 in 10^2.

Plasmas have many desirable properties in semiconductor device fabrication. In an argon plasma, the Ar^+ ions can be accelerated and used as tiny atomic-sized bullets to impinge a surface and dislodge atoms from another material. This is the sputtering process. The details of sputtering are discussed in Chapter 3. Instead of argon, if a reactive gas such as oxygen is used, in addition to oxygen ions, the plasma will also contain other neutral species such as O and O_3. These are highly unstable and highly reactive species that can accelerate surface chemical reactions. Oxygen plasmas are typically used for cleaning hydrocarbon contamination from semiconductor surfaces. Other gases such as SF_6 and CF_4 are used for etching silicon by creating fluorine (F) free radicals. The ions also play an important role in etching by bombarding the surface, like in sputtering, to create highly vertical etch profiles. The details of etching are discussed in Chapter 7.

2.3.1 Parallel Plate Configuration

The parallel plate is the simplest electrical plasma configuration. This is shown in Figure 2.19. The gap between two metal plates is evacuated and then filled with the desired plasma gas. Usually, the pump and the gas flow are kept continuously running to maintain a constant pressure and a constant density of plasma species, and also to prevent contaminants and byproducts from accumulating in the chamber. One plate is energized and the other plate is grounded. At some voltage, the gas will break down and will start conducting. The minimum voltage needed to break down the gas is a function of the product of the pressure (p) and the distance between the plates (d). At large values of pd, such as near atmospheric pressures, the breakdown voltage decreases as pd gets smaller. pd is proportional to the number of gas atoms that exists between the two plates and number of collisions the ions experience as they travel between the plates. As pd decreases, the number of collisions will decrease and less energy will be lost to the gas. Therefore, the voltage required to break down the gas will also be lower. As pd declines further, the voltage reaches a minimum value and then starts increasing again as shown in Fig 2.20. This occurs because below this value of pd there will be too few atoms between the plates to produce the required number of ionizing collisions to break down the gas. The breakdown voltage also depends on the gas type because the energy required to ionize the atoms or molecules also depends on the gas species.

The model that describes the gas breakdown is Paschen's law. It can be easily derived as follows: each electron emitted from the cathode is accelerated by the electric field between the plates, and upon each collision with a neutral gas atom creates a new electron–ion pair. This can be modeled as an avalanche multiplication process. If the avalanche multiplication coefficient is α, the number of

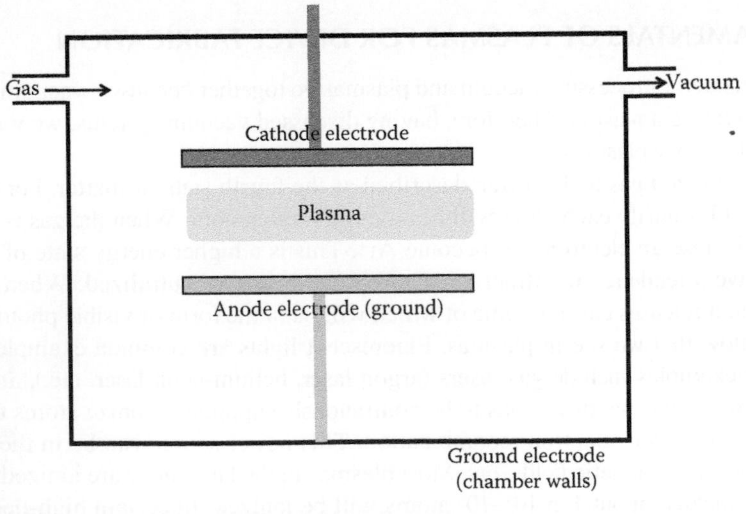

FIGURE 2.19 Configuration of a parallel plate plasma chamber.

additional electrons generated by the avalanche multiplication due to one electron traversing the distance d from the cathode to the anode plate will be $\left(e^{\alpha d}-1\right)$, where the −1 is to account for the original electron. Since electrons and ions are created in pairs, the number of ions created by one electron will also be $\left(e^{\alpha d}-1\right)$. These positive ions will accelerate in the opposite direction toward the cathode. Assuming we do not lose any ions to recombination, they will all reach the cathode electrode. These ions will collide with the cathode electrode and release additional electrons through a process known as secondary electron emission. This is quantified by the secondary electron emission coefficient γ that represents the probability of an electron being emitted by one incident ion. The value of this coefficient depends on the ion energy, electrode material and its surface condition.

In a self-sustaining plasma, for each electron that originally left the cathode, at least one electron has to be subsequently released by the ions that collide with the cathode. This can be mathematically stated as follows:

$$\gamma(e^{\alpha d}-1)=1. \tag{2.29}$$

This is known as the Townsend breakdown model. The multiplication coefficient α is a function of the electric field and gas pressure of the form

$$\alpha=\frac{pA}{e^{B\left(\frac{p}{E}\right)}} \tag{2.30}$$

where

A and B coefficients are specific to the gas type
E is the electric field

Since

$$E=\frac{V}{d}$$

where

V is the voltage
d is the distance between the electrodes, we can get

$$V=\frac{Bpd}{\ln\left(\frac{Apd}{\ln\left(1+\frac{1}{\gamma}\right)}\right)}. \tag{2.31}$$

Table 2.6 lists the A and B parameters for different gases, and Figure 2.20 shows the plot of Paschen's curve as a function of pd assuming a constant secondary ionization coefficient of $\gamma=0.05$ for all gases (though in practice the exact value of this coefficient depends on ion energy, cathode material and surface conditions). We can see that each curve exhibits a minimum breakdown voltage in the range of 0.1–10 Torr · cm, with the lighter gases requiring a larger value of pd compared to the heavier gases.

Once breakdown occurs, current flow proceeds as follows: positive ions will be attracted to the cathode and electrons will be repelled toward the anode, where the anode is usually connected to ground along with the chamber walls. The positive ions have to bombard the cathode with sufficient energy to cause electrons to be released. The released electrons will then be accelerated away from the cathode. This acceleration has to be sufficient for the electrons to collide with neutral gas atoms

TABLE 2.6
Gas Breakdown Parameters

Gas	A (Torr · cm)$^{-1}$	B (V/(Torr · cm))
N_2	10	310
Ar	12	180
He	3	34
H_2	5	130
CO_2	20	466
H_2O	13	290

Source: Fridman, A., *Plasma Chemistry*, Cambridge University Press, Cambridge, U.K., 2008.

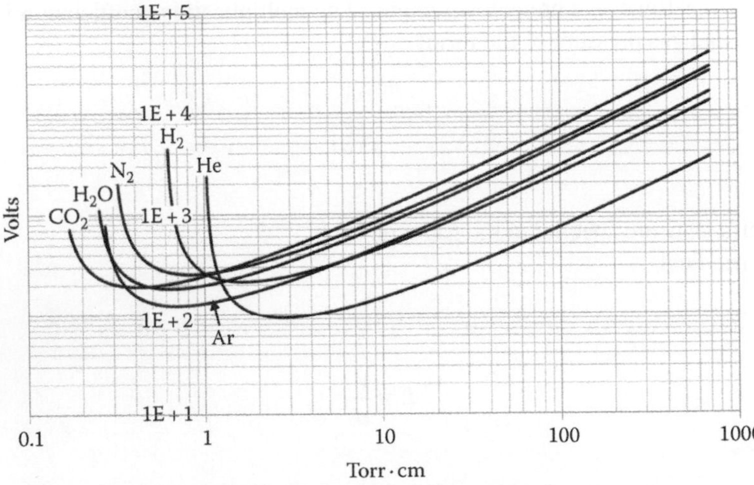

FIGURE 2.20 Paschen's curve for different atmospheric gases assuming a value of $\gamma = 0.05$ for all gases.

and cause ionization, releasing further ions and electrons in order to maintain a steady-state condition in the plasma. This is the onset of plasma discharge and is known as the Townsend discharge.

As current is increased, the plasma goes through other regions of operations. The important observation is that the *I–V* curve of a plasma is highly nonlinear—it contains positive and negative resistance regions. The dark Townsend discharge region is the initial breakdown model described by Equation 2.29 (Figure 2.20). As current is increased further, the voltage declines and enters the normal glow discharge region, which is visually the most identifiable region of operation. The glow is due to the ions recombining with electrons and emitting photons. The photon energies will be characteristic of the gas species present in the plasma. Argon emits a deep purple color, neon emits the well-known bright orange color, and oxygen emits a bluish-white color (see Figure 2.21). As current is increased further, the voltage remains relatively constant; the increase in current comes primarily from an enlargement of the cathode emission area. Once the plasma spreads across the entire cathode surface and there is no more surface left, further increase in current is accompanied by a sharp increase in voltage. This is the abnormal glow region and it is where most of the sputtering processes take place. Further increase in current can cause excessive heating of the cathode surface and can lead to arc discharges. This occurs due to thermionic emission of electrons (electrons emitted due to heating) rather than secondary emission (electrons emitted due to ion bombardment) A typical I-V characteristic of a plasma discharge is shown in Fig 2.21.

FIGURE 2.21 Emission from parallel plate oxygen plasma discharge as seen through a viewport.

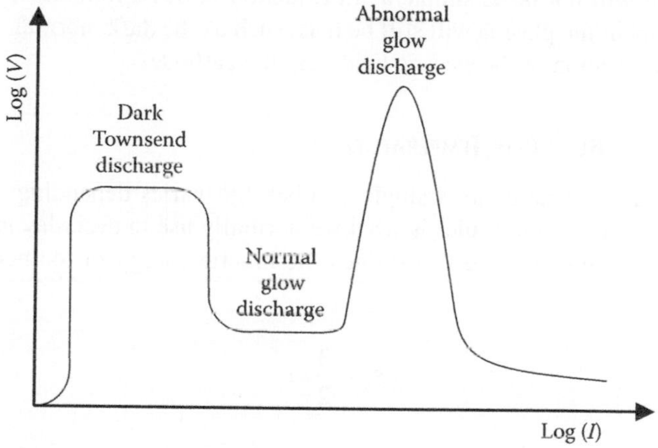

FIGURE 2.22 *I–V* characteristic of a plasma discharge.

It should be apparent from Figure 2.22 that supplying power to a plasma is more complicated than simply connecting a DC voltage source. The power supply has to traverse these nonlinear regions to reach the desired point of operation. Hence, plasma power supplies are designed differently than ordinary power supplies. Most are designed to supply a constant power rather than a constant voltage or constant current.

Additionally, except prior to the onset of the Townsend discharge, the electric field will not be uniform between the cathode and the anode. As current is increased, the electric field will become nonlinear as shown in Figure 2.23. Most of the voltage drop will appear near the cathode. This large voltage difference near the cathode, known as the sheath, develops partly to support the ion bombardment necessary to release the secondary electrons from the cathode. The center of the plasma will be at a relatively constant (but slightly positive) potential with nearly zero electric field.

If the cathode and anode parallel plates are not of equal sizes, or if a single plate is used as the cathode with the chamber walls as the grounded anode, the Townsend breakdown concept can still be used, except the field will not be uniform even at the onset of the Townsend discharge. The static

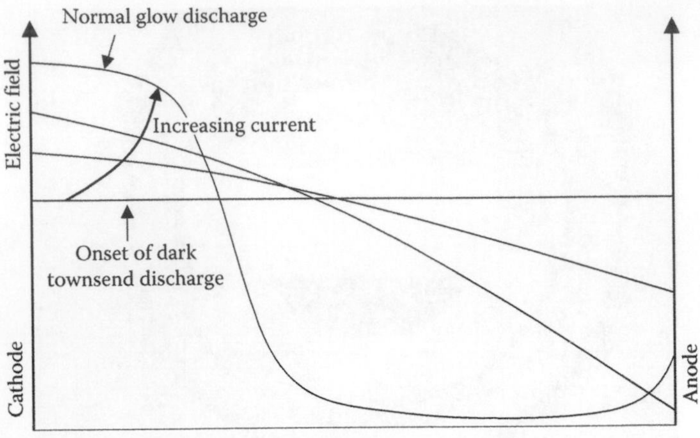

FIGURE 2.23 Electric field distribution in the plasma.

field distributions due to the nonplanar surfaces have to be numerically solved, but the resulting breakdown condition will not be as simple as in Equation 2.31. Nevertheless, the various conclusions that we noted about the plasma will still be true, such as the dark, normal, and abnormal glow regions and the concentration of the electric field near the cathode.

2.3.2 Electron and Bulk Gas Temperature

The temperature of the plasma is not a single number, but varies depending on the species. The temperature of the neural gas molecules is what we normally use in everyday language to describe how "hot" a gas is. The temperature determines the kinetic energy of the neutral gas molecules described by the following:

$$\varepsilon_g = \frac{3}{2}kT \tag{2.32}$$

where
ε_g is the energy of the neutral gas molecules
k is the Boltzmann constant
T is the absolute temperature

Under normal conditions, the electrons will be in equilibrium with the gas molecules and they both will have the same temperature. In a plasma, however, the electrons can be at a significantly higher temperature than the gas molecules because they are always under the influence of accelerating electric fields. However, this will not be a continuous acceleration. Electrons will accelerate over a distance equal to the mean free path λ before colliding with neutral molecules and giving up a small fraction of their energy $\Delta\varepsilon_e$ to the molecules, and then continue their acceleration. Assuming elastic (energy-conserving) non-ionizing collisions, this can be written as

$$\varepsilon_e' = \varepsilon_e - \Delta\varepsilon_e + qE\lambda \tag{2.33}$$

where
ε_e is the energy of the electron before the collision
$\Delta\varepsilon_e$ is the energy given up by the electron during the collision

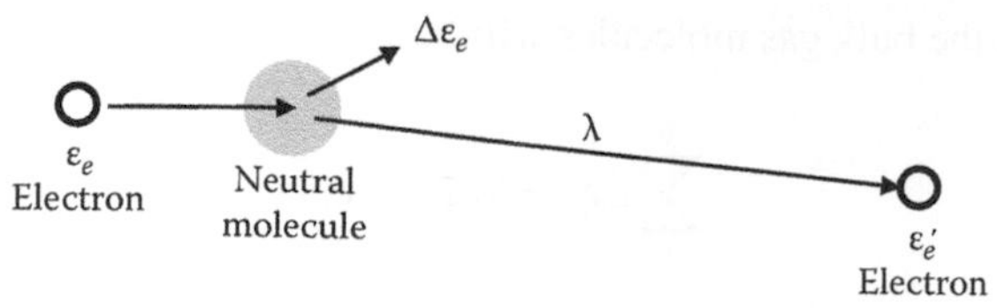

FIGURE 2.24 Collision of electrons with neutral molecules and their energy exchange.

E is the electric field
ε_e' is the energy of the electron after the collision and subsequent acceleration over a distance λ as illustrated in Figure 2.24

The energy lost by electrons $\Delta\varepsilon_e$ will be

$$\Delta\varepsilon_e = \frac{2mM}{(m+M)^2}(\varepsilon_e - \varepsilon_g) \tag{2.34}$$

where
m is the electron mass
M is the mass of the neutral molecule

The mass of an electron is significantly smaller than the molecular weight of all gases. For example, the atomic weight of argon is 73,000 times larger than the mass of an electron. Since $M \gg m$, Equation 2.34 simplifies to

$$\Delta\varepsilon_e = \frac{2m}{M}(\varepsilon_e - \varepsilon_g). \tag{2.35}$$

If we further assume $\varepsilon_e \gg \varepsilon_g$, this becomes

$$\Delta\varepsilon_e = \frac{2m}{M}\varepsilon_e. \tag{2.36}$$

Since $\frac{m}{M} \ll 1$, only a very small fraction of the energy will be lost by the electron during each collision. On the other hand, as we can verify from Equation 2.34, collisions between neutral gas molecules will incur large (50%) energy transfers because their masses will be equal. As a result, while the rate of energy transfer from electrons to neutral gas molecules is very slow and very inefficient, the energy transfer between neutral gas molecules is very fast and very efficient.

Furthermore, if the distance between the anode and cathode is L, each electron will undergo approximately $N = L/\lambda$ collisions. The total energy gained by the electron will be

$$\varepsilon_e = \sum_{n=0}^{n=N} qE\lambda\left(1 - \frac{m}{M}\right)^n \tag{2.37}$$

which can be expanded to

$$\varepsilon_e = qE\lambda\frac{1 - \left(1 - \frac{m}{M}\right)^N}{\frac{m}{M}}. \tag{2.38}$$

The total energy lost to the bulk gas molecules will be

$$\sum_{1}^{N} \Delta\varepsilon_e = NqE\lambda - \varepsilon_e. \tag{2.39}$$

For argon gas, at a pressure of 1 Torr, the mean free path is 50 μm. Assuming 100 V cathode potential and also assuming the field to be uniform (which is only approximately true since the field will be greater near the cathode), we can get $N = 2000$ for a cathode-to-anode distance of 10 cm. From this, we can estimate the electron energy as 98.6 eV. This means, the electrons will acquire an energy nearly equal to the applied potential energy of 100 eV. The equivalent temperature of the electrons can be evaluated from Equation 2.32 as

$$T_e = \frac{2}{3k}\varepsilon_e \tag{2.40}$$

which results in 758,000 K. This may seem like an unreasonably high temperature, but it is just the effective temperature of the electrons due to their kinetic energy, not the temperature of the gas itself. Nevertheless, this is the temperature that would be required if we were to create the same plasma by thermal excitation. With electrical excitation, the neutral gas species can remain at very low temperatures but the electron temperature can be raised by many orders of magnitude.

For the aforementioned example, the total energy lost to the gas molecules can be evaluated from Equation 2.39 as 1.4 eV. This energy will quickly thermalize because of the high rate of energy transfer between the gas molecules of equal weight. However, in a typical electrical plasma only a small fraction of the molecules will be ionized, of the order of 10^{-4} to 10^{-6}. As a result, each 1.4 eV will thermalize among approximately 10^5 neutral molecules, resulting in an average gain of only 1.4×10^{-5} eV for each gas molecule. If the total power discharged by the plasma is 100 W (for 1 A current), only 1.4×10^{-5} W will go toward heating up the gas molecules, and most of that heat will be transferred to the vacuum chamber walls. As a result, these plasmas are referred to as "cold" plasmas, where the electron temperature can be of the order of 10^6 K while the gas molecules remain at nearly room temperature.

Instead of 1 Torr, if we consider a lower pressure of 10 mT, the mean free path becomes 0.5 cm. For the same assumed geometry, the number of collisions between the anode and cathode becomes $N = 20$. The electron energy from Equation 2.38 then becomes almost equal to the applied potential energy of 100 eV. The energy lost to the gas bulk becomes 13 meV, and the thermalized energy per gas molecule becomes 1.3×10^{-7} eV, which is even more insignificant than before.

As already stated, since the electron temperature in a plasma is significantly larger than the neutral molecules and the ions, the electrons will diffuse further away from the plasma and occupy a larger volume than the ions. This imbalance in charge distribution will result in a slightly positive potential at the center of the plasma. This is referred to as the plasma potential V_p, and its magnitude indicates the difference in temperature between the ions and electrons. Since the anode is grounded, there will be a small potential drop V_p between the plasma and the anode, and an even greater potential drop will exist at the cathode terminal, as shown in Figure 2.25.

This type of plasma is also known as a diode plasma because it bears some resemblance to the original vacuum tube diodes. Reversing the polarity of the supply voltage will not make the anode and cathode reverse their roles unless they are also made of identical materials with identical geometries. As a result, the current flow can be highly asymmetrical due to differences in secondary electron emission coefficients, differences in temperatures (cathode will generally be hotter than the anode due to the ion bombardment that can lead to a higher secondary electron emission coefficient), or differences in electrode geometries. A triode plasma is also possible by introducing a third element designed to emit electrons into the plasma. A small number of electrons emitted by the third element

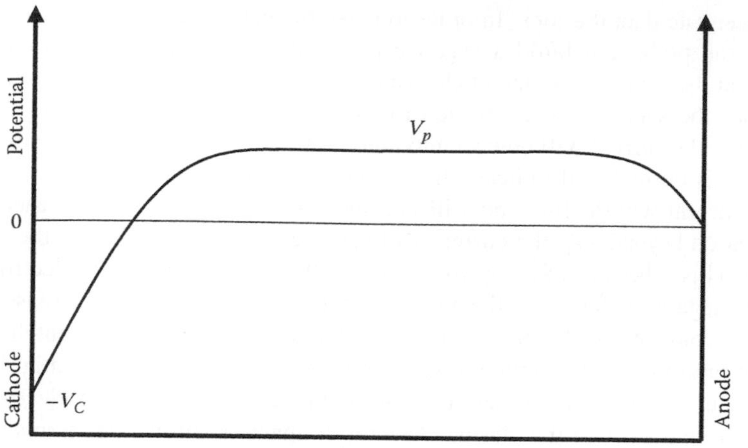

FIGURE 2.25 Potential distribution in a plasma with a DC excitation.

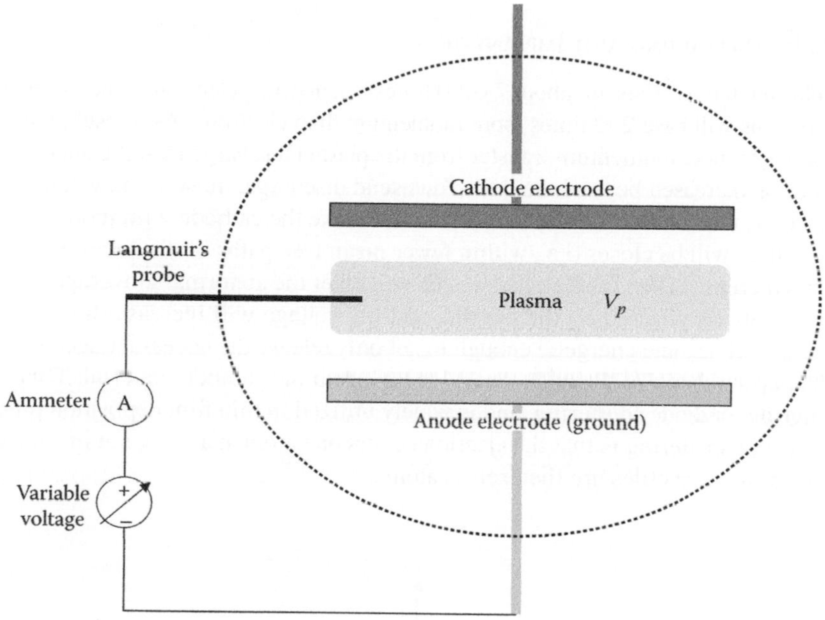

FIGURE 2.26 Langmuir's probe is a simple probe stuck into the plasma with a capability to adjust voltage and measure current.

can significantly increase the plasma density, so it acts as a control signal. This has some similarities to the triode vacuum tube, which was the predecessor to the semiconductor transistor.

2.3.3 Langmuir's Probe

The *I–V* characteristics of the excitation source, shown in Figure 2.21, reveals only limited information about the plasma. Further details such as electron and ion energies can be evaluated by inserting a probe into the plasma, known as Langmuir's probe, and measuring its *I–V* characteristics separately from the excitation source [19]. This is illustrated in Figure 2.26.

When nothing is connected to the probe (a floating probe), the probe will acquire a negative potential. This is because the electrons, due to their higher temperature, will be incident on the probe at a greater rate than the ions. In order to make the net current become zero, electrons will accumulate on the probe and build a negative potential $-V_f$ such that enough electrons will be repelled such that the net incident rate of electrons and the incident rate of ions are equal.

Now consider the same probe connected to an adjustable voltage source. If a large negative voltage is applied, the current will reach a maximum saturation value due to the positive ions striking the probe. This current will indicate the ion density at the probe's location. As the voltage is increased, the current will decline and will become zero when the potential reaches $-V_f$. As the voltage is increased beyond $-V_f$, the current will become positive even though the applied voltage is still negative. This is because, for any voltage larger than $-V_f$, the energetic electrons will be able to overcome the negative potential of the probe and enter the probe. Further increase in voltage will attract more electrons toward the probe and result in a rapid increase in current. The current will reach a positive saturation value as the probe potential reaches the plasma potential V_p. Both V_p and V_f indicate the electron and ion temperatures in the plasma. The saturation electron current will indicate the electron density in the plasma. As a result, many useful information about the plasma can be gathered from this simple technique. An example of a Langmuir's probe *I–V* characteristic is shown in Figure 2.27.

2.3.4 DC Ion Sputtering and Implantation

In an argon plasma, the Ar^+ ions are about 73,000 times heavier than electrons. Therefore, for the same energy, the Ar^+ ions will have 250 times more momentum than electrons. As a result, the cathode surface will experience more momentum transfer from the plasma discharge than the anode surface.

When power is increased beyond the dark Townsend discharge, most of the voltage drop will be near the cathode surface. This will make the Ar^+ ions strike the cathode with more energy than the anode because they will be closer (i.e., within fewer mean free paths of distance) from the cathode. When power is increased even further, the plasma will enter the abnormal glow region. The cathode surface will be entirely covered by the plasma, and the voltage will increase dramatically. In this regime, the ions will become energetic enough to not only release the necessary secondary electrons to sustain the plasma, but also dislodge neutral atoms from the cathode material. This is known as cathode sputtering, or diode sputtering, and is widely utilized in thin film deposition processes. The important aspect of sputtering is that the ejection occurs one atom at a time, not in chunks or pieces because the ejecting projectiles are themselves atoms.

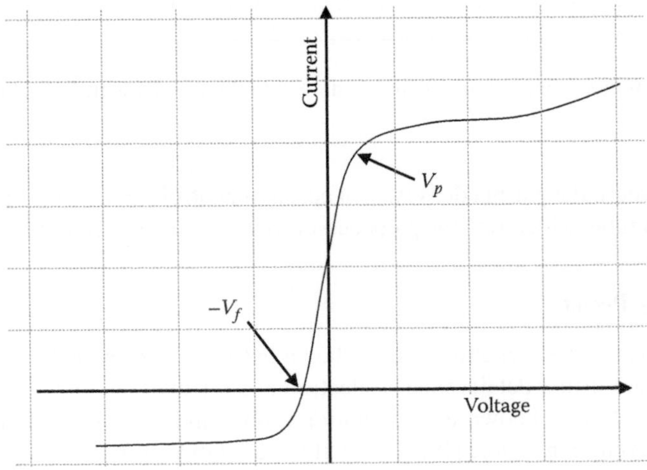

FIGURE 2.27 *I–V* characteristic of Langmuir's probe in a plasma.

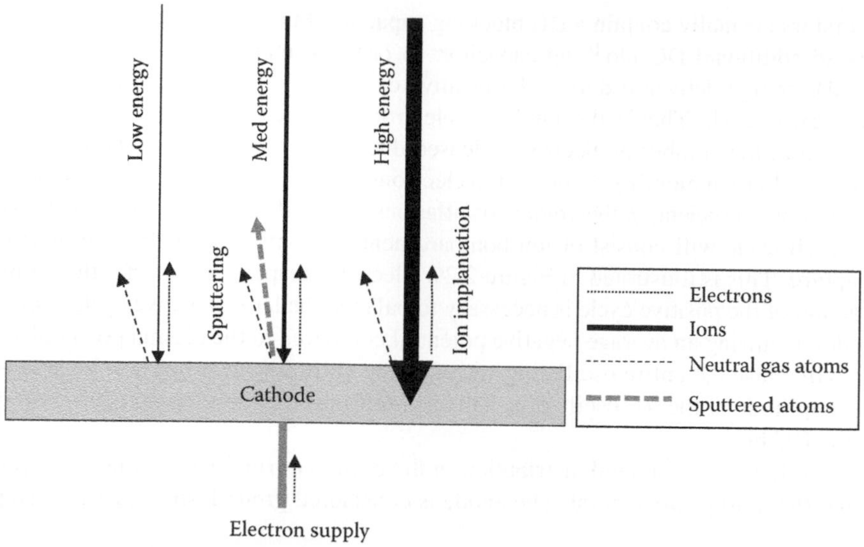

FIGURE 2.28 Three distinct regions of operation of a cathode discharge plasma.

If the discharge power is increased beyond the sputtering range, the ions can acquire enough energy to bypass sputtering and become buried in the cathode material. This is known as implantation. This is also a useful process and is used in semiconductor doping. Figure 2.28 illustrates all three of these modes—glow discharge, sputtering, and ion implantation.

In a typical sputter configuration, the material of interest (target) is attached or bonded to the cathode and the substrate is placed at some distance from the cathode. The ejected atoms from the cathode will land everywhere in the vicinity, including the substrate. The sputtered atoms are ejected within a small angular distribution to the normal of the cathode. Many models have been developed to predict the angular distribution of the sputtered atoms as a function of distance, diameter, pressure, and power. This will be discussed in more detail in Chapter 3.

In the sputtering regime, most of the electrical power will be dissipated as heat on the cathode plate due to ion bombardment, and very little due to ionization impacts and light emission. Therefore, the cathode plate has to be cooled by some fluid such as water to prevent overheating.

With a DC power supply, the target material has to be conductive. If the target is a dielectric, it will not conduct the current required to maintain the plasma. Therefore, DC sputtering is only suitable for sputtering conductive targets such as metals and certain semiconductors.

2.3.5 RF Plasma

If the cathode target is nonmetallic, as may be required for sputtering insulating materials such as SiO_2, a DC excitation will obviously not work because it will not be able to sustain a continuous current flow. In such cases, RF excitation can be used. The dielectric target that is sandwiched between the conductive cathode plate (typically copper) and the conductive plasma will behave as a capacitor and allow RF energy to pass through. To prevent electromagnetic interference, specific RF frequencies have been assigned for laboratory plasma excitations. One of the most commonly used plasma excitation frequencies is 13.56 MHz.

In a parallel plate plasma configuration, the electric field will switch directions at the RF frequency. However, ions being heavy particles, they will not accelerate and decelerate quickly enough to reposition from one terminal to the other at MHz frequencies. Instead, a steady-state condition will be reached where a fixed number of ions will be present at both terminals. Electrons on the other hand will easily move between the terminals during each cycle of the RF oscillation.

RF generators normally contain a DC blocking capacitor. Furthermore, an insulating target will also act as an additional DC blocking capacitor. In order to maintain charge neutrality at these capacitors, the charge delivered during the positive cycle has to be equal to the charge delivered during the negative cycle. That is, the number of electrons captured by the cathode during each cycle has to be equal to the number of electrons released due to ion bombardment. However, due to the significantly smaller ion mobility compared to electrons, as well as the low values of the secondary electron emission coefficient, γ, the voltage oscillations will shift downward such that a significant fraction of each cycle will consist of ion bombardment and only a small fraction will consist of electron capture. This is illustrated in Figure 2.29. Electrons respond a lot faster than ions, so only a small fraction of the positive cycle is necessary to balance the large negative cycle. This results in the electrodes acquiring an average negative potential compared to the plasma potential V_p. In most cases, this will cause the entire oscillating waveform to shift downward by a voltage nearly equal to half the amplitude of the RF oscillation. This is a self-bias that develops at each terminal and is known as the DC bias.

Figure 2.30 shows the potential distribution in the plasma during the most negative point in the RF cycle and the most positive point. The anode is considered ground, so it stays at zero potential.

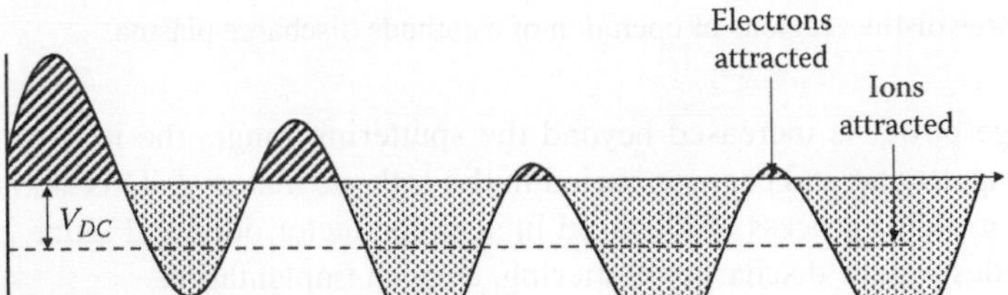

FIGURE 2.29 The evolution of DC bias on an RF plasma terminal.

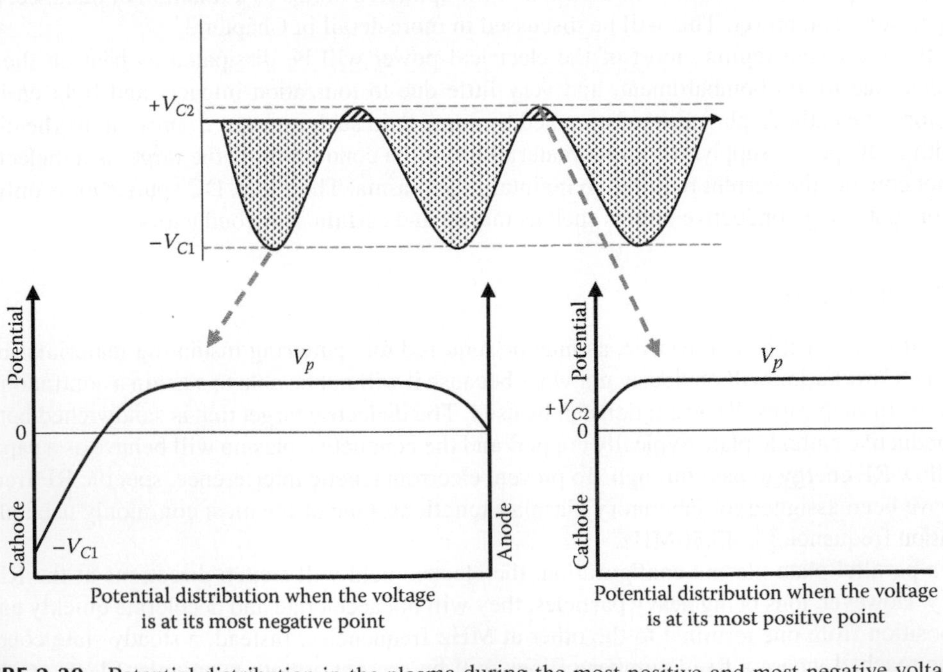

FIGURE 2.30 Potential distribution in the plasma during the most positive and most negative voltages of the RF cycle

During the most negative value $-V_{C1}$, the potential drop at the cathode sheath becomes large. When the voltage swings to $+V_{C2}$, the cathode sheath experiences a positive potential. It remains positive just long enough to collect enough electrons to neutralize the positive charges collected during the previous cycle. The time-averaged cathode voltage

$$V_{DC} = \frac{-V_{C1} + V_{C2}}{2} \tag{2.41}$$

will be negative valued because the magnitude of V_{C1} will be significantly larger than V_{C2}.

The DC bias is most commonly measured with a voltmeter attached to the cathode in series with an RF blocking inductor (RF choke). This measures the DC bias of the cathode with respect to the ground potential, and not necessarily with respect to the plasma potential V_p. However, the ion bombardment energy will be $q(V_p + V_{DC})$, not qV_{DC}. Therefore, the DC bias voltage is only part of the information necessary to evaluate the ion bombardment energy. A more accurate setup should include a probe to measure the plasma potential separately from the cathode voltage V_{DC}, as shown in Figure 2.31, although this is not always done in simple plasma configurations.

RF excitation also requires an impedance matching network to maximize the power delivered to the plasma. This is required with any RF or microwave transmitters; otherwise power could be reflected back into the RF generator causing inefficient power transfer and a significant thermal load in the generator. Typically, the output impedance of the RF generator and cables is 50 Ω, and the load must be designed to match this impedance. The load in this case is the plasma and

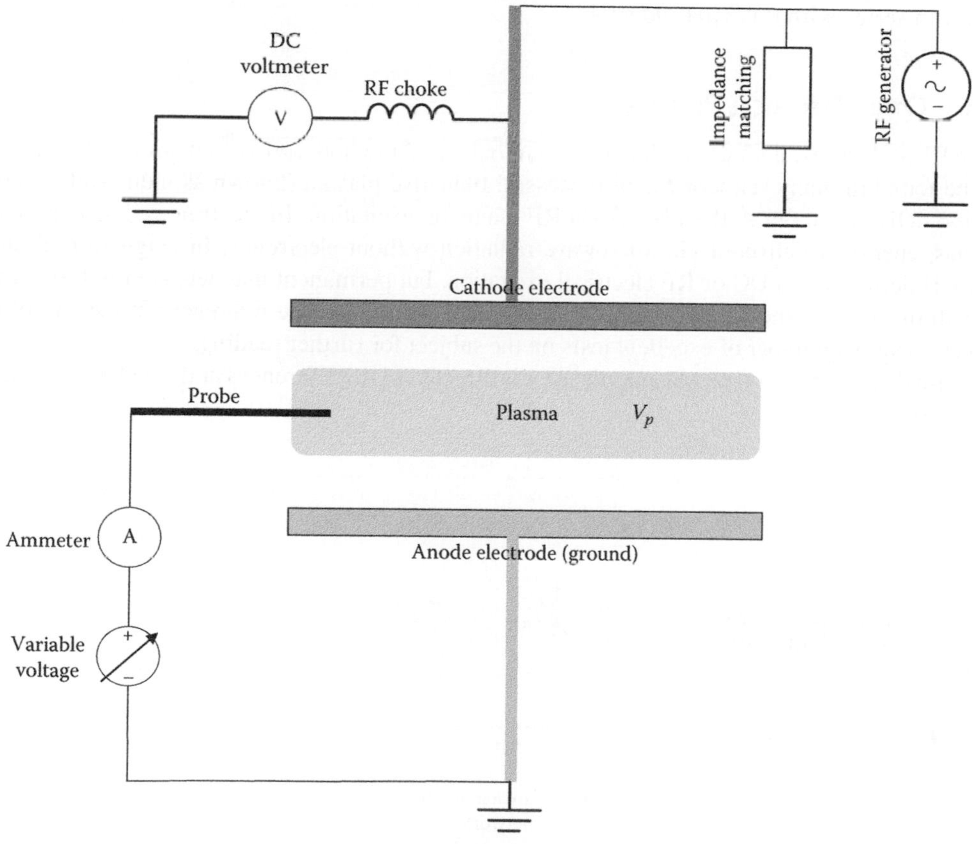

FIGURE 2.31 Measurement of DC bias in a plasma.

electrodes. However, the plasma impedance will not be a constant but will vary as a function of pressure, power, extent of ionization, gas type as well as the target material. Therefore, a tunable RF circuit network is necessary between the cathode and the RF generator to compensate for impedance mismatches in real time. In most plasma applications, the matching network automatically tracks and tunes the network by monitoring the reflected power in the cable. For example, since the plasma load is dominantly capacitive, its impedance will generally be of the form

$$Z_p = R_p - j\frac{1}{\omega C_p}. \tag{2.42}$$

With a parallel matching impedance network Z_n, as shown in Figure 2.32, the total impedance will be

$$Z_t = Z_p \| Z_n. \tag{2.43}$$

The matching impedance has to be inductive to cancel out the reactive part. This makes

$$Z_n = R_n + j\omega L_n. \tag{2.44}$$

For example, consider a plasma impedance of $Z_p = (5 - j25)\Omega = 25.5e^{-j78.7^\circ}$. In order to produce a total impedance of $Z_t = 25\ \Omega$, the required matching impedance has to be $Z_n = (12.8 + j15.2)\Omega$. The positive sign of the imaginary part indicates that the required network impedance is inductive. Assuming an RF frequency of 13.56 MHz, the required matching network becomes a 12.8 Ω resistor in series with 0.178 μH inductor.

2.3.6 Other Electrical Plasmas

While RF and DC excitations are the most commonly used methods, other excitation techniques can also be found in many semiconductor processes. Inductive plasma (known as inductively coupled plasma) delivers energy to the plasma via RF magnetic excitation. In electron cyclotron radiation plasmas, energy is delivered via microwave radiation without electrodes. In magnetron plasmas, energy is delivered via DC or RF electrical excitation, but permanent magnets are used to redirect the path of the ions and to focus the ions closer to the cathode. The references at the end of this chapter contain a number of excellent texts on the subject for further reading.

For further information on plasmas the user is directed to the references at the end of this chapter [16–18,20].

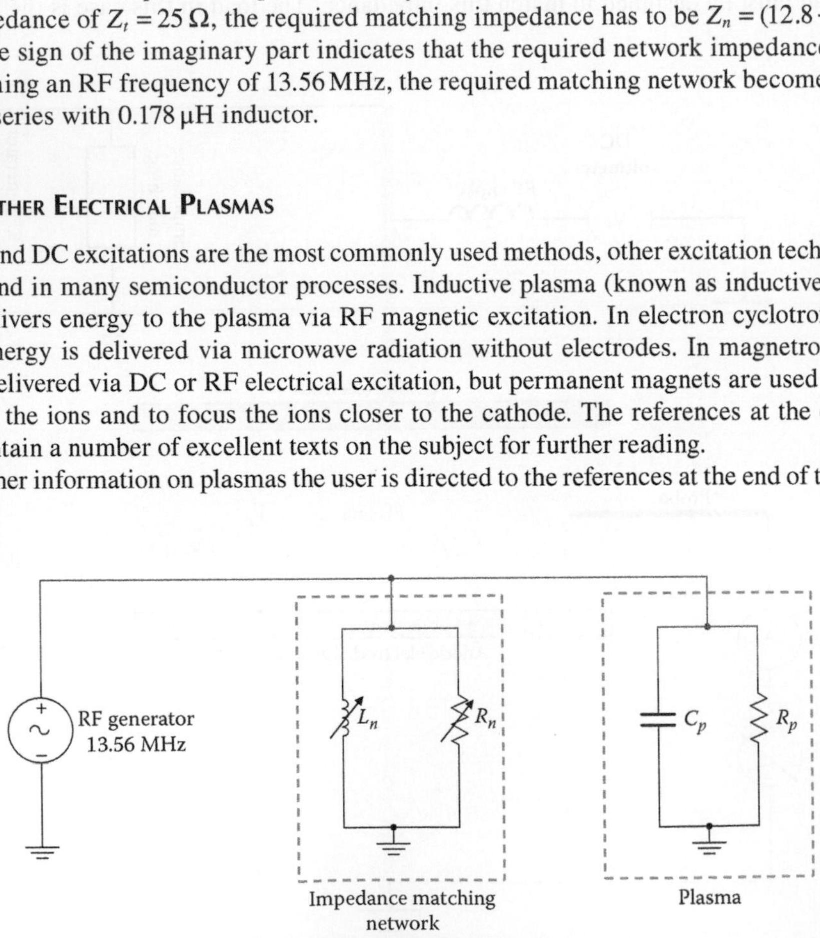

FIGURE 2.32 Impedance matching network for a capacitive plasma.

PROBLEMS

2.1 Consider a 50 L vacuum chamber connected by a 36 in. × ½ in. hose to a roughing pump. The roughing pump has a nominal pumping speed of 1 L/s at 760–10 Torr, and then falls by a factor of 10 for each factor of 10 drop in pressure. Assuming no leaks or desorptions, plot the pump-down curve and estimate the time necessary to reach 150 mT pressure in the chamber.

2.2 Consider a turbo molecular pump that has a pumping speed of 600 L/s for argon gas. A certain sputter deposition plasma process requires the pressure in the chamber to be 5 mT. Find the required argon flow rate in sccm to maintain this pressure.

2.3 A cryo pump has a pumping speed of 1200 L/s. A certain sputter deposition plasma process requires the pressure in the chamber to be 5 mT. Find the required argon flow rate in sccm to maintain this pressure. If the capacity of the pump for argon is 1000 standard liters, find how long the process can be run before the pump reaches its stated capacity.

2.4 Consider a 100 L vacuum chamber with a turbo molecular pump that has a pumping speed (for nitrogen) of 350 L/s and a maximum compression of 10^9. The foreline pressure is held constant at 10 mT. The water desorption follows the model given in Equation 2.22 with $Q_{d0} = 100$ sccm, $\alpha = 0.9$ and $\tau = 1$ s. Plot the pump down curve and estimate the time needed to reach 1 μT pressure.

2.5 A DC parallel plate plasma using oxygen as the feed gas has a cathode bias voltage of −200 V at a pressure of 250 mT. The distance between the plates is 5 cm. Assuming the electric field is uniform, calculate the energy lost by each electron to the neutral gas molecules. If the discharge power is 200 W and the ionization fraction is 10^{-6}, calculate the total power delivered by the electrons to neutral gas molecules.

LABORATORY EXERCISES

2.1 Measure and plot the pump down curve of your high-vacuum chamber. For the published pumping speeds of the roughing and high-vacuum pumps, plot the expected theoretical pump down curve including hoses and the chamber volume. Compare the two plots and estimate the desorption rates that is occurring in the system. Perform two separate tests, one after several hours of exposure to ambient air, and one with only a brief exposure to air. If equipped with a heater in the vacuum chamber, turn the heat on and estimate the increase in the desorption rate.

2.2 Using a portable CCD spectrometer through a vacuum viewport, collect the emission spectra of Ar, N_2, and O_2 plasmas. Compare the emission lines against the NIST Atomic Spectra Database (http://www.nist.gov/pml/data/asd.cfm) to verify the species in the plasma.

REFERENCES

1. Oerlikon Leybold Vacuum, *Fundamentals of Vacuum Technology*, Cologne, June 2007.
2. *Cryo-Torr® Pump Installation, Operation and Maintenance Manual*, Brooks Automation Inc., Chelmsford, MA, 2013.
3. J. F. O'Hanlon, *A User's Guide to Vacuum Technology*, 3rd edn., John Wiley & Sons, Inc., New York, 2003.
4. J. P. Dawson and J. D. Haygood, Cryopumping, *Cryogenics*, 5(2) (1965) 57–67.
5. A. Berman, Water vapor in vacuum systems, *Vacuum*, 47(4) (1996) 327–332.
6. M. Li and H. F. Dylla, Model for the outgassing of water from metal surfaces, *J. Vac. Sci. Technol. A*, 11 (1993) 1702.
7. A. Chambers, R. K. Fitch, and B. S. Halliday, *Basic Vacuum Technology*, 2nd edn., Institute of Physics Publishing, Bristol, U.K., 1998.

8. N. Yoshimura, *Vacuum Technology Practice for Scientific Instruments*, Springer, Berlin, Germany, 2008.
9. M. J. Holmes, Ultra-high vacuum technology: A review with particular reference to vacuum generators' contribution to the art, *Vacuum*, 33(7) (1983) 429–432.
10. A. P. Troup and N. T. M. Dennis, Six years of "dry pumping": A review of experience and issues, *J. Vac. Sci. Technol. A*, 9 (1991) 2048.
11. M. H. Hablanian, Design and performance of oil-free pumps, *Vacuum*, 41(7–9) (1990) 1814–1818.
12. M. H. Hablanian, *High-Vacuum Technology: A Practical Guide*, 2nd edn., Taylor & Francis, New York, 1997.
13. G. F. Weston, *Ultrahigh Vacuum Practice*, Butterworth & Co Ltd., London, U.K., 1985.
14. D. Hoffman, B. Singh, and J. H. Thomas, III, *Handbook of Vacuum Science and Technology*, Academic Press Inc., San Diego, CA, 1998.
15. A. Berman, *Total Pressure Measurements in Vacuum Technology*, Academic Press Inc., Orlando, FL, 1985.
16. A. Fridman, *Plasma Chemistry*, Cambridge University Press, Cambridge, U.K., 2008.
17. N. St J. Braithwaite, Introduction to gas discharges, *Plasma Sources Sci. Technol.*, 9 (2000) 517–527.
18. A. Bogaerts, E. Neyts, R. Gijbels, and J. van der Mullen, Gas discharge plasmas and their applications, *Spectrochim. Acta B*, 57 (2002) 609–658.
19. R. L. Merlino, Understanding Langmuir probe current–voltage characteristics, *Am. J. Phys.*, 75(12) (2007).
20. A. Ganguli and R. D. Tarey, Understanding plasma sources, *Curr. Sci.*, 83(3) (2002).

3 Physical and Chemical Vapor Deposition

Thin films are used for electrical, optical, chemical, and mechanical modification of a substrate surface. This is a vast and important subject area and extends well beyond device fabrication applications. Common everyday examples of thin films include titanium nitride coatings used in cutting tools, polytetrafluoroethylene (PTFE) coatings on nonstick cookware, antimicrobial coatings to prevent growth of bacteria, reflective coating on mirrors, and antireflective coatings on eyeglasses. In device fabrication, we are mostly concerned with the deposition of metals and dielectrics to achieve electrical, chemical, and optical modifications. The design of optical thin films is discussed in Chapter 4, but in this chapter, we will discuss the techniques used for making thin films. The techniques can be broadly classified into physical vapor deposition (PVD) and chemical vapor deposition (CVD). Within each of these categories, there are several different techniques and subcategories. We will not attempt to describe all of these in detail, but only focus on the methods most commonly used in typical research cleanrooms.

3.1 PHYSICAL VAPOR DEPOSITION

In PVD, a solid source material is vaporized atom by atom, or molecule by molecule, and deposited on the substrate at a controlled rate. The energy source used for vaporization is the main distinction between the different PVD techniques. In thermal evaporation, thermal energy is used for vaporizing the source. How this thermal energy is delivered can also be very different between the different deposition method. Heating with a resistive wire is the simplest approach, but it is also possible to deliver a more tightly focused energy using an electron beam. These are known as resistive evaporation and electron beam evaporation, respectively, as shown in Figure 3.1. One could also vaporize the source using energetic particles instead of thermal energy. This is similar to an abrasive blasting process except atomic projectiles are used instead of large particles. This is the sputter deposition process and can be thought of as a "cold" evaporation technique. Another approach is pulsed laser deposition (PLD), where energy is delivered via energetic laser pulses. All of these methods fall under the category of PVD. In CVD, the starting materials are in vapor or liquid form in a different chemical state than the final thin film. These starting materials (known as precursors) are delivered to the substrate surface at the appropriate pressure and temperature so that they may chemically react and form the desired film. In a thermal CVD, elevated temperatures are used to activate the reaction, and in plasma-enhanced CVD, a plasma is used to activate the reaction. Another variant of CVD is atomic layer deposition (ALD), where, instead of delivering the precursors simultaneously, they are delivered in alternating sequence to allow more control over the deposition properties.

3.1.1 Thermal Evaporation

Thermal evaporation is conceptually very simple—the temperature of a material is raised until it evaporates at a sufficiently high rate to produce deposition on adjacent surfaces. The temperature could be just a few hundred degrees in the case of volatile metals like indium or zinc, or several thousand degrees in the case of metals like molybdenum or platinum. The impingement rate of the background gases on the substrate has to be smaller than the impingement rate of the evaporating species to prevent the background gases from reacting and producing oxides and nitrides. Obviously, this means that evaporation has to be done in a high vacuum environment. The mean free path also

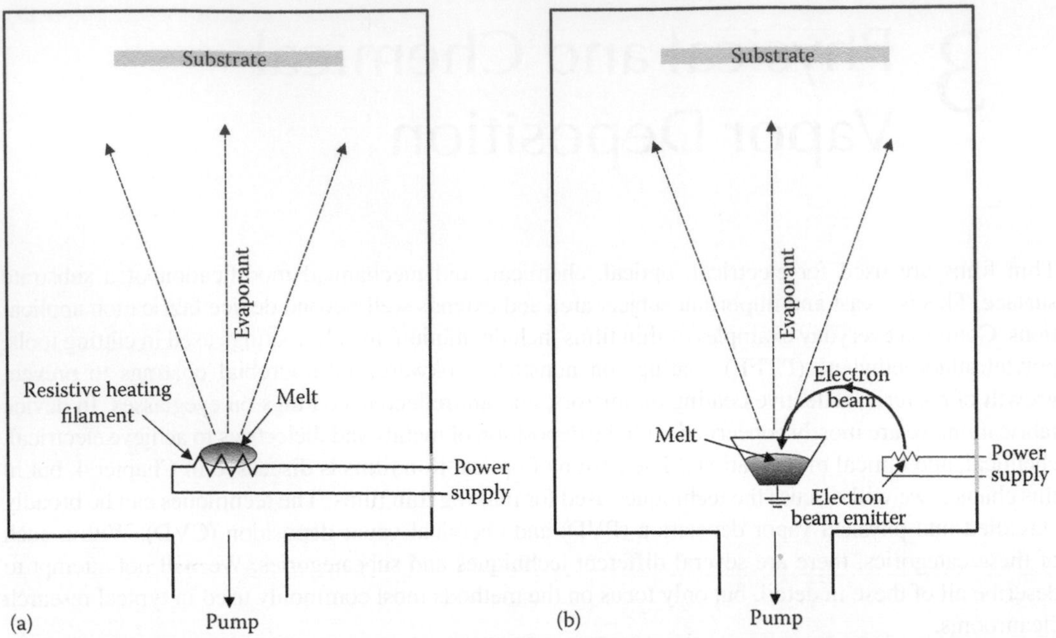

FIGURE 3.1 (a) Resistively heated evaporation and (b) electron-beam-heated evaporation.

has to be longer than the source-to-substrate distance (also known as the throw distance) to prevent collisions with the background gas molecules during transit. A greater throw distance also improves the deposition uniformity.

Every material is constantly evaporating, whether it is a solid or liquid, hot or cold. The evaporation rate is a strong function of the material's binding energy and its temperature. The number of atoms leaving the surface expressed in atoms/cm²/s can also be written in the units of pressure if the atomic weight and the temperature are known, as discussed in Chapter 2. This is the vapor pressure of the material and can be written as

$$Z_A = 3.5 \times 10^{22} \frac{P}{\sqrt{M \times T}} \tag{3.1}$$

where
Z_A is the rate in atoms/cm²/s
P is the vapor pressure in Torr
T is the vapor temperature in Kelvin
M is the molar mass

The evaporation rate can be converted to a thickness removal rate (or deposition rate in the case of condensation) using the following relationship:

$$r = \frac{Z_A}{\frac{\rho}{M} \times N_A} \tag{3.2}$$

where
ρ is the material density
M is the molar mass
N_A is the Avogadro number

The vapor pressure curves of several common elements are shown in Figures 3.2 and 3.3. For example, the vapor pressure of gold is 2 μT at 1000°C increasing to 30 mT at 1500°C. The vapor pressure of platinum is much lower, reaching only 1 μT at 1500°C. Though it is not a fundamental requirement, there is a correlation between melting temperature and vapor pressures. At a given temperature, elements with a low melting temperature such as aluminum have higher vapor pressures than the metals with a high melting temperature such as tungsten. However, this is not universally true. For example, chromium with a melting temperature of 1857°C has a vapor pressure of 100 mT at 1500°C (which is substantially below its melting temperature), but platinum with a melting temperature of 1772°C needs to be raised to 2300°C to reach 100 mT (which is substantially above its melting temperature). Therefore, it is important to consult a vapor pressure chart of different materials when working with thermal evaporation.

Boiling temperature is not a meaningful concept in vacuum evaporation. The definition of boiling point is the temperature at which the vapor pressure becomes equal to the atmospheric pressure, or 760 Torr. In vacuum, everything could be thought of as boiling.

The goal in thermal evaporation is to elevate the temperature of the material to create a high enough vapor pressure that can then condense on the substrate as a thin film. Melting is not a requirement. Adequate vapor pressure is the primary requirement. However, in the vast majority of cases the materials do melt. As shown in Figure 3.1, this makes it necessary to place the substrate above the evaporating species in the vacuum chamber.

3.1.1.1 Resistance Heating Method

In resistance-heated evaporation, the desired material in the form of pellets is placed on a tungsten or molybdenum sheet shaped like a boat (or wire basket). Then, a large DC current is flowed through the tungsten boat to raise its temperature (see Figure 3.1). The reason for tungsten or molybdenum is because of their high melting temperature and low vapor pressure so that they would not deform or evaporate. However, due to their low resistance, a very large current has to be supplied to create the required heating power. For a 0.5 Ω filament (when hot), 100 A of current will be necessary to produce 5 kW of heating at a supply voltage of only 50 V.

In resistance heating, the temperature will be highest at the heating filament, then the bulk of the evaporating species, and the lowest temperature will be at the evaporating surface of the material. This is the main drawback of the resistance heating method. The temperature profile is such that everything else is at a higher temperature than the surface that really matters. As a result, the process often runs too "hot," creating outgassing and evaporation of unwanted species in the chamber. Nevertheless, this method is widely used in the production of thin films because it is simpler than other forms of heating.

3.1.1.2 Electron Beam Evaporation

Thermal evaporation could be made more efficient with less collateral heating if we could heat the evaporating surface from the top rather than from the bottom. For example, consider heating the pellets with a lamp, laser, or electron beam. This would cause the evaporating surface, the surface that really matters, to have the highest temperature. The bulk of the evaporating species would have a lower temperature. The crucible could be water-cooled so nearly all but the evaporating surface would remain cool.

In e-beam heating, a hot tungsten filament generates a cloud of free electrons, like in an incandescent light bulb. The filament is held at a strong negative potential, of the order of −10 kV. This will result in the electrons being accelerated away from the filament (see Figures 3.1(b) and 3.4). Several beam forming plates near the filament redirect the electrons up and away from the filament, and a set of permanent magnets and electromagnets deflect the electron beam and focus them toward the center of the crucible. The electromagnets can be used to make small adjustments to the beam spot, as well as be programmed to perform sweeps to cover a larger surface area. The background pressure in the chamber needs to be low enough to produce a sufficiently long mean

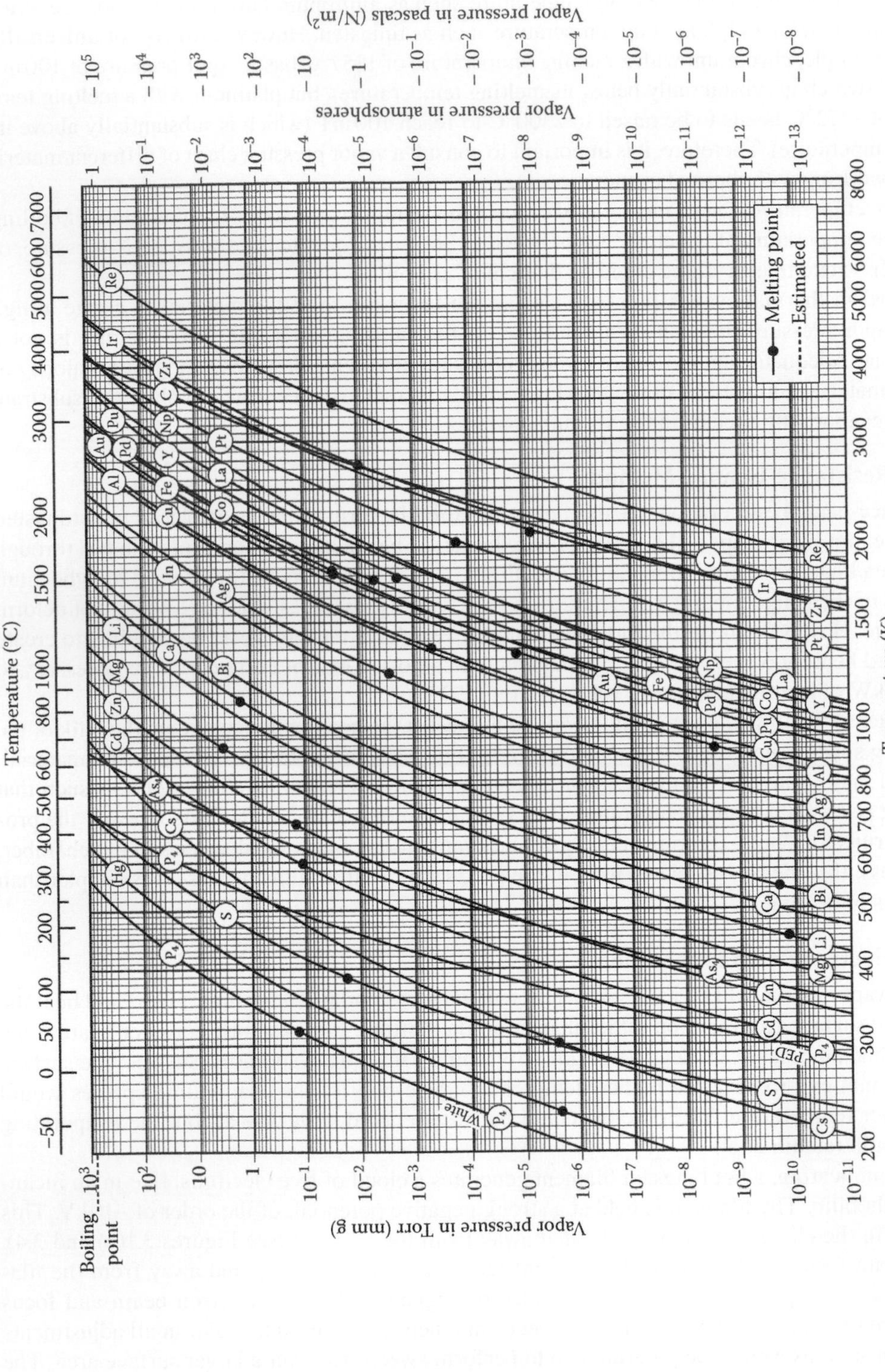

FIGURE 3.2 Vapor pressure curves of selected elements. (Phosphorous through Rhenium). (From Honig, R. E., *RCA Journal* 2, 195–204, 1957. Reproduced with permission from International Union for Vacuum Science, Technique and Applications.)

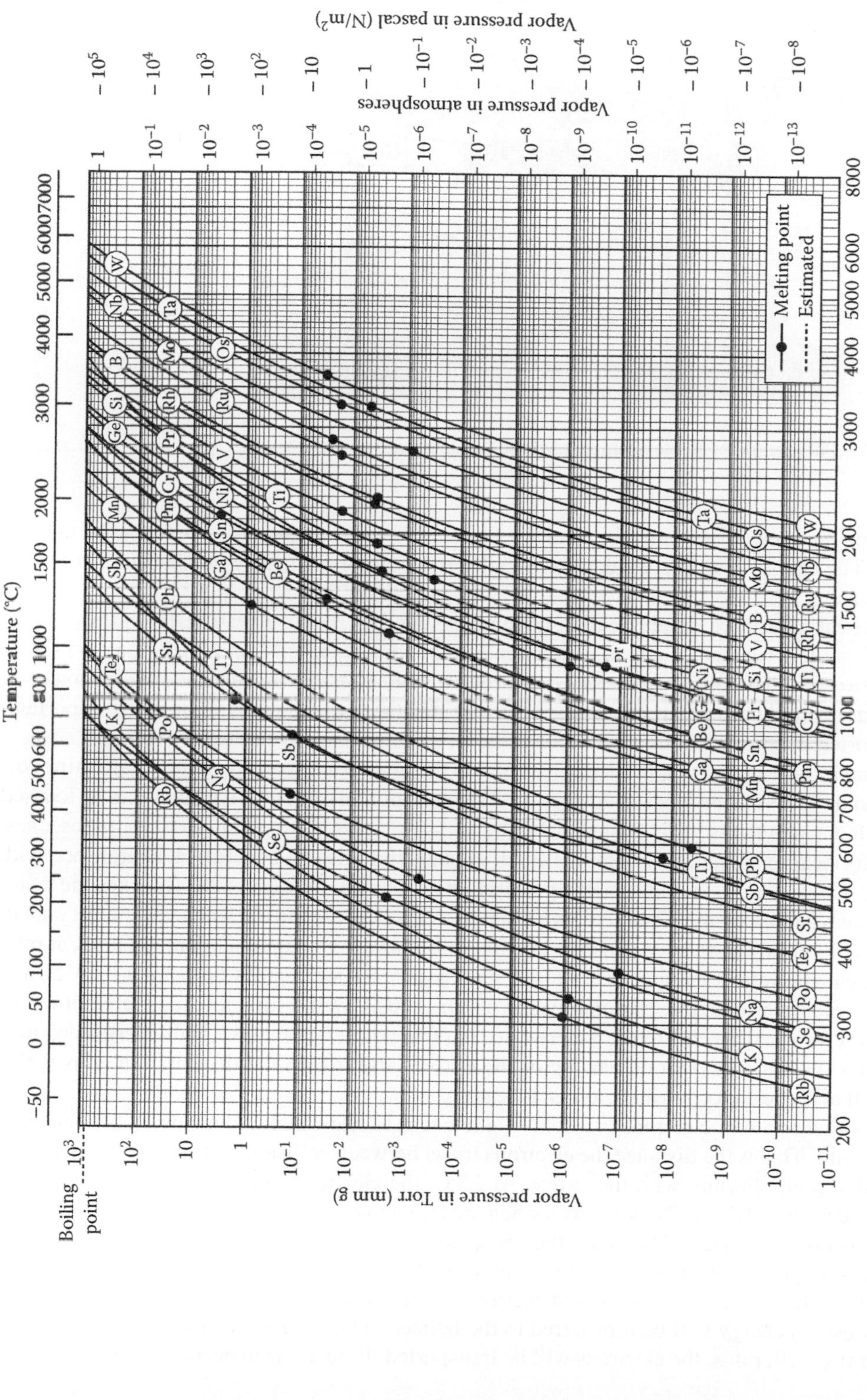

FIGURE 3.3 Vapor pressure curves of selected elements. (Rubidium through Tungsten). (From Honig, R. E., *RCA Journal* 2, 195–204, 1957. Reproduced with permission from International Union for Vacuum Science, Technique and Applications.)

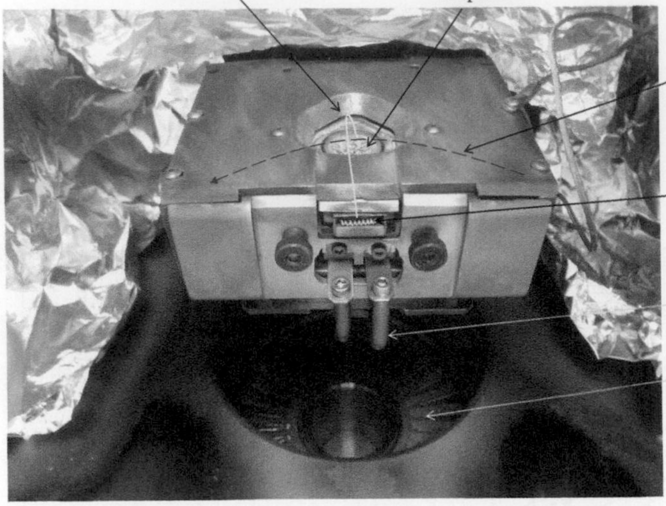

FIGURE 3.4 Photograph of an electron-beam-heated evaporation assembly.

free path for these electrons to travel from the filament to the crucible without encountering any collisions. The energy is ultimately delivered to the evaporating surface by bombardment of high-energy electrons.

E-beam heating is more expensive to install and operate than resistive heating. It requires a high-voltage DC source, separate filament and beam current control circuits, and the associated safety issues due to the high voltage. Special precautions also need to be made inside the chamber to prevent charge build ups and arcing. However, for applications where material purity is important, e-beam is often the best choice.

What makes e-beam heating particularly useful is the high-power density at the evaporating surface. Electron beams can be focused down to a very small spot. In principle, the beam can be focused down to a few nanometers in size using elaborate focusing fields, as done in electron microscopes. However, for heating applications 1 mm is small enough and this can be easily accomplished with simple magnets and coils. In addition, the penetration depth of the electron after it strikes the material is less than a micrometer. Therefore, the beam power is delivered to the source over a very small volume. The high-power density makes it possible to achieve very high melt temperatures at relatively low beam powers. If the spot size is 1 mm^2 and the penetration depth is 1 μm, for a 10 kV 50 mA beam, the power density at the evaporating site will be 500 MW/cm^3, for a total delivered power of only 500 W. Power is delivered to the material due to rapid deceleration of the high-energy electrons as they enter the material. Ohmic heating of the current after the initial deceleration is insignificantly small and does not contribute to much heating.

The mean free path of electrons in a solid is a few tens to a few hundred nanometers depending on temperature. This is the distance the electrons travel between collisions. The electrons in a solid are normally at equilibrium with the lattice. At 25°C, the electron temperature is also 25°C, and the average energy is $kT/q = 26$ meV. Even when the bulk temperature rises to 1500°C, the electron energy is still only 153 meV. Therefore, the energy of a 10 keV incident electron is several orders of magnitude larger. Upon entering the material, these high-energy electrons will undergo random collisions with the lattice with a net loss of energy after each collision. After a number of collisions, all of their excess energy will be transferred to the lattice, and the electrons will reach equilibrium with the lattice. After this, the electrons will be transported through the material as ordinary ohmic

FIGURE 3.5 Gold pellets in a tungsten crucible liner (left) and titanium pellets in a graphite crucible liner (right).

conduction. The penetration depth of high-energy electrons can be numerically modeled, but not really necessary for this application (some aspects of this is discussed in Chapter 6 under electron beam lithography). For 10 keV, this depth is of the order of a few hundred nanometers.

The crucibles are usually made of copper and are water-cooled. The evaporation species in the form of pellets are placed in the crucible pockets and heated with the electron beam. However, the thermal conductivity between the water-cooled copper pockets and the evaporation pellets will be high, which will require high beam powers to achieve the high temperatures necessary for evaporation. The thermal conductivity can be reduced by placing the pellets in a crucible liner made from a high-temperature material such as graphite, as shown in Figure 3.5. The liner allows high temperatures to be reached at lower beam powers. The liner material has to be electrically conductive to allow the electrons to flow through, but also act as an adequate thermal barrier.

In some cases, graphite can interact with the evaporating species and form compounds. Gold for example is prone to creating hotspots and spitting in a graphite crucible. A tungsten or molybdenum crucible liner can alleviate this problem, although its thermal conductivity will be greater than graphite and will require higher beam powers. One could also use a combination—a tungsten crucible with an outer graphite sleeve or spacer.

3.1.1.3 Thermal Evaporation Rate from the Source

Consider aluminum as the evaporation source. At 500°C, its vapor pressure will be about 1 pT, which is a very small value. Using the atomic weight of aluminum of 27 g/mol, this vapor pressure can be converted to atomic removal rate using Equation 3.1 resulting in

$$Z_A = 3.5\times10^{22}\frac{10^{-12}}{\sqrt{27\times(500+273)}} = 2.4\times10^{8}\ \text{atoms/cm}^2\text{/s}. \tag{3.3}$$

The density of aluminum at 500°C is around 2.6 g/cm³. Therefore, the removal rate in thickness/time using Equation 3.2 is

$$r = \frac{2.4\times10^{8}}{\frac{2.6}{27}\times6.02\times10^{23}} \approx 4.2\times10^{-15}\ \text{cm/s}. \tag{3.4}$$

This is an extremely small number, so it is more meaningful to describe it in terms of monolayers/time rather than thickness/time. The monolayer thickness of aluminum is

$$a = \left(\frac{M}{\rho N_A}\right)^{\frac{1}{3}} = \left(\frac{27}{2.4 \times 6.02 \times 10^{23}}\right)^{\frac{1}{3}} = 2.6 \times 10^{-8} \text{ cm} = 2.6 \text{ Å}. \tag{3.5}$$

Therefore, the evaporation rate of aluminum at 500°C can be calculated as 1 monolayer every 71 days!

Now raise the temperature of aluminum to 1200°C. The vapor pressure will rise to 10 mT. The melting temperature is 660°C; therefore, it will be a liquid at 1200°C. Repeating the earlier calculation, using a density of 2.4 g/cm³, we can find

$$Z_A = 3.5 \times 10^{22} \frac{0.01}{\sqrt{27 \times (1200 + 273)}} = 1.75 \times 10^{18} \text{ atoms/cm}^2\text{/s} \tag{3.6}$$

$$r = \frac{1.75 \times 10^{18}}{\frac{2.4}{27} \times 6.02 \times 10^{23}} = 3.5 \times 10^{-5} \text{ cm/s}. \tag{3.7}$$

This is 350 nm/s, which is a significant removal rate. In terms of monolayers, this will be

$$\frac{3.5 \times 10^{-5}}{\left(\frac{27}{2.2 \times 6.02 \times 10^{23}}\right)^{\frac{1}{3}}} = 1290 \text{ monolayers/s}. \tag{3.8}$$

3.1.1.4 Deposition Rate and Distribution

The evaporation rate from the source will obviously not be the same as the deposition rate because of the angular distribution of the flux and the distance between the source and substrate. The distribution of the evaporating flux density (atoms/cm²/s) can be fitted to a $\cos^n(\theta)$ relationship. The simplest model is $\cos(\theta)$, known as the Knudson distribution, and is the kinetic gas analog of the Lambert's law of optical radiation [1]. This makes the evaporating flux density

$$R(\theta) = R(0)\cos(\theta) \tag{3.9}$$

where $R(0)$ is the flux along $\theta = 0$. If we also assume that the source-to-substrate distance D is much larger than the source diameter such that the source can be treated as a point source, the total flux R_T (atoms/cm²) can be written as

$$R_T = \int_0^{\pi/2} (2\pi D\sin(\theta))(Dd\theta)(R(0)\cos(\theta)) = \pi D^2 R(0). \tag{3.10}$$

From this, we can get

$$R(0) = \frac{R_T}{\pi D^2} \tag{3.11}$$

$$R_T = \pi D^2 R(0). \tag{3.12}$$

This assumes evaporation from a point source, which is not always the case, especially with resistive heating method. Even in e-beam heating, the high vapor pressure near the evaporating source can lead to short mean free paths and increased scattering and make the evaporation source appear as if it has a large effective area [2]. Nevertheless, for most practical applications, the cosine law produces reasonably accurate results.

Consider gold at a temperature of 1750°C (melting temperature is 1064°C). The vapor pressure at 1750°C is 1 Torr. Therefore, using the atomic weight of 197 g/mol,

$$Z_A = 3.5\times10^{22}\frac{1}{\sqrt{197\times(1750+273)}} = 5.5\times10^{19}\ \text{atoms/cm}^2/\text{s}. \tag{3.13}$$

Furthermore, if we assume that the evaporating surface has an area of 10 mm^2,

$$R_T = Z_A A = 5.5\times10^{18}\ \text{atoms/s}. \tag{3.14}$$

If the source-to-substrate distance is 30 cm, the flux density at the substrate will be

$$\frac{5.6\times10^{18}}{\pi\times30^2} = 2\times10^{15}\ \text{atoms/cm}^2/\text{s}. \tag{3.15}$$

The standard density of gold is 19.3 g/cm^3. Therefore, the film growth rate will be

$$r = \frac{2\times10^{15}}{\frac{19.3}{197}\times6.02\times10^{23}} \approx 3.3\times10^{-8}\ \text{cm/s or } 3.3\ \text{Å/s}. \tag{3.16}$$

This is the rate at $\theta=0$. At other angles, the rate will drop as $\cos(\theta)$. If the substrate is a 6 inch wafer, based on the cosine distribution, we can expect a variation of 3% from center to the edge.

An interesting point to note is that, even though the vapor pressure at the evaporating surface is 1 Torr, the vapor pressure at the substrate will be significantly lower:

$$P = \frac{Z_A\sqrt{M\times T}}{3.5\times10^{22}} = \frac{2\times10^{15}\sqrt{197\times(1750+273)}}{3.5\times10^{22}} = 35\ \mu\text{T}. \tag{3.17}$$

This difference in pressure arises due to the geometry of the vapor flux and the molecular flow of the species. The expanding area of the vapor results in a decrease in the flux density and vapor pressure.

Gold has to be heated well above its melting temperature to reach the 1 Torr vapor pressure and achieve 3.3 Å/s at the substrate for the aforementioned geometries. Similarly, aluminum has to be heated to 1400°C (which is also well above its melting temperature) to reach 1 Torr vapor pressure and a corresponding growth rate of 9.6 Å/s. With chromium, the temperature required to get the same 1 Torr vapor pressure is 1700°C, which is significantly lower than its melting temperature of 1857°C. Chromium typically remains a solid during thermal evaporation. This is known as sublimation. Many dielectrics, including MgF_2 and SiO_2, as well as some metals such as chromium evaporate by sublimation.

3.1.1.5 E-Beam Evaporation of Dielectrics

A material has to conduct electrons in order to be evaporated by an electron beam. Therefore, it may seem odd that dielectrics such as SiO_2, TiO_2, and Al_2O_3 can be evaporated just as easily as metals. This is possible due to internal dielectric breakdown and leakage currents that develop through grain boundaries of the insulator at elevated temperatures. For this to happen, the e-beam power and temperature has to be raised gradually. A rapid increase can cause the dielectric to charge up and explosively spit the material. Also, the current discharge will not be uniform across the entire surface. There will be some areas where the charge will build up to high levels and deflect the electron beam, and other areas where the current will reach high densities and become too hot. All of these can result in instabilities in the deposition rate. One technique to avoid these problems is to sweep the electron beam over a large surface area. This reduces the current density and allows each site to discharge and recover after being bombarded by electrons.

3.1.1.6 Reactive Thermal Evaporation

The background pressure during normal thermal evaporation is in the range of μT, made up mainly of outgassing species from heated surfaces in the chamber. In reactive evaporation, a background gas such as oxygen or nitrogen is intentionally bled into the chamber to cause a chemical reaction and produce a compound film on the substrate. With a small gas flow, this pressure can be increased to the mT range. The level of pressure needed for reactive evaporation depends on the deposition rate, desired stoichiometry, and reaction rate. Examples of reactive evaporation include VO_2 films from metallic vanadium, SiO_2 from silicon, and MgO from magnesium. Reactive evaporation can also be performed with a plasma that increases the reactivity due to the free radicals generated in the plasma [3].

3.1.1.7 Thermal Evaporation of Alloys and Compounds

Whether or not a compound can be evaporated depends on its thermal stability at elevated temperatures. This will be dictated by the bond strength of the compound. Stable compounds like SiO_2 and Al_2O_3 can be easily evaporated. Since these bonds do not dissociate during evaporation, the resulting film will be at nearly the same stoichiometry as the evaporating source. However, a small loss of the volatile component is usually inevitable, such as the loss of oxygen when evaporating SiO_2. TiO_2 is particularly known for this because it exhibits many stable oxidation states [4]. This can be prevented, or controlled to some extent, by bleeding a small flow of background oxygen when evaporating these oxides, just like a reactive evaporation. However, many compounds will completely dissociate at high temperatures and evaporate as separate species. For example, if we attempt to evaporate GaP, which is a semiconductor, it will dissociate into gallium and phosphorous. Since the vapor pressure of phosphorous is significantly greater than gallium, the source will release phosphorous and retain gallium. As temperature is increased further, gallium will evaporate and deposit on the substrate. Therefore, only the most thermally stable compounds can be deposited by evaporation.

3.1.1.8 Ion-Assisted Deposition

Since evaporated atoms condense on the substrate gently and at very low energies, they tend to have higher porosity and poor adhesion and experience faster atmospheric degradation compared to more energetic depositions like sputtering. The porous nature also leads to lower refractive indices and the absorption of environmental moisture. Moisture can cause shifts in refractive index as a function of temperature. The porous nature can also induce a significant amount of intrinsic stress in the film. Since optical coating designs tend to utilize a large number of stacked thin films, film stress can become a major consideration.

One way to mitigate these problems is to increase the substrate temperature during deposition. However, many substrates cannot withstand high temperatures, and high temperatures can also result in outgassing, making the deposition process more difficult. Ion assistance in conjunction with evaporation can significantly improve the film characteristics without substrate heating, and in many cases, it can be used in combination with substrate heating. Ion bombardment can densify the films and reduce porosity, resulting in films with increased mechanical and environmental durability, lower scatter, lower optical absorption, and lower stress. As a result, ion-assisted deposition (IAD) is widely used in the production of optical films [5].

In an IAD setup, a separate ion source is installed inside the vacuum chamber. This ionizes a gas and directs a stream of energetic ions at the substrate, as illustrated in Figure 3.6. Ion sources

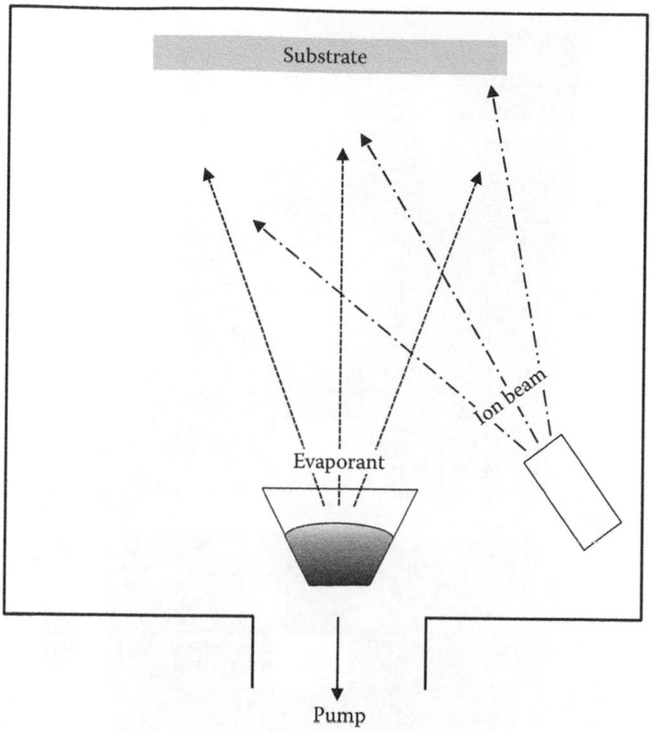

FIGURE 3.6 Configuration of ion-assisted deposition (IAD).

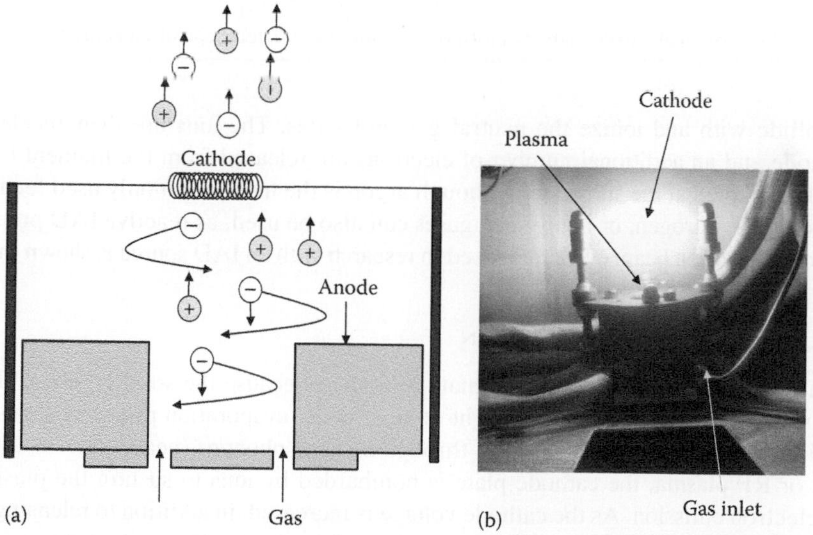

FIGURE 3.7 End-Hall ion source configuration. (a) Schematic and (b) end-Hall source in operation.

come in many different configurations, but they all contain a cathode, anode, and a neutralizer. One common variant, known as the end-Hall ion source, uses a hot tungsten filament that acts as the cathode (Figure 3.7). The electrons emitted from the cathode accelerate toward the anode in a spiraling path due to the permanent magnets installed between the anode and the cathode, and this results in an increased interaction length between the gas atoms and the electrons. These energetic

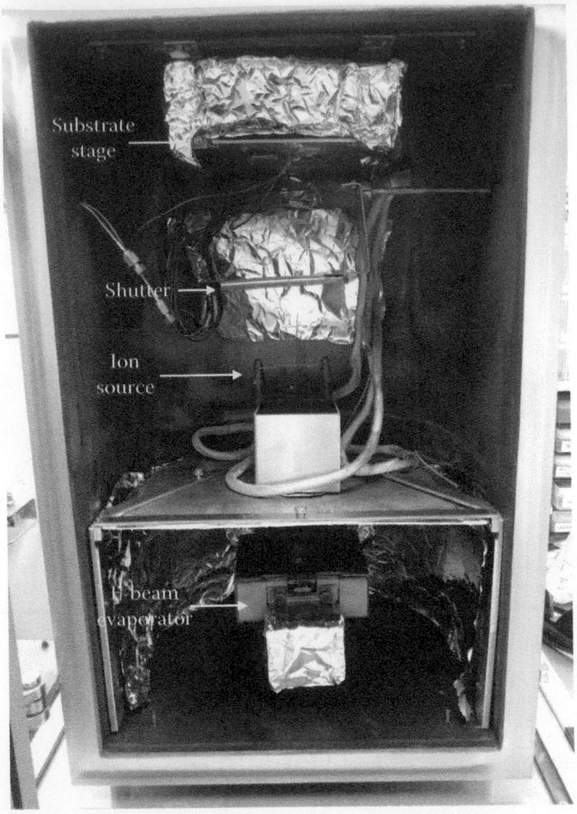

FIGURE 3.8 Photo of an electron beam evaporator with an ion-assisted deposition source used for research.

electrons collide with and ionize the neutral gas molecules. The ions are then accelerated away from the anode and an additional number of electrons are released from the filament to neutralize the ions being directed at the substrate. Although argon is the most commonly used feed gas for the ion source, oxygen, nitrogen, or many such gases can also be used, as reactive IAD processes [6].

A photo of an electron beam evaporator used in research with an IAD source is shown in Figure 3.8.

3.1.2 Sputter Removal and Deposition

Sputter removal can be considered a sublimation process because the solid is turned directly to a vapor without melting. It can also be thought of as a "cold" evaporation process because energy is delivered via external projectiles instead of through internal phonons (heat).

In a DC or RF plasma, the cathode plate is bombarded by ions to sustain the plasma through secondary electron emission. As the cathode voltage is increased, in addition to releasing secondary electrons, the ions will also eject neutral species from the cathode. These ejected species will land everywhere in the vicinity of the cathode including the substrate. This is the sputtering process. Since sputtering is a physical removal of material from the target, it is a PVD method just like evaporation. The material to be sputtered is referred to as the target. The material being deposited on is the substrate. The target surface undergoes sputter removal, and the substrate surface undergoes sputter deposition.

A typical sputtering cathode removed from a vacuum chamber is shown in Figure 3.9. The cathode plate is typically a water-cooled copper plate with an electrical connection on the water side and vacuum on the other side. The target material is thermally and electrically bonded to this

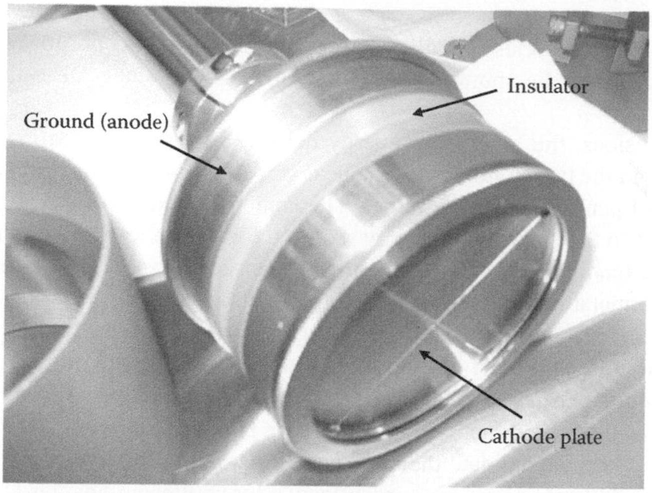

FIGURE 3.9 Magnetron sputtering cathode.

FIGURE 3.10 Photo of a sputtering system with two magnetron cathodes used in research.

copper plate. The ions in the plasma strike the target instead of the copper cathode plate, ejecting the target atoms. Water cooling is necessary because a large portion of the power (60%–80% or more) supplied to the plasma is dissipated as heat on the cathode due to the ion bombardment [7,8]. Kinetic energies of the sputtered atoms, photon emission, and electron heating constitute only a smaller portion of the supplied power. A photo of a vacuum chamber with two magnetron sputtering cathodes and a rotating substrate stage is shown in Figure 3.10.

3.1.2.1 Sputter Removal Mechanism

Upon collision, the incident ions dissipate their kinetic energy to the target's surface. If the ion energy is larger than the binding energy of the target atoms, it can dislodge and eject a target atom through a primary or secondary collision. Momentum and energy are always conserved, and kinetic energy can also be conserved if the collisions are elastic. At higher ion energies, the primary collision can set off a cascading series of collisions that can propagate deeper into the target. This can still cause surface atoms to be sputtered if a reflecting cascade reaches the target's surface with sufficient energy.

For primary collisions, the amount of energy delivered by the ions will be a function of the masses of the ions and the target atoms. The nuclear stopping power of the target atoms determines how far the ions will penetrate into the target. The angle of incidence determines the momentum transfer from the ion to the target. The energy required for the removal of an atom from the target will be related to the binding energy of the atoms in the target. Most of these effects are well understood and can be computationally modeled and experimentally verified.

Although any gas can be used in the plasma, argon is most commonly used gas for sputtering. The first reason is that it is an inert gas. Sputtering must be done with an inert gas plasma unless we intentionally want the target atoms to chemically react with the gas to form compounds during deposition. Second, sputter yield increases when the ion mass is closely matched to the target mass, which makes argon a better candidate than helium and neon for a large number of target elements. The heavier xenon and krypton ions are more effective for sputtering heavier atoms. Finally, argon is an inexpensive gas since it is the third most abundant gas in the atmosphere after nitrogen and oxygen. It is easy to extract, purify, and compress. Therefore, argon is a good compromise in overall performance for nearly all inert sputtering applications.

3.1.2.2 Sputter Yield

The sputter yield S is the number of ejected target atoms for each incident ion. It can be calculated using ion–matter interaction models, and tables of experimental data have been compiled for various sputter gases, target materials, and ion energies. Sputter yield is generally found to vary as a function of ion mass, ion energy, angle of incidence, target atomic mass, and target surface quality. Sputter yield has been found to increase from left to right on the periodic table, which roughly correlates inversely with binding energy (heat of sublimation). The sputter yield is also related to the atomic masses of the ion and the target atom. Maximum energy transfer occurs when the ion mass is close to the atomic mass of the target. Table 3.1 shows tabulated values for sputter yield for a few elemental targets as a function of argon ion energies. A more extensive compilation of sputter yields for elements using different ions can be found in Reference 9.

There is a threshold energy below which no sputtering takes place. Beyond this threshold, yield increases with increasing energy and then reaches a plateau. For larger energies, sputter yield starts to decline because ions penetrate too far below the target's surface to cause sputtering. This is the ion implantation regime, and is used for doping intentional impurities into semiconductors. The ion energy in this case is used for controlling the landing depth of the dopant species.

TABLE 3.1
Sputter Yield Values for a Few Common Target Materials from Argon Ions

Target	100 eV	200 eV	300 eV	600 eV
Ag	0.63	1.58	2.20	3.40
Al	0.11	0.35	0.65	1.24
Au	0.32	1.07	1.65	2.43
Cu	0.48	1.10	1.59	2.30
Ti	0.08	0.22	0.33	0.58
W	0.07	0.29	0.40	0.62

Source: Laegreid, N. and Wehner, G.K., *J. Appl., Phys.*, 32, 365, 1961.

Sputter yields can also be calculated from fairly simple empirical models. Figure 3.11 shows the calculated sputter yields of a few common metals using Ar^+ ions as a function of the ion energy using the empirical model described in Reference 10. We can see that the sputter yield of gold with Ar^+ ions shows a threshold energy of about 50 eV and reaches a peak of 12 at 90 keV. Titanium has a threshold around 100 eV and reaches a peak of 2 around 35 kV. Figure 3.12 shows the same calculations done with He^+ ions, where it can be seen that the yields are about an order of magnitude smaller than with Ar^+. Figure 3.13 shows the sputter yields for the same elements using Xe^+ as the ions, where an increased sputter yield can be seen for all elements, and a higher implantation threshold.

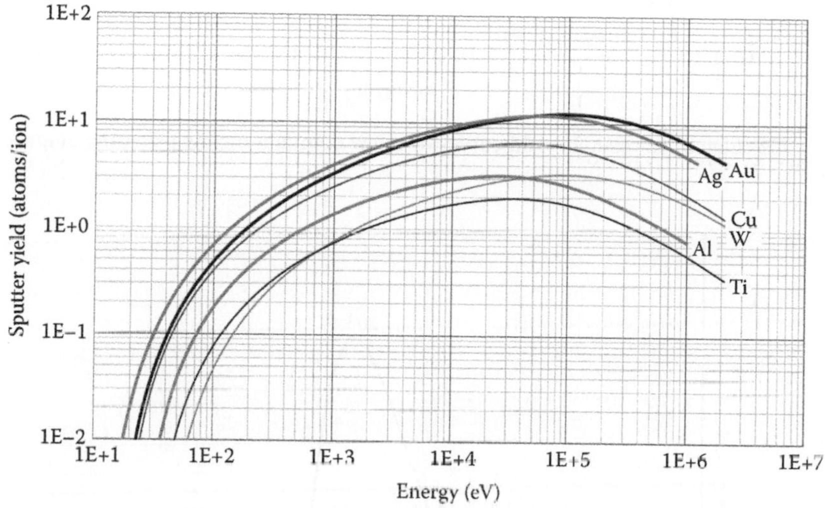

FIGURE 3.11 Calculated Ar^+ sputter yield curves of a few common metals using the empirical model in Reference 10.

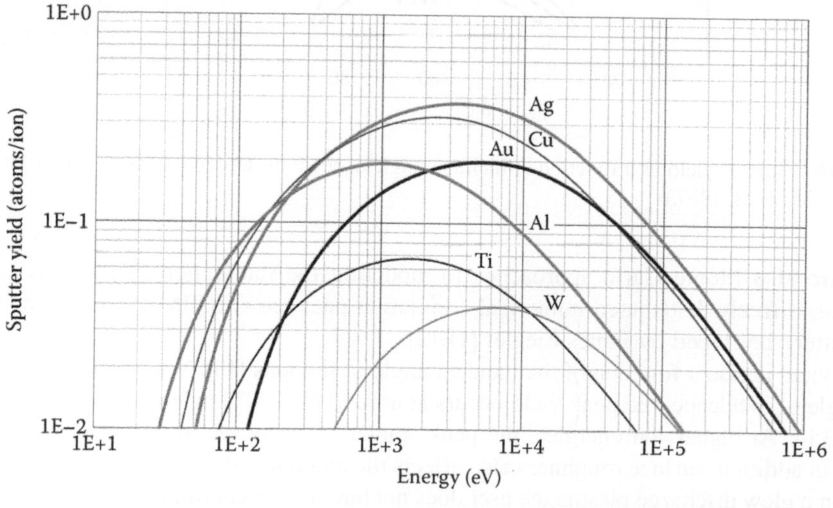

FIGURE 3.12 Calculated He^+ sputter yield curves of a few common metals using the empirical model in Reference 10.

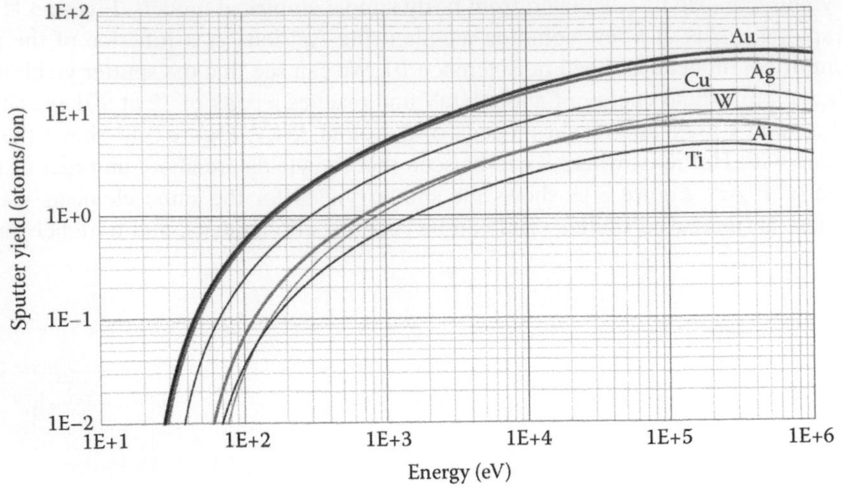

FIGURE 3.13 Calculated Xe^+ sputter yield curves of a few common metals using the empirical model in Reference 10.

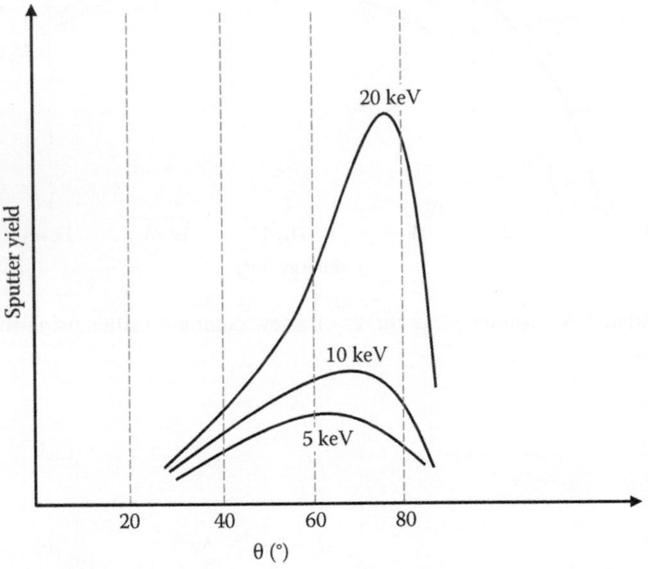

FIGURE 3.14 Sputter yield as a function of incident angle of the ions. (Adapted from Oliva-Florio, A. et al., *Phys. Rev. B*, 35, 2198, 1987.)

There are also Monte Carlo approaches to modeling the interaction of the ions with atoms. A widely used freely distributed numerical simulation package is SRIM (Stopping and Range of Ions in Matter) developed by James Ziegler [11,12].

Sputter yield is also a function of the incident angle of the ions [13]. When sputter yield is plotted against angle of incidence, the peak yield occurs at around 60°–70° angle of incidence, as illustrated in Figure 3.14. At higher ion energies, the peak sputter yield moves toward the normal incidence angle [14]. In addition, surface roughness also affects the angular distribution of sputter yields [15,16]. However, in a glow discharge plasma the user does not have direct control over the angle of incidence of the ions. The exception is in ion beam sputtering, where a beam of ions is directed at a target's surface from a separate ion source to induce sputtering.

3.1.2.3 Magnetron Sputtering

In a conventional DC plasma, electrons are released from the cathode through secondary electron emission and are accelerated toward the anode. These electrons create more electrons and ions through impact ionization, and the ions return to the cathode to release secondary electrons to sustain the plasma. Since the secondary electron emission coefficient γ is typically much smaller than 1, many impact ionization events will be needed to sustain the plasma. Therefore, the ions will be spread across many mean free paths of distance from the cathode. The ions will have to undergo as many collisions as electrons, and they will be less energetic when they reach the cathode, having lost some of their energy to nonionizing collisions with the gas. This results in a low sputter rate at the cathode and heating of the plasma gas and the substrate. A magnetron plasma overcomes some of these problems by placing cylindrical permanent magnets behind the cathode to alter the trajectory of the electrons so that they traverse a circuitous route instead of moving parallel to the electric field lines. Ions, on the other hand, will move parallel to the electric field and will not respond to the magnetic fields because their velocities are much smaller than electrons. This can be viewed as electrons having a shorter effective mean free path than ions in a direction normal to the cathode. As a result, more ions will be generated closer to the cathode, resulting in a greater ion density, and they will reach the cathode with fewer or no collisions, while electrons continue to collide and ionize a large number of atoms. The higher energy and higher density of the arriving ions results in a higher sputter rate. The greater ion density also allows the magnetron cathode to sustain a plasma at much lower pressures than a simple cathode plate. Magnetrons can be operated at 1–10 mT, whereas simple plasmas may require 100–500 mT. Today, magnetrons are widely used in sputtering applications, and they come in a large number of different configurations for DC as well as RF excitations [7,8,17].

The placement of magnets on a circular magnetron cathode is shown in Figure 3.15. Two ring magnets are concentrically placed to create a radially symmetric magnetic field on the target surface. Between the two magnets where the magnetic field is parallel to the cathode plate, the plasma density will be significantly greater than elsewhere. This also results in a significantly higher sputter rate in those areas, and the target will show a characteristic racetrack-shaped erosion pattern, as shown in Figure 3.16. This uneven erosion can sometimes be a problem in production environments where maximum target utilization is an important consideration.

Ferromagnetic targets such as iron and nickel do not work well with magnetron cathodes because they will shield the magnetic fields from reaching the plasma. These target materials could still be used with a magnetron if they are sliced into very thin sheets to allow the magnetic fields to penetrate the material.

3.1.2.4 Sputter Removal Rate

The removal rate (atoms/s) is the product of the ion current multiplied by the sputter yield. This can be written as

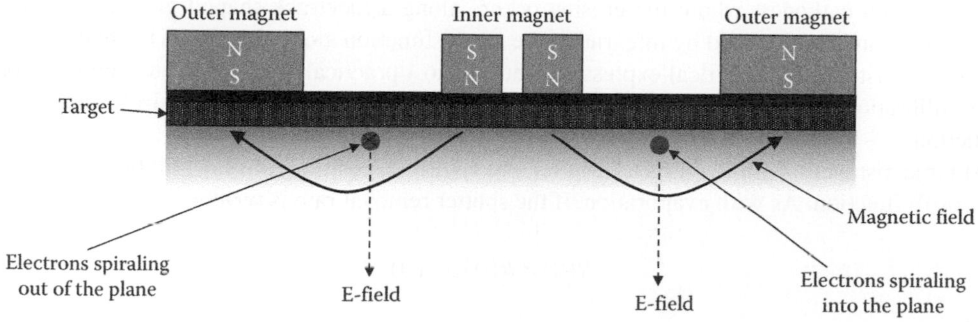

FIGURE 3.15 Circular magnetron cathode with two concentric magnets. The electrons will move in a spiraling path as indicated in the figure away from the cathode instead of moving parallel to the E-field.

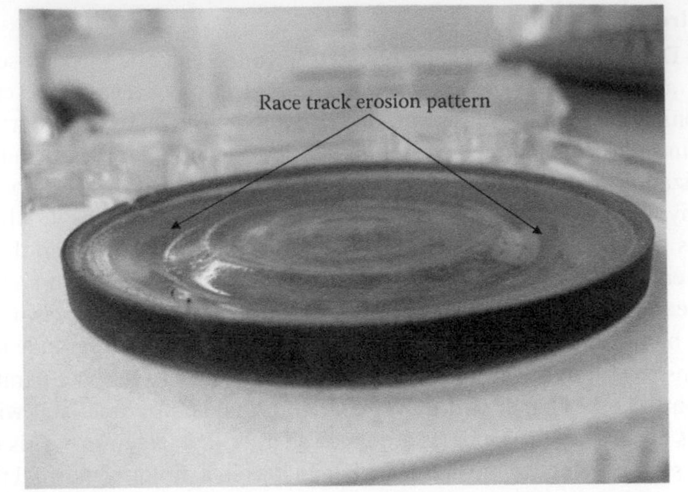

FIGURE 3.16 Erosion pattern on a magnetron target.

$$R_T = \frac{I_p}{q} S. \tag{3.18}$$

For example, consider a tungsten target in an argon plasma, with a DC power of 200 W and a DC voltage of 400 V. In a magnetron plasma, we can assume that the cathode voltage is nearly equal to the ion energy. The sputter yield for 400 eV Ar^+ on tungsten can be looked up from Figure 3.11 to be 0.35. The ion current is 200/400 = 0.5 A. Therefore, the sputter removal rate is

$$R_T = \frac{0.5}{1.602 \times 10^{-19}} \times 0.35 = 1.1 \times 10^{18} \text{ atoms/s} \tag{3.19}$$

This is the total removal rate from the entire target surface, so in order to find the deposition rate on a substrate a certain distance away, we would need to know the geometry of the system.

3.1.2.5 Sputter Deposition Rate

For a normally incident ion, the ejected atoms will have an angular distribution that can be fitted to a $\cos^n(\theta)$ relationship [18]. The value of n depends on the target material, orientation (if it is a crystal), and ion energy. The simplest model is $\cos(\theta)$, which can give a rough order of magnitude result. For magnetron cathodes, where the erosion occurs along a racetrack-shaped groove, the angular distribution can be calculated by integrating the $\cos(\theta)$ function along the ring. This makes it more difficult to write down analytical expressions, but in most practical situations one simply performs a few calibration runs to determine the thickness distribution rather than rely entirely on analytical predictions.

At large distances from the target, however, the sputter deposition rate can still be approximated as a $\cos(\theta)$ function. As with evaporation, if the sputter removal rate is written as

$$R(\theta) = R(0)\cos(\theta) \tag{3.20}$$

assuming that the target-to-substrate distance D is much larger than the target diameter, the total removal rate R_T can be written as

$$R_T = \int_0^{\pi/2} \left(2\pi D \sin(\theta)\right)\left(D d\theta\right)\left(R(0)\cos(\theta)\right) = \pi D^2 R(0) \tag{3.21}$$

where $R(0)$ is the deposition rate at $\theta = 0$. From this, we can get

$$R(0) = \frac{R_T}{\pi D^2}. \tag{3.22}$$

Continuing the previous example with tungsten, the deposition rate at a 30 cm distance from the target will be

$$R(0) = \frac{R_T}{\pi D^2} = \frac{1.1 \times 10^{18}}{\pi 30^2} = 3.9 \times 10^{14} \text{ atoms/cm}^2\text{/s}. \tag{3.23}$$

If the density and atomic number of tungsten are known, this rate can be converted to a film growth rate. The standard density of tungsten is 19.25 g/cm^3, and the atomic weight is 183.8 g/mol. Therefore, the film growth rate will be

$$r = \frac{Z_A}{\frac{\rho}{M} \times N_A} = \frac{3.9 \times 10^{14}}{\frac{19.25}{183.8} \times 6.02 \times 10^{23}} \approx 0.6 \times 10^{-8} \text{ cm/s or } 0.6 \text{ Å/s}. \tag{3.24}$$

This is the rate normal to the target surface at $\theta = 0$. At other angles, the rate will drop as $\cos(\theta)$.

3.1.2.6 Dependence of Sputter Deposition Rate on Pressure

If the plasma power is held constant and the gas pressure is changed, both the ion current and voltage will change such that the product remains a constant. An increase in pressure will produce an increase in the ion density in the plasma. As a result, the ion current will increase and the voltage will decline. Even though this reduction in voltage will result in a lower sputter yield, due to the increase in ion current and the nonlinear relationship between sputter yield and voltage, there will be an overall increase in the sputter removal rate.

As pressure continues to increase, the voltage will continue to drop and the decline in sputter yield will start to dominate the sputter removal rate. Increased scattering from the gas will also enlarge the angular distribution of the sputtered atoms. Both of these effects will result in a decline in the sputter deposition rate. Therefore, if power is held constant and pressure varied, the sputter removal rate will initially increase and then decrease as shown in Figure 3.17. The optimum operating point is the pressure at which the deposition rate reaches a maximum.

3.1.2.7 Energy of the Sputtered Atoms

The energy of the ejected species will be a function of the incident ion's energy and its interaction with the target. Due to the numerous interaction pathways in the target, the ejected atoms will contain a continuous energy spectrum. For most sputtered surfaces, the atoms have energies in the range of 1–10 eV. For example, the energy spectrum of copper atoms sputtered from 500 eV and 5 kV Ar^+ ions is shown in Figure 3.18. This was calculated using the SRIM software that uses a Monte Carlo model [11]. It is interesting to note that the energy distribution remains nearly the same between 500 eV and 5 kV, with the only difference between the plots is the area under the curve (which corresponds to sputter yield). Nevertheless, the important point is that, sputtered atoms are ejected with energies far greater than the energies encountered in thermal evaporation. It is insightful to calculate the equivalent temperature of a sputtered atom. For the 10 eV atom, this temperature is equivalent to

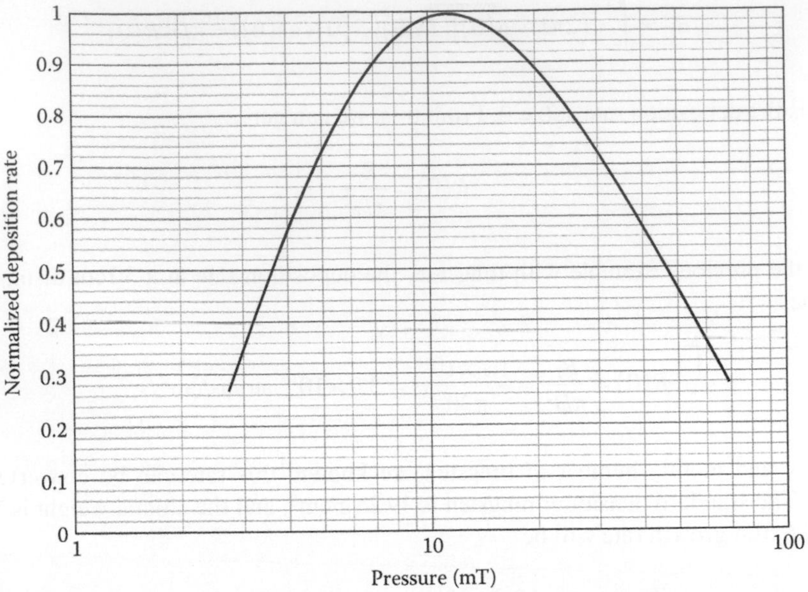

FIGURE 3.17 Magnetron sputter rate as a function of pressure at a constant power.

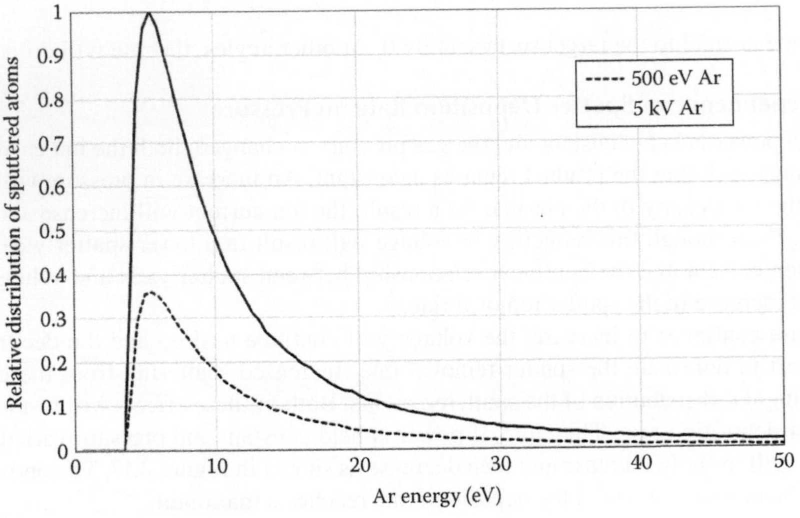

FIGURE 3.18 Energy spectrum of sputtered copper from 500 eV and 5 kV argon ions, as simulated by SRIM software. (From http://www.srim.org/.)

$$T = \frac{2E}{3k} = \frac{2 \times 10}{3 \times 8.62 \times 10^{-5}} = 7.7 \times 10^{4}\ \text{K}. \tag{3.25}$$

3.1.2.8 Sputter Up versus Sputter Down

The target can be mounted facing down toward the substrate, or facing up. Gravity does not influence the ions or sputtered atoms because their energies are too high. Sputter down is more flexible because it allows differently shaped substrates to be placed on a flat surface rather than having to mount them upside down with retaining clips. Sputter up is beneficial for avoiding large particulates and contaminants (which are influenced by gravity) from falling down on the substrate. The casings and other surfaces can become heavily coated and may peel off at some point during the deposition. The target can also once in a while eject large chunks of atoms due to thermal stress or defects on the target surface. However, sputter-up can also cause debris to fall on the target face which can then get sputtered, leading to contamination of the deposition film.

3.1.2.9 Compound Sputtering

When compounds are sputtered, ion bombardment will cause the compounds to be ejected as atoms instead of as whole molecules. The atoms will also have different sputter yields due to their different masses and diameters. If one type of atom has a higher yield than the other, it will be preferentially ejected over the other types. This may seem like a serious problem when sputtering compound materials, but it has been found that a certain steady-state composition and depth profile are quickly reached where the sputter removal of both elements occurs at the original stoichiometric composition [19]. This should not be surprising because the only way to consume a target to its entirety is by depleting its constituent elements in their original proportions.

Consider a target with a binary compound *AB*. If *A* has a higher sputter yield than *B*, the initial sputtering process will remove more of *A* than *B*. This will make the target surface become enriched in *B* and deficient in *A*. As a function of depth, the *B*/*A* ratio will be highest near the surface, falling to the original stoichiometric ratio after roughly the range of the incident bombarding ions. The majority of *B* will get sputtered from the surface where its concentration is greatest and the incident ions are also more energetic. The majority of *A* will be sputtered from slightly below the surface, where the ions will be less energetic due to secondary collisions. This will result is a stoichiometric removal of the atoms. More importantly, the same depth profile will be maintained as the target erodes to completion. This is depicted in Figure 3.19. As an example, consider SiO_2. This target often develops a brown tint due to the enrichment of silicon on the surface. However, the net erosion of the target will occur as stoichiometrically correct SiO_2.

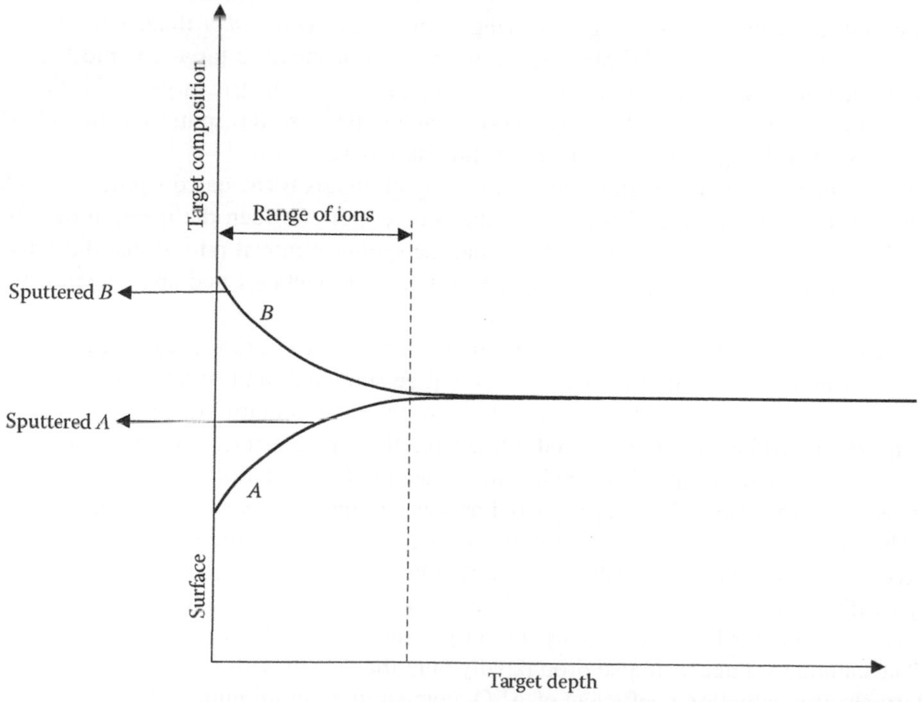

FIGURE 3.19 Target depth profile of a binary compound where the sputter yield of *A* is greater than *B*.

Nevertheless, stoichiometric removal does not always imply stoichiometric deposition. In the case of SiO_2, not all oxygen atoms will recombine with the silicon atoms on the substrate to reform SiO_2. Some oxygen will be lost to the pumping system due to its higher volatility compared to silicon. A common technique used to compensate for this effect is to flow a small amount of oxygen with the argon feed gas to the plasma. This allows the depositing film to become replenished with new oxygen atoms to make up for the loss of oxygen.

Compared to metals, the bond strength and energy of vaporization of dielectrics are high, which makes their overall sputter rates very low. Models for the calculation of the sputter yield of ionic compounds such as oxide and nitrides are significantly less mature than that of elements.

3.1.2.10 Co-Sputtering

Co-sputtering is the process of sputtering different target materials from different cathodes such that they deposit on the substrate at the same time. It forms a simple mixture of the sputtered species. Although the same mixtures can be sputtered from a single alloy target, co-sputtering allows one to change the composition of the film without having to purchase a new target every time a different composition is required. Examples include SiGe films made by co-sputtering silicon and germanium and conductive optical films such as Zn-In-Sn-O produced by co-sputtering ZnO and ITO (indium-tin-oxide). Reactive co-sputtering can also be done by introducing a reactive gas. Silicon and titanium can be co-sputtered with oxygen gas to produce a mixture of SiO_2 and TiO_2. SiO_2 has a refractive index of 1.48 and TiO_2 has a refractive index of 2.5. By co-sputtering, one could realize a range of refractive index value between these extremes, as well as create graded refractive index profiles for optical thin-film applications.

3.1.2.11 Reactive Sputtering

An elemental target can be sputtered with a mixture of argon and another reactive gas (such as oxygen or nitrogen) to create compounds. Sputtering a metallic target to produce a dielectric thin film has a number of attractive qualities—dielectrics generally have very low sputter yields but metals have a high sputter yield; dielectrics have poor thermal conductivity and the target can easily crack due to thermal stress during sputtering; sputtering of metallic targets can be done with a DC plasma, whereas dielectrics require an RF plasma; the production of metallic targets is much easier and can be obtained in greater purity and at lower cost compared to dielectric targets [17]. The reactive sputtering of aluminum to form Al_2O_3 is a good example because the sputtering rate of Al_2O_3 is anemically low, but the sputter rate of metallic aluminum is very fast.

However, a major difficulty with reactive sputtering of metals is the narrow process window for achieving optimum conditions. This is due to the competition between the reaction rate that produces a dielectric film on the target's surface, and the sputter removal rate of that dielectric. This balance is greatly influenced by the vastly different cathode voltages and sputter rates when the target is fully metallic vs. when it is fully dielectric.

Considering the reactive sputtering of aluminum, referring to Figure 3.20, as the oxygen flow is increased from zero, initially the metal atoms will be sputtered, and some of these metal atoms will become oxidized at the substrate. The cathode voltage will remain high and any oxides forming on the target will be quickly sputtered off, leaving the target surface mostly metallic. In other words, the sputter removal rate of the oxide will be equal to the oxidation rate of the target. Despite the increase in oxygen flow, the oxygen partial pressure in the chamber will not show a noticeable increase because all the oxygen will be consumed (gettered) by the sputtered metal atoms.

As oxygen flow is increased further, a point will be reached where all the sputtered metal will be fully oxidized. This is the optimum point of operation for reactive sputtering. However, if the oxygen is increased further, even briefly, the target surface will build up a continuous layer of oxide. The cathode voltage will also drop rapidly (and the current will increase) due to the higher secondary electron emission coefficient of Al_2O_3 compared to aluminum, and the sputter rate will also drop precipitously. The sputtered species will be Al_2O_3 rather than aluminum. At this point,

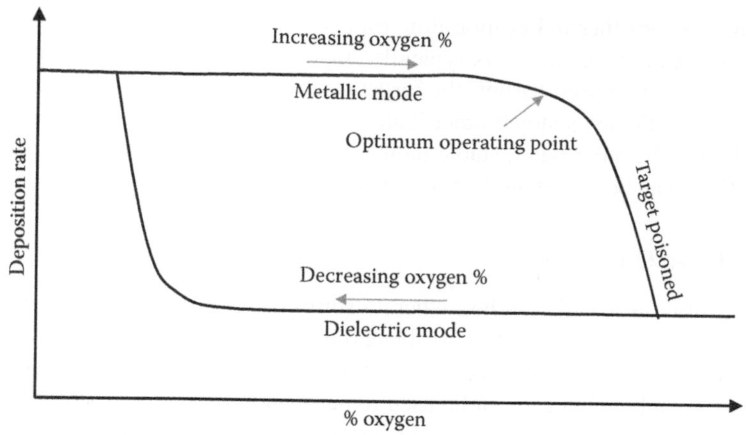

FIGURE 3.20 Hysteresis curve of the deposition rate with oxygen flow in reactive metal deposition.

the oxygen flowing into the chamber will show a rapid rise in the oxygen partial pressure having depleted all the sources of pure metal, and the target is said to have been "poisoned." In the case of DC plasma, this can produce arcing and instabilities due to the dielectric breakdown of the thin layer of Al_2O_3 on the target face. In the case of an RF plasma, a steady plasma will be sustained but with a very low sputter rate and a low DC bias.

Reducing the oxygen flow will reverse this condition, except the process will not follow the same trajectory back to the original condition. This is a well-known hysteresis effect in reactive sputtering [20,21], and it makes automation difficult for maintaining operation at the optimum point. Due to the low sputter removal rate of the oxide, the oxidation rate of the metallic target exceeds the removal rate of the oxide, and the oxygen level has to be reduced significantly below the previous transition point where it became "poisoned." A typical hysteresis curve is shown in Figure 3.20.

3.1.2.12 Thermal Evaporation versus Sputtering

Typical background pressures encountered during magnetron sputtering is about 1–10 mT. The mean free path at this pressure is about 1 cm. During thermal evaporation, the pressure is much lower, of the order of 1 μT, except in reactive evaporation or IAD. At 1 μT, the mean free path is 100 m. As a result, thermal evaporation can be considered to be almost a line-of-sight deposition. The evaporating species arrive at the substrate without encountering any collisions. In sputtering, depending on the distance between the target and substrate, there could be some collisions. The line-of-sight deposition in thermal evaporation has important benefits when films need to be patterned by lift-off lithography.

In evaporation, the arriving species will have energies close to the thermal energies of the source. The highest temperature encountered during thermal evaporation is of the order of 2500°C. The average thermal energy at 2500°C is 234 meV. In sputtering, the energies of the arriving species are of the order of 10 eV, which is several orders of magnitude higher. Therefore, thermal evaporation is a low-energy deposition compared to sputtering. The atoms land on the substrate much more gently in thermal evaporation, whereas in sputtering they collide with the substrate at high energies causing damage and heating. On the other hand, the high deposition energy in sputtering can be advantageous for compacting and densifying the film. In thermal evaporation, the film will have a lower density than in sputtering, and voids and gaps in the film can form more easily. The porous nature of the evaporated films can be also exploited to make unique thin films, such as nanostructured thin film for applications in sensors and for ultralow refractive index applications [22,23]. These are some of the inherent differences between sputtering and thermal evaporation.

Another important difference between sputtering and thermal evaporation is their ability to deposit compounds. Neither method can maintain a perfect stoichiometry, but sputter deposition has many advantages over thermal evaporation, especially for alloys and mixtures. The difference in sputter yield between two elements is typically smaller than the difference in their vapor pressures. Furthermore, as discussed earlier, the target's surface composition becomes conditioned to sputter the constituent elements stoichiometrically, although some loss of more volatile species is always inevitable. On the other hand, stable molecules like SiO_2 may not dissociate under thermal evaporation, so these can be evaporated or sputtered equally well.

3.1.3 Pulsed Laser Deposition

Pulsed laser deposition is a PVD method where a series of high-energy pulses from a ultraviolet (UV) laser beam are focused onto a target to ablate the atoms off the surface. The ablated atoms are then collected on a substrate surface located a certain distance away. No background gas is necessary for this process, although a gas can be used to perform a reactive deposition or to compensate for loss of species from a compound target. In that sense, PLD can be thought of as a sputtering process with a greater flexibility to use with any background gas, or without any gas at all. The laser and the focusing optics are all located outside the vacuum chamber, so it also has the advantage of needing minimal installation inside the chamber [24]. A typical layout of a PLD system is shown in Figure 3.21.

By far the greatest advantage of PLD is the stoichiometric removal of the target atoms, even more so than sputtering. This arises due to the short duration of the ablation process and the high laser fluence. The target surface temperature reaches an extremely high value, on the order of 5000 K, for a few nanoseconds. Because this is substantially higher than the ablation threshold of all elements, and because it lasts for a much shorter duration than the time necessary to establish a steady-state, all constituent elements are removed equally regardless of their volatility. As a result, PLD is used mostly for producing thin films of complex compounds that are too unstable to evaporate, or have widely different sputter yields and volatility to use with sputter deposition methods. Examples of films produced with PLD include yttrium barium copper oxide (YBCO) in superconductors and copper zinc tin sulfur (CZTS) in solar cells. The biggest drawback of PLD is the poor film uniformity and slow deposition rates. Since ablation occurs from a single small point on the target and the ablated plume has a narrow angular distribution, the deposition typically covers only a small area on the substrate. Moving the substrate to a greater distance will improve uniformity, but the rate will also drop very fast, and this distance is ultimately limited by the size of the vacuum chamber.

UV lasers are the most commonly used excitation sources in PLD systems, such as KrF (248 nm), ArF (193 nm), and F_2 (157 nm). The UV wavelength allows most of the energy to be absorbed within a short penetration depth of the target, which is necessary for the atomic removal of the target. Longer wavelengths will deposit the laser energy deeper inside the target and could result in clusters

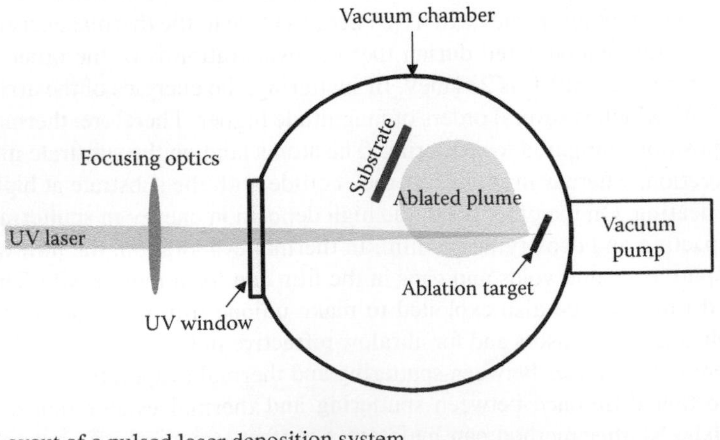

FIGURE 3.21 Layout of a pulsed laser deposition system.

of atoms being ejected. Even with UV lasers, ejection of micron-sized particles and molten liquid droplets can still occur, and it continues to be a problem with PLD.

As mentioned before, PLD allows a wide range of gases and pressures to be used. It is possible to use the gas to decelerate the ablated species, or filter out certain energies. Since pressure controls the scattering rate of the ablated atoms with the background gas atoms, the speed of the depositing atoms on the substrate can be controlled by adjusting the gas pressure. Similar techniques have also been used to create nanoparticles [25]. PLD is also more suitable for reactive deposition of certain metals that require a high level of background pressure. One example is the synthesis of VO_2, which when produced from pure metallic vanadium requires a narrow process window with a high level of oxygen pressure and high substrate temperatures [26]. The pressure can be too high for magnetron sputtering or e-beam evaporation, but PLD is more flexible to allow a large range of pressures to be used during deposition.

3.2 CHEMICAL VAPOR DEPOSITION

One of the characteristics of PVD is that the source material is in the same chemical state as the final thin film. Reactive sputtering and reactive evaporation can be considered minor exceptions to this definition, where a gas is used to modify the chemical state of the film. In CVD, all of the starting materials are volatile and are in a different chemical state than the intended film material. The film is produced from gaseous precursors as a result of a chemical reaction that occur at the surface of the substrate. Besides device fabrication, CVD is widely used in many industrial processes, such as titanium nitride and tungsten carbide coatings on cutting tools, silicon carbide diffusion barrier coatings in turbine engines, and other thermal barrier and wear-resistant coatings as well as in the production of ceramics [27,28].

The deposition rate in CVD will be directly related to the gas pressure. Higher pressures will provide more reactive species and result in a greater reaction rate. This is another major difference between CVD and PVD. The pressures used in CVD are generally in the range of hundreds of mT to several Torrs in the viscous flow regime, whereas in PVD the pressures are rarely higher than a few mT.

CVD can be used for making dielectric or metal films. Crystalline films can also be made under specific conditions on lattice-matched substrates, and is known as epitaxy. A characteristic of CVD is that the films produced are generally conformal (good step coverage)—that is, they exhibit nearly equal deposition rates on horizontal and vertical surfaces. This is a desirable property in applications such as passivation, planarization, and insulation, as well as in filling holes (known as vias) with metals in integrated circuit processes. The conformal property arises due to the isotropic diffusion of the reactive gases, whereas in PVD the deposition species follow a line-of-sight path. The stoichiometry of a compound can also be easily controlled in CVD over a very large range. This allows one to tune the material properties such as density, refractive index, and stress to meet different application requirements. The major drawbacks of CVD are the requirement for high substrate temperatures and the use of potentially hazardous gases. With PVD, substrate heating or the use of gases are not fundamental requirements, except when necessary to condition the film during the deposition.

SiO_2, Si_3N_4, SiC, amorphous silicon, refractory metals, and diamond are examples of films that can be made with CVD. The gas precursor for silicon is most often SiH_4 (silane) or SiH_2Cl_2 (dichlorosilane). Examples of metal precursors are WF_6 (tungsten hexafluoride), $(CH_3)_3Al$ (trimethyl aluminum), $CuCl_2$ (copper chloride), etc. A few commonly used CVD reactions are as follows:

- $SiH_4 + 2N_2O \rightarrow SiO_2 + 2N_2 + 2H_2$
- $SiH_4 + O_2 \rightarrow SiO_2 + 2H_2$
- $3SiH_4 + 4NH_3 \rightarrow Si_3N_4 + 12H_2$
- $3SiH_2Cl_2 + 4NH_3 \rightarrow Si_3N_4 + 9H_2 + 3Cl_2$
- $WF_6 + 3H_2 \rightarrow W + 6HF$

- $SiH_4 \rightarrow Si + 2H_2$
- $CH_4 \rightarrow C + 2H_2$
- $Al(CH_3)_3 \rightarrow Al +$ hydrocarbons

The precursors used in CVD can be generally categorized into halides, hydrides, and metal organics. In the aforementioned list, WF_6 and SiH_2Cl_2 are halides, SiH_4 and CH_4 are hydrides, and $Al(CH_3)_3$ is a metal organic.

One of the goals in CVD is to allow the primary reaction that produces the film to occur on the substrate and not in the gas phase. Gas phase reactions are undesirable because the by-products can precipitate on the substrate and produce a film similar to PVD. Depending on the size of the precipitates, this can result in large grains, pin holes, and film roughness. Gas phase reactions can be reduced by choosing the pressure and temperature such that the reactions only occur on the heated substrates.

For the reaction to be thermodynamically favorable, the parameter known as the Gibb's free energy has to be negative. For example, the reaction $SiH_4 + 2N_2O \rightarrow SiO_2 + 2N_2 + 2H_2$ has a Gibb's free energy $\Delta G = -1121$ kJ/mol at room temperature, becoming $\Delta G = -1184$ kJ/mol at 1000°C. These values were obtained from the FactSage thermodynamic software [29]. Similarly, $SiH_4 + O_2 \rightarrow SiO_2 + 2H_2$ has $\Delta G = -913$ kJ/mol at room temperature and has $\Delta G = -832$ kJ/mol at 1000°C. However, the reaction kinetics is much more difficult to predict, and would depend on the exact reaction pathway. The activation energy E_a, which dictates the reaction rate, is often measured rather than calculated. Most of these reactions require an elevated temperature, in the range of 500°C–1000°C in order to proceed at a reasonable rate. For the nonchemist, it may seem odd that silane (SiH4) and oxygen require high temperatures to react, given that silane is a highly pyrophoric gas that reacts with oxygen in the air spontaneously and violently. This has to do with the pressure. At atmospheric pressures, even if the probability of reaction is very low at room temperature, there are a large number of oxygen molecules that some of them will react. The energy released during the reaction will be sufficient to cause a local increase in temperature to accelerate the next reaction, eventually leading to a rapid increase in reaction rate.

There are two main types of CVD reactors: cold wall reactors and hot wall reactors. In a cold wall reactor, only the substrates inside the reactor are heated, either inductively or radiatively, while the walls are kept cool. This prevents deposition from taking place on the walls because such depositions could eventually accumulate and flake off leading to debris and contaminations on the substrates. However, the cold walls can also lead to undesirable temperature gradients in the chamber. As a result, cold walls are most suitable for applications requiring thick-film depositions. Hot wall reactors are more common in semiconductor applications. Here, the walls and the substrate are heated to the same temperature. Depositions will take place on all surfaces, including the chamber walls, but uniformity will be significantly better than in cold wall reactors.

The design of a CVD system is driven mostly by the rate-determining step for the reaction. In the simplest model, there are two rate-determining steps: (1) diffusion rate of the reactants from the free flowing gas to the substrate surface through the stagnant boundary layer and (2) surface reaction rate. The smaller of the two will limit the overall reaction rate. This can be thought of as two serial processes each with a certain conductance (or resistance). The smallest conductor (largest resistor) will limit the overall flow through the system, as shown in Figure 3.22.

For uniform deposition, it is desirable to have a laminar gas flow instead of a turbulent flow. Turbulent flows can cause differences in gas concentration inside the chamber and lead to non-uniformities in the deposition. The flow characteristics of a viscous fluid are determined by its Reynolds number, which is a dimensionless parameter specified as

$$Re = \frac{\rho \upsilon d}{\eta} \tag{3.26}$$

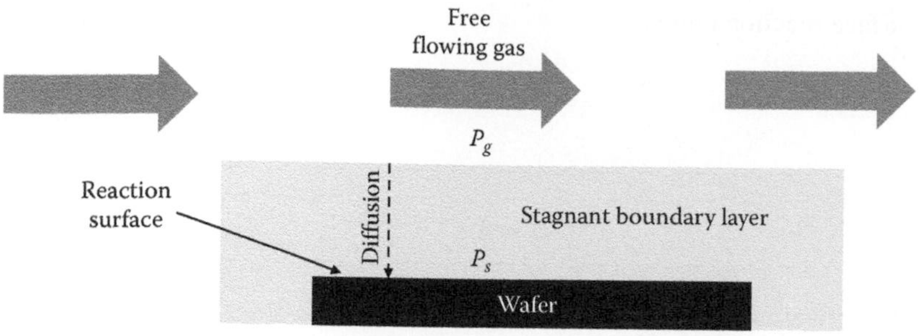

FIGURE 3.22 Gas flow characteristics in a chemical vapor deposition process.

where
- ρ is the density (which is directly related to pressure)
- υ is the flow velocity
- d is the diameter of the reaction tube (or chamber)
- η is the viscosity

Reynolds number is a representation of the ratio between the viscous force and the inertial force in a flow. When the flow is dominated by viscous forces, laminar flow results. When it is dominated by inertial forces, turbulent flow results. Generally, $Re < 1000$ is desired to ensure laminar flow. The gas velocity υ will have to be maintained at a value necessary to supply the feed gas for the reaction, and the dimensions of the reactor will be mostly dictated by the substrate sizes. As a result, ρ (or, equivalently pressure) is the only quantity that can be changed to ensure laminar flow. Most CVD reactors operate at Reynolds numbers of around 100.

With laminar flow, there will be a thin stagnant boundary layer on all surfaces, and transport through this layer occurs by diffusion. A difference in partial pressure will develop between the free flowing gas stream and the stationary substrate surface. This flow rate can be written as

$$J_{12} = h_g(P_1 - P_2) \tag{3.27}$$

where
- P_1 is the partial pressure of the reactant species in the free flowing gas
- P_2 is the partial pressure near the substrate surface

The adsorption process is

$$J_{23} = k_a(P_2 - P_3) \tag{3.28}$$

with the pre-exponential factor following the well-known Arrhenius equation

$$k_a = A_a e^{-(E_{aa}/kT)} \tag{3.29}$$

where E_{aa} is the activation energy for the adsorption process. Simultaneously, there will be a desorption process:

$$J_{32} = k_d(P_3 - P_2) \tag{3.30}$$

with

$$k_d = A_d e^{-(E_{ad}/kT)} \tag{3.31}$$

where E_{ad} is the activation energy for the desorption process.

Finally, the surface reaction rate is

$$J_{30} = k_s P_3 \tag{3.32}$$

with

$$k_s = A_s e^{-(E_{as}/kT)} \tag{3.33}$$

where E_{as} is the activation energy of the reaction.

The combination of all three factors can be visualized using a circuit model where $1/h_g$ is the diffusion resistance, $\frac{1}{k_a - k_d}$ is the resistance due to the net adsorption, and $1/k_s$ is the reaction resistance. The final current flow J_{30} represents the film growth rate (see Figure 3.23).

We can consider two limits:

1. If $\frac{1}{h_g} \gg \frac{1}{k_a - k_d} + \frac{1}{k_s}$ (diffusion resistance is large), then $J_{30} \approx h_g P_1$. In this case, the reaction rate is determined only by the diffusion resistance. This is the diffusion-limited case.
2. If $\frac{1}{h_g} \ll \frac{1}{k_a - k_d} + \frac{1}{k_S}$, then $J_{30} \approx \frac{P_1}{\frac{1}{k_a - k_d} + \frac{1}{k_S}}$. In this case, the adsorption and reaction resistances are greater than the diffusion resistance, and the reaction rate is determined mostly by the surface properties. This is the surface-limited case.

The boundary layer thickness δ increases as the Reynolds number gets smaller. This is expressed as [30]

$$\delta = \frac{5x}{\sqrt{Re}} \tag{3.34}$$

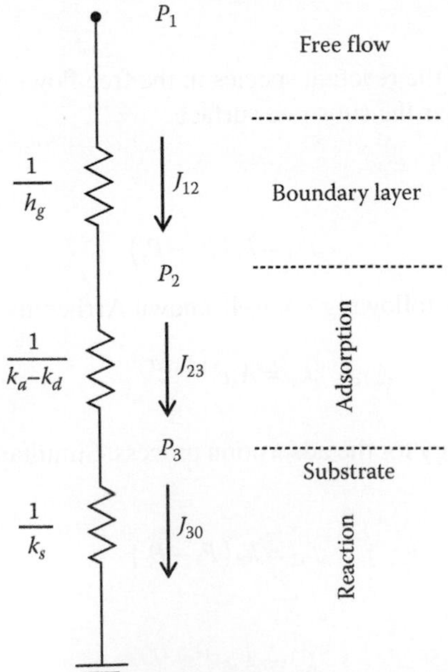

FIGURE 3.23 Circuit description of the rate-determining steps of the chemical vapor deposition reaction.

where x is the distance from the leading edge of the flow container. The diffusion coefficient varies inversely with pressure as

$$D = \frac{2}{3pd^2}\sqrt{\frac{k^3T^3}{\pi^3 m}} \tag{3.35}$$

where
p is the pressure
d is the diameter of the gas molecules
m is the molecular mass

The viscosity is independent of pressure and is expressed as

$$\mu = \frac{2}{3d^2}\sqrt{\frac{kT}{\pi^3}}. \tag{3.36}$$

Combining Equations 3.26 and 3.34 through 3.36, the diffusion parameter h_g that varies as D/δ can be found to be

$$h_g \propto \frac{D}{\delta} \propto \frac{T^{\frac{3}{4}}}{P^{\frac{1}{2}}}. \tag{3.37}$$

Although this equation is not quantitative, it still provides some guidance on how pressure and temperature play a role in the gas transport through the boundary layer. We can see that the boundary layer diffusion parameter h_g will get larger at lower pressures and higher temperatures. Therefore, pressure and temperature will determine which of the two rate-limiting factors that we discussed earlier will dominate the process.

Most commonly encountered CVD systems are

- Atmospheric pressure CVD (APCVD)
- Low-pressure CVD (LPCVD)
- Plasma-enhanced CVD (PECVD)

3.2.1 Atmospheric Pressure Chemical Vapor Deposition

This is typically used as a purge-and-flow system at near atmospheric pressures. Due to the high pressure, from Equation 3.37, h_g will be small, and the primary rate-determining step will be the diffusion rate through the boundary layer (case 1 of the two limiting cases discussed earlier). Typically, no vacuum systems are used in APCVD. The chamber is initially purged with argon or nitrogen to remove residual gases. Then highly diluted reactive gases are flowed over the heated substrates at atmospheric pressures. The dilution is necessary to reduce the gas phase reactions that can occur before the gases reach the substrates. Deposition rates are high, generally greater than 1000 Å/min. Since the reaction is limited by the diffusion rate, the substrates have to be placed adequately far apart to prevent bottlenecks in the gas transport. Obstructions can create turbulent flows and this will result in variations in deposition rates. Also, some reactions in the gas phase are unavoidable due to the higher pressures and slower diffusion (which results in longer gas phase residency times). This results in a large number of particles being produced, which have to be continuously removed with an exhaust, but some of them will inevitably become embedded in the film.

In a tube reactor configuration the substrates have to be placed horizontally and parallel to the gas flow to ensure a uniform flow free of turbulence. An alternative configuration to achieve higher throughputs is a linear nozzle injector. The gases are injected at high speeds from separate ports within the injector with nitrogen or argon curtains to reduce mixing in the gas phase. The high injection speed reduces the residency time and therefore gas phase reactions. The heated substrates typically move horizontally under the injectors.

Due to the challenges in film uniformity, one rarely finds APCVD systems in typical device fabrication cleanrooms. However, they are attractive for some applications such as the manufacture of solar cells where the devices are made on large flat panels and low cost and high throughputs are more important than high quality and uniformity [31].

3.2.2 Low-Pressure Chemical Vapor Deposition

In LPCVD, the pressure is in the range of 1 Torr or lower. From Equation 3.37, we can verify that h_g will get larger, so the primary rate-determining step will be the surface rates (adsorption, desorption, and surface reaction rates). This is case 2 of the two limiting cases we discussed. The primary factors that affect deposition rate will be temperature and pressure. Due to the low pressures, the deposition rates will be lower, typically in the range of 10–100 Å/min. Most importantly, maintaining uniform temperature and pressure is easier than maintaining uniform flow characteristics. This is the main advantage of LPCVD over APCVD. Since the deposition rate is independent of the gas flow characteristics, it is possible to stack the wafers vertically very close to each other in a tube furnace to improve manufacturing throughput. However, microscopic features on the substrate may still encounter diffusion barriers that will limit the step coverage over those structures. For this reason, a high desorption rate k_d is beneficial, as this would increase the overall surface resistance and minimize any effects from h_g. This is referred to as a "high exchange flux" condition, where only a small fraction of the species arriving at the substrate participate in the reaction, while a large portion is desorbed from the surface [32].

Since the gas flow pattern is less critical in LPCVD, the reactor can be configured as a horizontal or vertical tube, depending on how the wafers are stacked. Figure 3.24 shows a horizontal tube setup. Due to the high exchange flux condition, a large portion of the reaction species go unused and will end up in the exhaust stream. Due to the hazardous nature of most LPCVD precursors, an exhaust scrubber is required before the gases can be vented out. It is also possible to recycle the gases, but such a system is rarely used due to concerns of contamination.

The high temperature and conformal nature of the deposition makes LPCVD ideal for making dense pinhole-free thin films. As a result, LPCVD is widely used in electronics, photonics, and MEMS. The most common films are Si_3N_4 (silicon nitride) and SiO_2 at temperatures ranging from 500°C to 800°C. Although some LPCVD depositions, such as tungsten, can be done at lower temperatures, the main

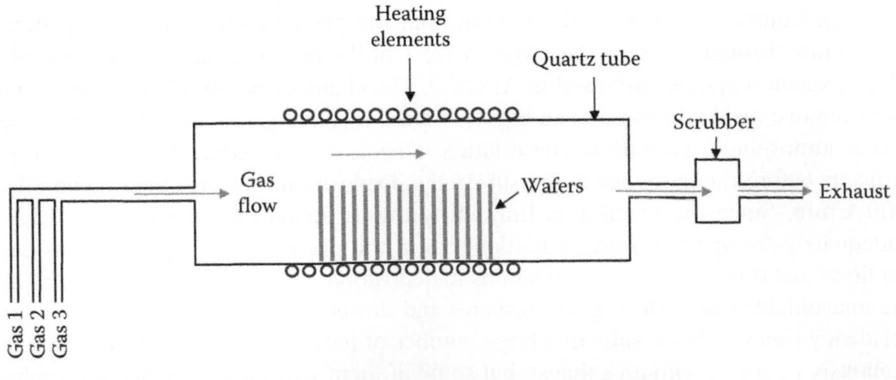

FIGURE 3.24 Low-pressure chemical vapor deposition horizontal tube reactor.

drawback of this method is the high temperature. This eliminates certain substrates and temperature-sensitive devices, including CMOS electronics in the later stages of the fabrication process.

3.2.3 Plasma-Enhanced Chemical Vapor Deposition

For CMOS electronics, there was a need to develop films similar in quality to LPCVD but at lower substrate temperatures. For example, when making metal interconnects on an integrated circuit, each metal layer has to be interleaved between insulating dielectric layers, and exposing the nearly complete chips to LPCVD temperatures would have a detrimental impact on their thermal budget. In PECVD, the pressure is maintained at similar values as LPCVD (<1 Torr), but the substrate temperature is lowered to around 300°C. At this temperature, the LPCVD reaction rate would be nearly zero. Therefore, RF plasma excitation is used to reduce the activation barrier (E_{as} in Equation 3.29). The plasma decomposes the gases into free radicals, lowers the activation energy, and significantly accelerates the reaction rate.

Since this is still a surface reaction rate-limited process, the gas flow characteristics will not significantly affect the uniformity, just as it was with LPCVD. However, the film growth rate will be a strong function of the plasma density and the proximity of the wafers to the plasma. This places restrictions on the chamber configurations and substrates placement. The simplest RF plasma is a parallel plate configuration, which will require the substrates to be placed horizontally similar to the APCVD system. Figure 3.25 shows an example of a parallel plate PECVD system. The wafers are placed horizontally on the heated plate in close proximity to the cathode. The cathode plate also contains a gas shower head to uniformly distribute the gas, and the gas is drawn symmetrically around the hot plate by four pump inlets. Horizontal tube configurations similar to LPCVD can also be used, but special electrode configurations will be needed to ensure a uniform plasma density in the entire tube volume.

The RF discharge powers are generally kept low, at the minimum power necessary to overcome the activation energy E_{as}. Excessive discharge powers can increase the probability of reaction in the gas phase and can produce particulates and PVD-like properties. Also, at high RF powers, the reaction point will move from being surface reaction-limited to transport-limited, and nonuniformities due to gas flow patterns can also start to emerge.

A drawback of PECVD films compared to LPCVD is the inclusion of high levels of byproducts and other gases. Hydride-based PECVD reactions generally contain excess hydrogen in the films. This arises due to the low temperature that prevents the gases from fully desorbing during the

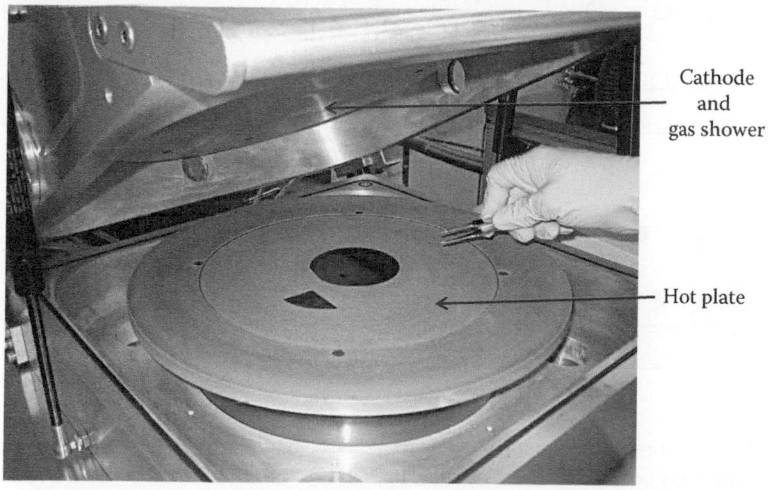

FIGURE 3.25 Plasma-enhanced chemical vapor deposition system.

reactions. As a result, achieving stoichiometrically correct films becomes difficult, which can affect their physical properties and resistance to chemical etching.

The ion bombardment of the plasma can densify the film, which can be beneficial, but it can also lead to a compressively stressed film. Ion damage to other components on the substrate can also be a concern in PECVD. Despite these disadvantages, PECVD is widely used for making the dielectric layers for metal interconnects in CMOS manufacturing [33].

For more comprehensive reviews of CVD processes, the reader is directed to the references listed at the end of this chapter [29,31,34–36].

3.2.4 Atomic Layer Deposition

ALD is relatively new and is quickly growing to become a mainstream deposition method. It is a variant of the conventional CVD process, except the precursors are flowed sequentially rather than simultaneously. Assuming two precursors containing the species *A* and *B* to produce a compound *AB*, the precursor containing *A* is flowed first, and the molecules are allowed to attach to the substrate by adsorption. The chamber is then purged with an inert gas, and the precursor containing *B* is flowed next. The *B* precursor molecules will react with the adsorbed *A* precursor molecules and produce the film *AB* (see Figure 3.26). Since there is only a finite number of *A* precursor molecules adsorbed on the surface, the reaction is self-limiting, that is, it will come to a stop once all the adsorbed *A* precursor molecules have reacted. The film produced during one cycle will typically be one monolayer thick, or less, depending on the extent of coverage and the sticking coefficient of the precursors on the substrate. This cycle is repeated many times to produce the desired film thickness. Because adsorption is a nondirectional process driven by diffusion, all surfaces will be coated equally with excellent step coverage, as shown by the example in Figure 3.27. ALD is best suited for applications that need very thin, very-high-quality, pinhole-free films. The gate oxide of MOS transistors is the most coveted application for ALD. While the supremacy of silicon originally relied on its ability to form high quality thermal oxides as the gate dielectric of MOS transistors, ALD has made it possible to make gate dielectrics from other materials with even higher dielectric constants, such as Al_2O_3 and HfO_2.

Al_2O_3 and TiO_2 are some of the most developed processes used for ALD. Al_2O_3 is created using trimethylaluminum ($Al(CH_3)_3$) and H_2O as the two precursor gases. TiO_2 is created by $TiCl_4$ and H_2O. The reactions proceed as follows [38]:

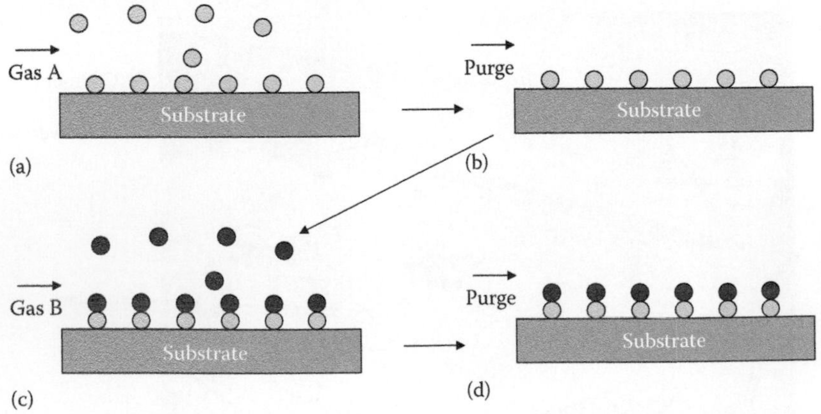

FIGURE 3.26 Pulse and purge description of an atomic layer deposition process using two gases precursors A and B. (a) Precursor gas containing A is flowed and allowed to attach to the substrate; (b) purge gas removes all traces of precursor A from the chamber; (c) precursor gas containing B is flowed and allowed to react with the adsorbed A precursor molecules; and (d) purge gas removes all traces of precursor B from the chamber.

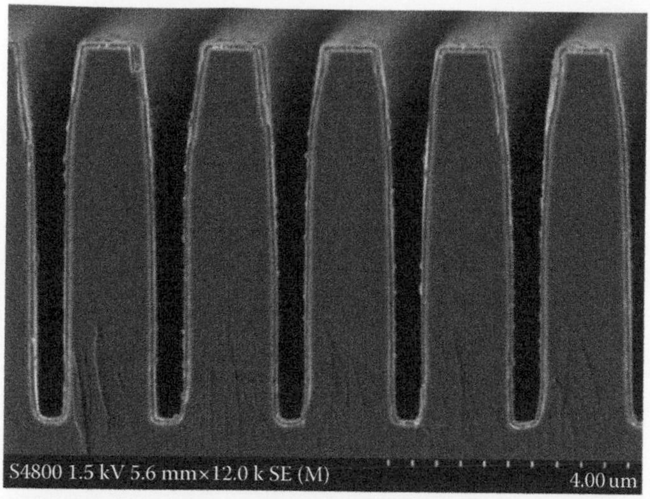

FIGURE 3.27 Conformal deposition of ALD films on a high aspect ratio structure. (Reprinted with permission from Pore, V., Hatanpää, T., Ritala, M., and Leskelä, M., Atomic layer deposition of metal tellurides and selenides using alkylsilyl compounds of tellurium and selenium, *J. Am. Chem. Soc.*, 131(10), 3478. Copyright 2009 American Chemical Society.)

$$2Al(CH_3)_3 + 3H_2O \rightarrow Al_2O_3 + 3CH_4$$

$$TiCl_4 + 2H_2O \rightarrow TiO_2 + 4HCl$$

The process temperature can be anywhere from room temperature to 300°C. The temperature not only affects the reaction rate, but also the adsorption and desorption rates. Therefore, higher temperatures do not necessarily lead to higher deposition rates.

While binary compounds are the most developed for ALD, single element metals can also be deposited by reducing a metal-containing precursor. These processes are generally more difficult than binaries and require plasma activation of hydrogen. Nevertheless, processes for a number of metals such as tantalum, silicon, germanium, and tungsten have been developed, and the list continues to grow. Although ALD is a very promising technique, its major disadvantage is the slow rate of growth due to the pulse and purge cycles.

3.3 THIN-FILM MEASUREMENTS

3.3.1 Thickness Measurement with a Quartz Crystal Microbalance

Measuring the deposition rate of incident atoms in real time is an important requirement in thin-film deposition. The quartz crystal microbalance (QCM) is the most widely used method for in situ measurement of thin films. This is based on measuring the mass loading effect on a piezoelectric crystal. One may also be familiar with quartz crystal oscillators as precise timekeeping components in electronic circuits and watches. QCMs are also used in a large number of chemical and biological applications to measure small changes in mass [39,40].

Quartz is the crystalline form of SiO_2. It is also a piezoelectric material. This means that when a voltage is applied across the crystal faces, it will induce a strain. The crystal orientation of the quartz substrate is selected such that a transverse strain is induced when a voltage is applied across the two parallel faces of the substrate. If the voltage is oscillating, a transverse acoustic wave will be induced in the crystal. The acoustic wave will reflect off both substrate boundaries and setup a standing wave. This resonance frequency will be a function of the crystal thickness and the

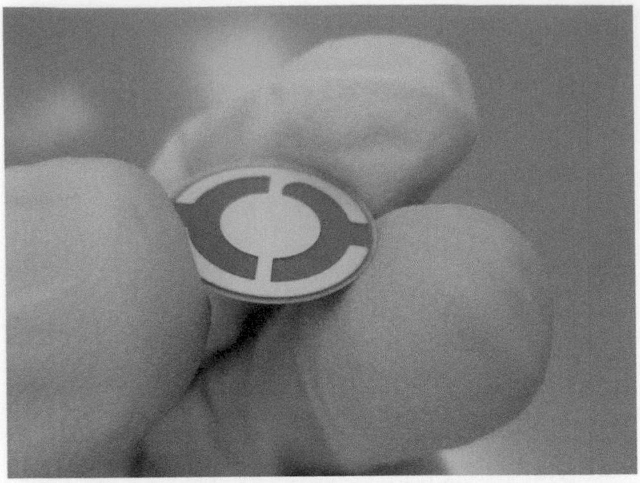

FIGURE 3.28 Quartz crystal microbalance.

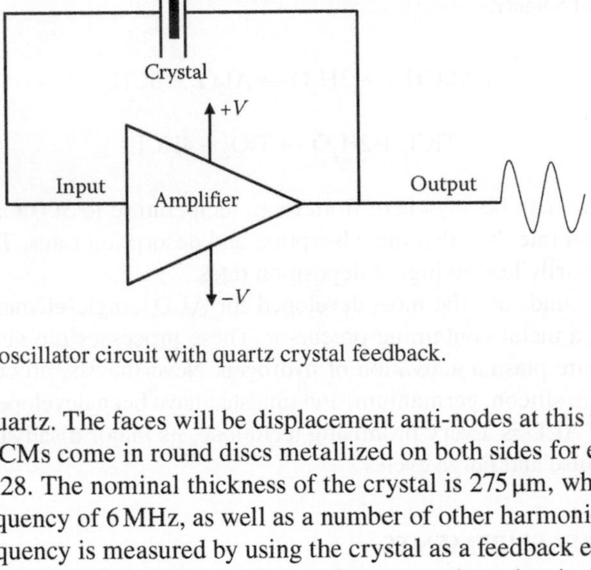

FIGURE 3.29 Active oscillator circuit with quartz crystal feedback.

acoustic velocity in quartz. The faces will be displacement anti-nodes at this resonance. Most commercially available QCMs come in round discs metallized on both sides for electrical connections, as shown in Figure 3.28. The nominal thickness of the crystal is 275 μm, which produces a fundamental resonance frequency of 6 MHz, as well as a number of other harmonics.

The resonance frequency is measured by using the crystal as a feedback element of an amplifier circuit (see Figure 3.29). The electrical impedance of the crystal reaches its lowest value when the frequency of the electrical signal is exactly equal to the resonance frequency of the acoustic signal. When used as a feedback element, the amplifier will enter a self-sustained oscillation at this frequency, which can then be measured with a frequency counter.

Since the face of the crystal is an anti-node of the standing wave, a thin film that becomes attached to the face will act as an additional inertia that will reduce the vibration frequency. The relationship is described in the Sauerbrey equation:

$$\frac{\Delta f}{f} = \frac{2f}{Z_q} m \tag{3.38}$$

where

f is the resonance frequency
Δf is the frequency shift

Z_q is the acoustic impedance of quartz
m is the mass per unit area of the film

If the density of the film is known, the mass per unit area can be converted to a thickness value. This equation treats the film as an infinitesimally thin mass attached to the end of the quartz crystal face. As a result, it only works for very thin films that produce frequency changes of 2% or less. For thicker films, the acoustic impedance of the film must also be taken into consideration. This is done in the so-called Z-match method. It is an extension of the Sauerbrey equation and includes the acoustic impedance of the film material [39]. The acoustic impedance of the film is specified as a ratio compared to that of quartz, and is known as the z-ratio. This is the most widely used method in thin-film deposition systems. In order to translate the change in oscillation frequency Δf to a film thickness, the model requires two parameters for each film: its density and z-ratio.

In a typical application, a fresh crystal starts with a frequency of 6 MHz. As films are deposited, the frequency will drop. Generally, the sensitivity of the crystal starts to degrade when its frequency falls significantly below it nominal value, such as 5 MHz. When this happens, the crystal should be discarded and a new crystal installed. Quartz crystals are inexpensive and can be replaced often.

To ensure correct readings, the QCM is installed inside the vacuum chamber in the path of the incident deposition flux, as close as possible to the substrate location. A typical example of a QCM installation in an evaporation chamber is shown in Figure 3.30.

As with any measurement system, there are a number of factors that produce inaccurate results from a QCM that are worth discussing.

3.3.1.1 Temperature Sensitivity

The crystal angle (known as the AT-cut) is selected to minimize the sensitivity of QCM frequency to temperature changes. However, this sensitivity is designed to be a minimum only at room temperature. The temperature sensitivity is worse at elevated temperatures that may be encountered in thin-film deposition systems. In thermal evaporation systems, the crystal is exposed to a significant amount of radiative heating. Therefore, the crystals are normally installed in a water cooled fixture to maintain a constant temperature. Nevertheless, the front face of the crystal can still experience temperature swings due to the low thermal conductivity of quartz. A rise in

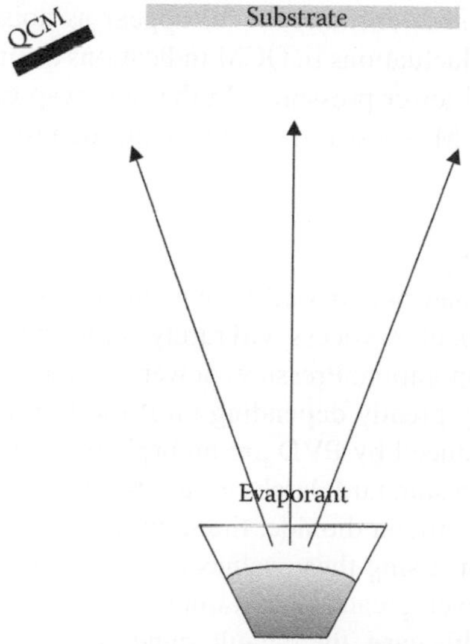

FIGURE 3.30 Installation location of a quartz crystal microbalance inside an evaporation vacuum chamber.

temperature will increase the resonance frequency and register as a negative rate of film growth, and a drop in temperature will reduce the frequency and register as a positive rate of film growth. In a typical deposition cycle, the crystal will under-indicate during the ramp up and over-indicate during the ramp down. This aspect may not make a huge difference in thick-film applications where the ramp-up and ramp-down portions are relatively short compared to the total deposition time, but it will significantly affect the results when trying to grow very thin films in the range of 10–50 Å.

3.3.1.2 Tooling Factor

To get accurate indications, the crystal should be located as close to the substrate as possible. However, due to the finite substrate sizes, some difference in thickness between the QCM and the substrate is unavoidable. This difference should be a fixed number and can be calibrated out during trial runs. The ratio $t_{substrate}/t_{QCM}$ is known as the tooling factor (TF). Since the substrate is normally placed at the center of the deposition flux and the QCM at the edge, the TF is usually greater than 1. Typical numbers are 1.0–1.2 depending on the geometry of the system.

In thermal evaporation, the deposition rate follows a cosine distribution fairly accurately so the TF should also follow a cosine value and should remain constant over a wide range of operations. This is not the case with sputter deposition. The distribution of the deposition flux is a function of pressure and power, so the TF will also change with these parameters.

3.3.1.3 Film Stress

QCM relies on having stress-free films with good adhesion to the crystal face. A compressively stressed or tensile-stressed film will produce a lateral force on the crystal face and change its curvature, especially after a significant accumulation of film material. This will affect the crystal resonance and produce unreliable readings. Excessively stressed films can also create internal defects and voids. These areas will change the way the acoustic fields propagate through the film and affect the readings. It is also possible for the films to entirely delaminate from the QCM and show a large frequency jump, which will be indicated as a large negative deposition rate.

3.3.1.4 Deposition Energy

The energy of the incident atoms can also affect the QCM operation. In sputter deposition, the QCM will be bombarded with high-energy atoms that will transfer momentum to the crystal and produce random acoustic excitations. These excitations will appear as noise in the QCM resonance. As a result, it is normal to see large fluctuations in QCM indications during sputter deposition, especially at higher discharge powers and lower pressures. In thermal evaporation, the readings will contain a lot less noise. As a result, QCM is less reliable in sputter deposition systems than in evaporation systems.

3.3.1.5 Density and *z*-Ratio

Although the density of most materials is well known and documented, the film being grown in a particular chamber using a particular process will rarely conform to these values. Sputter deposition produces a denser film than evaporation. Pressure, power, and substrate temperatures can also affect the density. They can also vary greatly depending on if the film is polycrystalline or amorphous. The vast majority of films produced by PVD are amorphous or polycrystalline. For some materials, the density values quoted in standard databases are for their crystalline form. For example, the density of silicon, germanium, silicon dioxide, titanium dioxide, etc., found in databases are most likely for their crystalline form. Using these values in a QCM measurement in a PVD system can lead to errors. The *z*-ratio has even greater uncertainties. The acoustic impedances of many materials are simply not known. In this case, the default value of 1.0 is used, which essentially assumes the acoustic impedance of the film to be the same as that of quartz.

3.3.2 Thickness Measurement with a Stylus Profiler

While QCM is the most commonly used method for monitoring the thickness during deposition, it has a number of shortcomings as described in the previous section. One always needs a reliable technique to verify the film thickness after the deposition so that the QCM can be calibrated. Probably the most reliable and repeatable thickness measurement tool is the contact stylus profiler. It measures a step height variation by dragging a fine tip across the sample and measuring its deflection. If all sources of vibrations can be reduced, the accuracy and repeatability of this method can be as little as 1 nm.

In order to measure the thickness of a deposited film, a sharp step is required. This is most easily done by placing an ink dot with a fine-tipped marker on a companion substrate, performing the deposition and dissolving the ink away in a solvent such as acetone. This would expose the substrate and the sharp transition step at the film/substrate interface can be used to measure the film height with a profilometer (Figure 3.31).

3.3.3 Measurement of Optical Properties

One could measure the optical transmission and reflection of the film in real-time through transparent viewports in the deposition chamber. An ellipsometer can also be used. But these techniques require the film to be transparent at the interrogation wavelength and would also require the viewports to remain clean and free of deposition buildup.

Similar to the stylus profiler, one could also use a companion sample to measure the optical reflection and transmission as a function of wavelength after the deposition has been completed. This can then be fitted to known dispersion models of materials to extract the refractive index and film thickness. This method is particularly useful for optical thin-film applications where the optical thickness may be slightly different from the physical thickness measured from the stylus profiler.

3.3.4 Thin-Film Stress

Film stress is the tendency of a film to want to shrink or expand. If it wants to shrink but it is prevented to do so due to its attachment to the substrate, the molecules in the film will be held in a stretched position. This produces a film that is under tensile stress. If it wants to expand, but it is

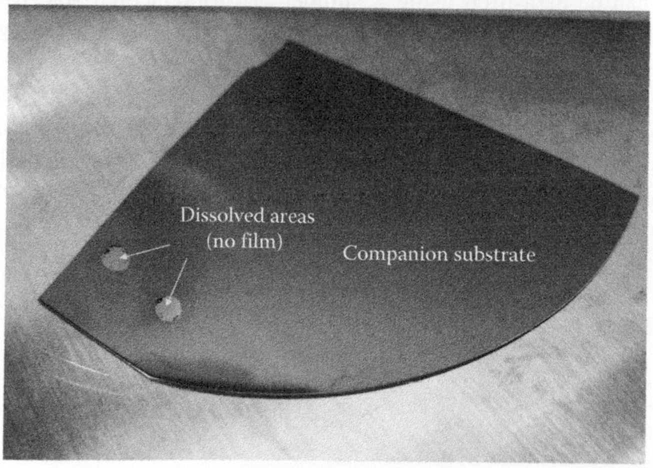

FIGURE 3.31 An ink spot can be used to create a step height for measurement with a profilometer.

prevented from doing so, the molecules will be held in a compressed position. This is a film with a compressive stress.

When the compressive or tensile forces exceed the adhesion force between the film and substrate, the film can peel, crack, buckle, or delaminate. Even if the film does not separate, it will change the curvature of the underlying substrate. In applications with tight mechanical tolerances, this can become a problem. Therefore, understanding and controlling stress is an important aspect of thin-film engineering.

3.3.4.1 Origins of Film Stress

The deposition process itself introduces stresses in a film. For example, in sputter deposition the high energy of the incident species will compact the film into a compressed state. The distances between the atoms will be smaller than their equilibrium distances. After the deposition, the film will want to expand to reduce its internal energy, which is when it will warp the substrate or delaminate. In thermal evaporation, the opposite can happen. Due to the very low energies, the film can have a larger atomic distance than its equilibrium distances [41]. After the deposition, atoms may migrate closer together causing the film to become stressed in a tensile state. In CVD, the films can be tensile or compressive depending on the reaction pathways. Reactions with byproducts that desorb from the substrate often tend to form tensile films. Inclusion reactions produce compressive films. For example, Si_3N_4 from SiH_4 and NH_3 produces tensile films due to the desorption of hydrogen, but SiO_2 by direct oxidation of silicon produces compressive films due to the inclusion of oxygen. In PECVD, film stress can be controlled to some degree by utilizing the ion bombardment effect. All of these stresses are known as intrinsic stress or residual stress.

Extrinsic stress is caused by external means, such as changes in temperature. Any mismatch in the coefficient of thermal expansion between the film and the substrate will induce a stress when the temperature changes. Absorption of moisture or other vapors into the film can also induce extrinsic stress. Extrinsic stresses are usually smaller in magnitude than intrinsic stresses.

3.3.4.2 Measurement of Stress

Stress is defined in the same units as pressure. This is the amount of force that exists within a cross-sectional area of the film. Imagine a stretched wire. This will be under tensile stress. If we take the force that keeps the wire stretched and divide it by its cross-sectional area, we will get the stress value. The most commonly used unit for stress is Pa, where $1\,\text{Pa} = 1\,\text{N/m}^2$. However, the magnitudes commonly encountered in thin films are in MPa or GPa. Tensile stresses are designated with positive values and compressive stresses with negative values.

Stress in a film can be measured using a number of different methods. Raman spectroscopy can often reveal stresses in a film. Acoustic methods and infrared spectroscopy can also be used. However, the most widely used method is a direct measurement of the substrate curvature induced by the film stress. By measuring the curvature before and after the thin-film deposition, the stress can be accurately evaluated. It may seem surprising that a thin film of the order of 1000 Å can bend a 0.5 mm thick substrate, but the curvature we are speaking of is extremely small and imperceptible to the eye. For example, the radius of curvature before deposition might be 100 m (yes, meters), and after deposition it might be 80 m. This small change in curvature indicates a tensile stressed film.

The relationship between curvature and stress is given by Stoney's equation:

$$\sigma = \left(\frac{E}{1-\upsilon}\right)\left(\frac{t_s^2}{6t_f}\right)\left(\frac{1}{r_a}-\frac{1}{r_b}\right) \tag{3.39}$$

where

E is the Young's modulus of the substrate
υ is the Poisson's ratio of the substrate
t_s is the substrate thickness

t_f is the film thickness
r_a is the radius of curvature after deposition
r_b is the radius of curvature before deposition

This model assumes that the strain energy is stored entirely in the substrate. This is the reason for why the mechanical properties of the film are not contained in the equation.

Since silicon is a very common substrate, it is useful to note its values. For (100) wafers along the (110) direction, $E_{Si} = 171\,\text{GPa}$ and $\upsilon_{Si} = 0.06$, and along the (100) direction it is $E_{Si} = 130\,\text{GPa}$ and $\upsilon_{Si} = 0.28$ [42]. For example, consider a 5000 Å film on a 500 µm silicon substrate. If the pre- and post-deposition radii of curvatures were −50 m and −30 m, respectively, along the (110) direction, the film stress can be calculated to be −153 MPa (compressive).

The radius of curvature can be measured very accurately using a number of different methods. One method is to use a stylus profiler over a very long scan length of at least 1 inch. It is important to scan at exactly the same location before and after the deposition because the intrinsic substrate curvature may not be symmetric. Figure 3.32 shows an example of a profile scan before and after the film deposition. Another approach is to use a noncontact optical interference method. This method can produce a 2D map of the surface topography that can then be compared and subtracted after the deposition. This method can produce a stress map of an entire wafer.

Another useful number is the stress value multiplied by the film thickness. This quantity will be in the units of force per length and represents the shear force imparted by the film on the substrate per unit length. As the film thickness increases, the stress value may remain the same, but the shear force will increase. At a certain shear force, the film may separate from the substrate. Therefore, delaminations are more likely to occur when the films grow thicker, even if the stress values are the same.

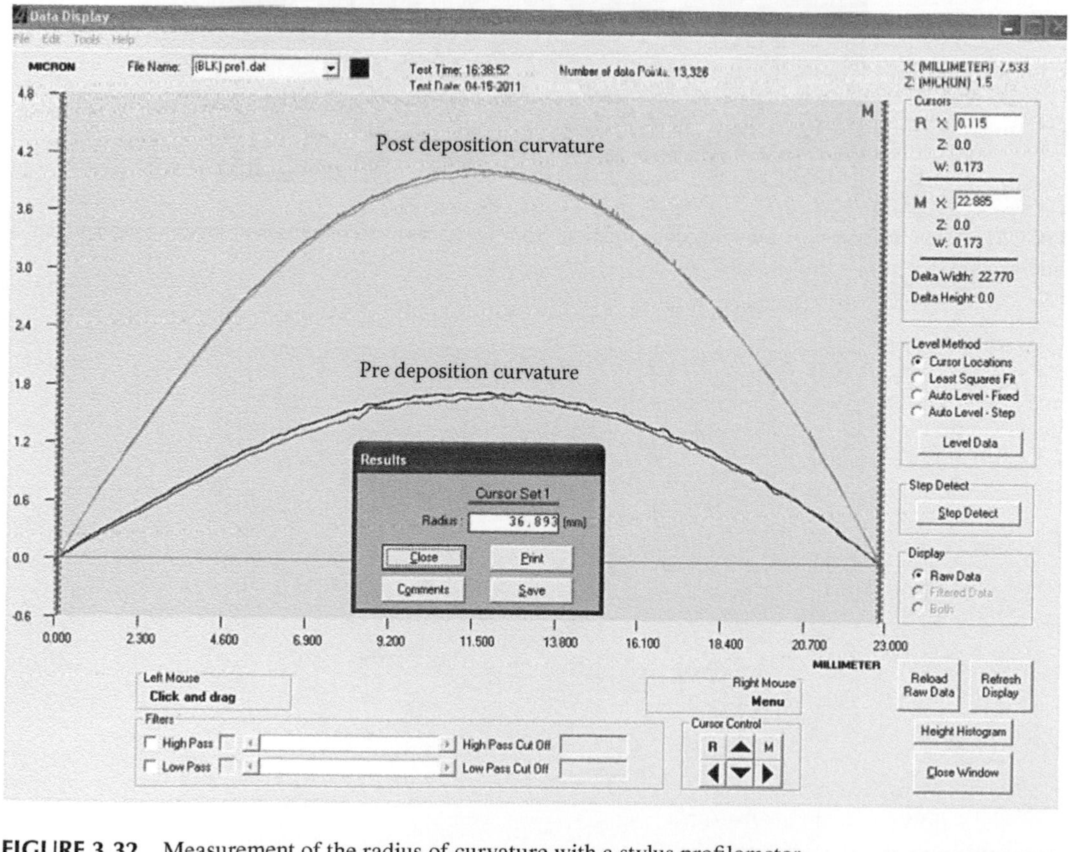

FIGURE 3.32 Measurement of the radius of curvature with a stylus profilometer.

3.3.4.3 Compressive Stress

Compressive stress is noted with a negative stress value. It causes the substrate to become more convex-shaped (as viewed from the film side). Too much compressive stress will tend to lift the film off the substrate in order to make it more convex. This will show up as blistering and buckling. To the naked eye, these buckling delaminations may only appear as a discoloration or as a surface texturing, but under a microscope the buckles should be easy to identify. An example of buckling delaminations of a sputtered silicon film is shown in Figure 3.33.

3.3.4.4 Tensile Stress

Tensile stress has a positive stress value. It causes the substrate to become more concave as viewed from the film side. Excessive tensile stress can cause cracks in the film or it can separate from the substrate at the edges and curl inward to make a more concave film. Very thick films produced by thermal evaporation can display this character. Figure 3.34 shows an example of curling from the substrate edges due to excessive tensile stress from a thick germanium film made by electron beam evaporation.

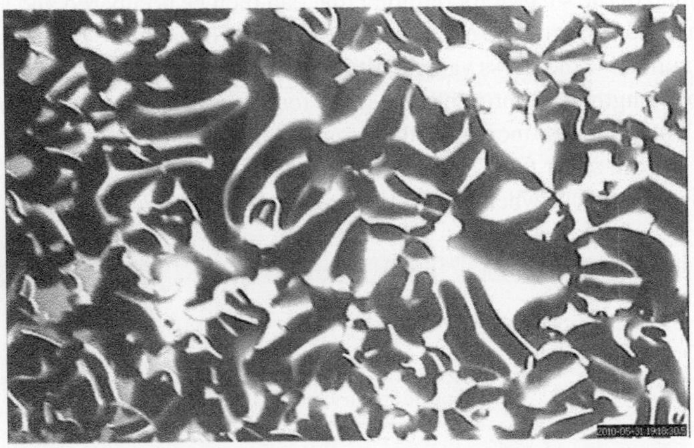

FIGURE 3.33 Example of bucking delamination from an excessive compressive stress viewed under a microscope.

FIGURE 3.34 Example of peeling due to excessive tensile stress.

3.3.4.5 Stress Reduction

It is possible to reduce intrinsic stress by increasing the energy of the atoms during deposition so that they relax to a low-stress configuration. The most common method to accomplish this is by raising the temperature of the substrate during deposition, in the range of 300°C–500°C. However, in some cases the extrinsic stress could become large when the substrate is cooled down after the deposition. Furthermore, not all substrates can withstand high substrate temperatures. An alternative method is ion bombardment of the thin film during the deposition, as with IAD. The ions carry significant energy (several hundred eV), so they are equivalent to a significantly higher temperature than any substrate heater can accomplish. However, IAD is generally only effective with tensile films. With increasing ion energy, the tensile stress can be reduced, or even be turned into a compressively stressed film.

3.4 THIN-FILM MATERIALS

In this section, a brief summary of the thin-film materials most commonly used in many laboratories is described.

3.4.1 Titanium

Titanium is most commonly deposited by evaporation or sputtering. It is one of the most reactive metals, and therefore requires a very low base pressure and extremely low levels of desorption during deposition. Otherwise, one will end up with an oxidized form of titanium. During evaporation, it is common to notice a decline in the background pressure as titanium starts to evaporate (which is known as the gettering effect). With most other materials, the pressure will actually increase during deposition due to an overall increase in the chamber temperature and the associated outgassing. Titanium is commonly used as an adhesion layer for metals such as gold, silver, and copper. Evaporation works slightly better than sputtering because it is easier to monitor the desorption rates during deposition.

3.4.2 Chromium

Chromium is also reactive like titanium, and it requires a very low base pressure before deposition. It also exhibits some gettering effect during deposition. It can be sputtered or evaporated. Under evaporation, it sublimates rather than melt. With electron beam evaporation, the beam spot should be continuously scanned to prevent evaporating from a single spot because fresh material will not flow in like with other materials that melt. Chromium is also commonly used as an adhesion metal like titanium.

3.4.3 Aluminum

Aluminum is also reactive and requires a low base pressure. It can be sputtered or evaporated. Although it melts at a very low temperature (660°C), it needs to be raised significantly above its melting temperature to create a reasonably high vapor pressure. As a result, it tends to wick into many surfaces and profusely wet the crucible liners and create contaminations.

3.4.4 Copper

The reactivity of copper is not very high, so it can be easily sputtered or evaporated. Copper has a very high electrical conductivity and is widely used as an interconnect metal in electronics.

3.4.5 Gold

Gold is very inert and does not react with background gases under normal conditions. It is easy to evaporate or sputter. However, due to its high cost, evaporation is more economical than sputtering because it can be bought in small pellets as needed for each project. Also due to its high cost, the protective liners used in vacuum chambers (typically aluminum foil) can be sent to reclaim and extract the gold. In most cases, an adhesion layer, such as titanium or chromium is necessary before depositing gold. It is often used as the final metallization in devices because it will not oxidize or corrode.

3.4.6 Silver

Silver evaporates and sputters wells, and also has fairly low reactivity. It is a soft metal like gold and requires an adhesion layer. When deposited at very small thicknesses (<10 nm) directly on silicon or silicon dioxide surfaces, it forms distinct droplets due to its poor wetting properties. Thin silver films have reasonable transparency in the visible spectrum, so they are used as infrared blocking films on architectural windows.

3.4.7 Platinum

Like gold, platinum is an expensive metal, so although it can be sputtered, it is more economical to evaporate it from small pellets. It is also not reactive. Platinum has a very large work function so it can be used as an ohmic contact in p-type semiconductor materials, as well as for making Schottky diodes. It has poor adhesion so it must be used with a metal like titanium. Platinum requires high powers to evaporate because it needs to be raised to a very high temperature in order to produce a reasonable vapor pressure. It melts at 1768°C, but needs to be raised to about 2500°C to produce about 100 mT of vapor pressure.

3.4.8 Nickel

Being a ferromagnetic metal, nickel cannot be sputtered with a magnetron cathode because it would shield the magnetic fields and prevent it from confining the plasma. It can, however, be sputtered if it is sliced into a very thin sheet and then bonded to a copper target. It can be easily evaporated, although with an e-beam evaporator, one needs to be aware of its effects on the focusing magnets. Initially, the beam may not focus properly, but once the nickel reaches its curie temperature of 355°C, its magnetism will disappear and it will behave normally.

3.4.9 Tungsten

Tungsten has one of the highest melting temperatures, therefore, it cannot be easily evaporated. Fortunately, it has a reasonably high sputter yield, so it is most often deposited by sputtering. Tungsten can also be deposited with CVD using WF_6 as the precursor gas. This is used for via-filling applications in integrated circuits.

3.4.10 Molybdenum

Molybdenum is very similar to tungsten and is difficult to evaporate due to its low vapor pressure and high melting temperature, but it can be easily sputtered. It has an interesting property that it etches fast in hydrogen peroxide, forming an oxide that is readily water soluble.

3.4.11 Vanadium

This metal is very similar to titanium, and exhibits a pronounced gettering effect. It can be sputtered or evaporated easily, but the background pressure has to be low to prevent oxidation.

3.4.12 Silicon

Silicon can be sputtered or evaporated, but it is highly reactive so it requires a very low base pressure and low desorption rates. PVD silicon films are not very stable, because the unfilled silicon bonds can oxidize over time. It can also be produced by CVD reactions. One of the common variant is the hydrogenated amorphous silicon that is produced by the pyrolysis of silane gas in a CVD reactor. The hydrogen terminated silicon is very stable, and it is widely used in applications requiring amorphous silicon (such as solar cells).

3.4.13 Germanium

Germanium is similar to silicon, but it is less reactive and can be evaporated or sputtered, or deposited by CVD.

3.4.14 Aluminum Oxide

Al_2O_3 has a very low sputter rate, and for this reason it is used as a shielding material in many plasma fixtures. Sputtering from an Al_2O_3 target is very slow, but it can be deposited by reactive sputtering from an aluminum target. It can also be evaporated fairly easily, directly from a dielectric source or reactively from aluminum. ALD and CVD processes can also be used to produce Al_2O_3.

3.4.15 Magnesium Fluoride

MgF_2 finds many applications in optical coatings because it is one of the few materials with a refractive index lower than SiO_2. It can be sputtered or evaporated easily. It produces very high vapor pressures at low temperatures, so it is really easy to evaporate. However, the evaporated film tends to be highly stressed, so it needs to be used in conjunction with IAD or substrate heating when depositing large thicknesses (which is needed in optical applications due to its low refractive index). The loss of fluorine during deposition is difficult to compensate because one cannot easily flow fluorine as a background gas, so a common technique is to use a small oxygen background to convert any metallic Mg to MgO to keep the optical film losses low (albeit at a slightly increased refractive index).

3.4.16 Silicon Dioxide

SiO_2 is the ubiquitous "oxide" and can be sputtered, evaporated, or deposited by CVD, among many other techniques. Sputter rates tend to be very slow, but evaporation can be fast. SiO_2 is so stable that oxygen loss during deposition is usually not a problem, but a small oxygen background can help, and does not seem to cause any adverse effects to the film.

3.4.17 Titanium Dioxide

TiO_2 is a popular optical film because it has one of the highest refractive indices in the visible spectrum (about 2.5). It sputters very slowly, and the targets tend to crack very easily due to thermal gradients during sputtering. Evaporation of TiO_2 is also not straightforward because it easily decomposes to form one of many other stable oxide states, such as Ti_2O_3 and Ti_2O_5. Normally, one evaporates Ti_2O_3 pellets under an oxygen ambient and substrate heating to create TiO_2 on the substrate.

3.4.18 Niobium Oxide

Nb_2O_5 is also a dielectric of interest in optical thin-film applications because of its high refractive index of 2.3 in the visible spectrum. It is an attractive alternative to TiO_2 albeit at a slightly lower refractive index. It can be easily sputtered with a small oxygen background gas.

3.4.19 Zinc Sulfide

ZnS is also a material of significant interest in optical applications. It has a reasonably high refractive index of 2.3 in the visible spectrum. It is also used as an infrared thin film because of its extended transparency in the mid-wave and long-wave infrared spectrum. ZnS is most easily deposited by sputtering. It can also be evaporated but the zinc contamination of the vacuum chamber should be considered before evaporating ZnS.

3.4.20 Vanadium Oxide

VO_2 is an interesting material because it exhibits an insulator-to-metal phase transition at 68°C. It is usually deposited from a metallic vanadium source by sputtering, evaporation, or PLD in an oxygen ambient at elevated substrate temperatures. Like TiO_2, this material exhibits mixed valence states. VO_x is a mixed oxide of vanadium that is much easier to deposit than VO_2. It has a large temperature coefficient of resistance, and is used as the sensor element in thermal imaging cameras.

PROBLEMS

3.1 Consider thermal evaporation of platinum on a 4 in. diameter substrate in a vacuum chamber with a throw distance of 45 cm. If the desired deposition rate is 3 Å/s, assuming the melted source area is 5 mm^2, calculate the approximate temperature of the metal. Determine if the metal will be melting or sublimating. Calculate the expected center-to-edge variation in film thickness on the 4 in. substrate.

3.2 If aluminum is sputtered with argon at 200 W discharge power and 200 V cathode voltage, calculate the average target removal rate in Å/s.

3.3 Referring to Figure 3.15, determine what would happen to the electron trajectories if the inner magnet is flipped so that its N pole is facing up.

3.4 Using the SRIM software, calculate the average sputter yield and the energy distribution of the sputtered atoms for copper and tungsten for normally incident Ar^+ ions with an energy of 500 eV.

LABORATORY EXERCISES

3.1 Perform a thermal evaporation on the largest possible substrate that can be accommodated, and take thickness measurements at various radial positions. Examine if the uniformity fits the expected cos(θ) distribution.

3.2 Install two small (approx. 1 inch × 1 inch) silicon substrates in a thermal evaporation chamber. The first one should be mounted perpendicular to the deposition flux (which is the normal configuration), and the second one should be at a very steep angle of incidence, such as 85° or greater. An angle block may have to be utilized for the latter. Then deposit approximately 1000 Å of titanium. After the deposition, the film on the first sample should visually appear as a normal metallic film, and the second sample should appear dark in color. Examine the cross-sections of the films under a scanning electron microscope. The first sample should not show any structure but the second sample should contain tilted nano-columns. This is known as glancing angle deposition [22]. Investigate how the angle of incidence of the deposition flux plays a role in this growth mechanism.

3.3 For the above two samples, measure the substrate curvature with a stylus profilometer before and after the deposition. The profile scans must be taken at exactly the same locations. Apply Stoney's equation to calculate the film stress for both cases, and determine if it is compressive or tensile.

REFERENCES

1. L. Holland and W. Steckelmacher, The distribution of thin films condensed on surfaces by the vacuum evaporation method, *Vacuum*, 2(4) (October 1952) 346–364.
2. E. B. Graper, Distribution and apparent source geometry of electron-beam-heated evaporation sources, *J. Vac. Sci. Technol.*, 10 (1973) 100.
3. W. Heitmann, Reactive evaporation in ionized gases, *Appl. Opt.*, 10 (1971) 2414–2418.
4. S.-C. Chiao, B. G. Bovard, and H. A. Macleod, Repeatability of the composition of titanium oxide films produced by evaporation of Ti_2O_3, *Appl. Opt.*, 37(22) (August 1998) 5284–5290.
5. P. J. Martin, H. A. Macleod, R. P. Netterfield, C. G. Pacey, and W. G. Sainty, Ion-beam-assisted deposition of thin films, *Appl. Opt.*, 22(1) (1983) 178–184.
6. H. R. Kaufman, R. S. Robinson, and R. I. Seddon, End-hall ion source, *J. Vac. Sci. Technol. A*, 5 (1987) 2081.
7. S. Swann, Magnetron sputtering, *Phys. Technol.*, 19(2) (1988) 67.
8. R. K. Waits, Planar magnetron sputtering, *J. Vac. Sci. Technol.*, 15 (1978) 179.
9. N. Laegreid and G. K. Wehner, Sputtering yields of metals for Ar^+ and Ne^+ ions with energies from 50 to 600 ev, *J. Appl. Phys.*, 32 (1961) 365.
10. Y. Yamamura and H. Tawara, Energy dependence of ion-induced sputtering yields from monatomic solids at normal incidence, *At. Data Nucl. Data Tables*, 62 (1996) 149–253.
11. James F. Ziegler, M. D. Ziegler, J. P. Biersack, SRIM—The stopping and range of ions in matter, *Nucl. Instrum. Methods Phys. Res. B*, 268(11–12) (2010) 1818–1823, 2010. http://www.srim.org/.
12. J. P. Biersack and W. Eckstein, Sputtering studies with the Monte Carlo program TRIM.SP, *Appl. Phys. A*, 34 (1984) 73–94.
13. A. Oliva-Florio, R. A. Baragiola, M. M. Jakas, E. V. Alonso, and J. Ferrón, Noble-gas ion sputtering yield of gold and copper: Dependence on the energy and angle of incidence of the projectiles, *Phys. Rev. B*, 35 (1987) 2198.
14. P. Wetz, W. Krüger, A. Scharmann, and K.-H. Schartner, The angular dependence of the sputtering yield of Ag, Ni and Fe by high energy noble gas ions, *Radiat. Meas.*, 27(4) (July 1997) 569–574.
15. A. Hu and A. Hassanein, How surface roughness affects the angular dependence of the sputtering yield, *Nucl. Inst. Methods Phys. Res. B*, 281 (June 15, 2012) 15–20.
16. M. Küstner, W. Eckstein, V. Dose, and J. Roth, The influence of surface roughness on the angular dependence of the sputter yield, *Nucl. Instrum. Methods Phys. Res., Sect. B*, 145(3) (1998) 320–331.
17. I. Safi, Recent aspects concerning DC reactive magnetron sputtering of thin films: A review, *Surf. Coat. Technol.*, 127 (2000) 203–219.
18. S. Swann, Film thickness distribution in magnetron sputtering, *Vacuum*, 38(8–10) (1988) 791–794.
19. J.-C. Pivin, An overview of ion sputtering physics and practical implications, *J. Mater. Sci.*, 18(5) (May 1983) 1267–1290.
20. J. Heller, Reactive sputtering of metals in oxidizing atmospheres, *Thin Solid Films*, 17(2) (August 1973) 163–176.
21. J. Musila, P. Barocha, J. Vlcek, K. H. Nam, and J. G. Han, Reactive magnetron sputtering of thin films: Present status and trends, *Thin Solid Films*, 475 (2005) 208–218.
22. M. M. Hawkeye and M. J. Brett, Glancing angle deposition: Fabrication, properties, and applications of micro- and nanostructured thin films, *J. Vac. Sci. Technol. A*, 25 (2007) 1317.
23. J.-Q. Xi, M. F. Schubert, J. K. Kim, E. F. Schubert, M. Chen, S.-Y. Lin, W. Liu, and J. A. Smart, Optical thin-film materials with low refractive index for broadband elimination of Fresnel reflection, *Nat. Photon.*, 1 (2007) 176–179.
24. H.-U. Krebs, M. Weisheit, J. Faupel, E. Süske, T. Scharf, C. Fuhse, M. Störmer et al., Pulsed laser deposition (PLD)—A versatile thin film technique, *Adv. Solid State Phys.*, 43 (2003) 505–518.
25. T. Donnelly, S. Krishnamurthy, K. Carney, N. McEvoy, and J. G. Lunney, Pulsed laser deposition of nanoparticle films of Au, *Appl. Surf. Sci.*, 254(4) (December 15, 2007) 1303–1306.
26. T.-W. Chiu, K. Tonooka, and N. Kikuchi, Influence of oxygen pressure on the structural, electrical and optical properties of VO_2 thin films deposited on ZnO/glass substrates by pulsed laser deposition, *Thin Solid Films*, 518(24) (October 1, 2010) 7441–7444.
27. L. F. Pochet, P. Howard, and S. Safaie, CVD coatings: From cutting tools to aerospace applications and its future potential, *Surf. Coat. Technol.*, 94–95 (October 1997) 70–75.
28. R. F. Bunshah (ed.), *Handbook of Deposition Technologies for Films and Coatings*, Noyes Publications, Park Ridge, NJ, 1994.
29. http://www.factsage.com/.

30. Y. Xu and X.-T. Yan, *Chemical Vapour Deposition: An Integrated Engineering Design for Advanced Materials*, Springer-Verlag London Limited, London, U.K., 2010.
31. L. Woods and P. Meyers, Atmospheric pressure chemical vapor deposition and jet vapor deposition of CdTe for high efficiency thin film PV devices, Technical report, National Renewable Energy Laboratory, Golden, CO, August 2002.
32. M. J. Cooke, A review of LPCVD metallization for semiconductor devices, *Vacuum*, 35 (1985) 67–73.
33. K. Mackenzie, Silicon nitride for MEMS applications: LPCVD and PECVD process comparison, MEMS and Sensors Whitepaper Series, Plasma-Therm LLC, St. Petersburg, FL January 2014.
34. H. O. Pierson, *Handbook of Chemical Vapor Deposition (CVD): Principles, Technology, and Applications*, Noyes Publications, Park Ridge, NJ, 1999.
35. A. C. Jones and M. L. Hitchman, *Chemical Vapour Deposition: Precursors, Processes and Applications*, Royal Society of Chemistry, Cambridge, U.K., 2009.
36. J.-H. Park and T. S. Sudarshan, *Chemical Vapor Deposition*, ASM International, Materials Park, OH, 2001.
37. V. Pore, T. Hatanpää, M. Ritala, and M. Leskelä, Atomic layer deposition of metal tellurides and selenides using alkylsilyl compounds of tellurium and selenium, *J. Am. Chem. Soc.*, 131(10) (2009) 3478–3480.
38. S. M. George, Atomic layer deposition: An overview, *Chem. Rev.*, 110 (2010) 111–131.
39. C. K. O'Sullivan and G. G. Guilbault, Commercial quartz crystal microbalances—Theory and applications, *Biosens. Bioelectron.*, 14 (1999) 663–670.
40. C. Lu and A. W. Czanderna (eds.), *Applications of Piezoelectric Quartz Crystal Microbalances* (Methods and Phenomena, Volume 7), Elsevier Science Publishing Co., New York, 1984, pp. 1–393.
41. E. Klokholm, Intrinsic stress in evaporated metal films, *J. Vac. Sci. Technol.*, 6 (1969) 138.
42. G. C. A. M. Janssen, M. M. Abdalla, F. van Keulen, B. R. Pujada, and B. van Venrooy, Celebrating the 100th anniversary of the Stoney equation for film stress: Developments from polycrystalline steel strips to single crystal silicon wafers, *Thin Solid Films*, 517(6) (January 30, 2009) 1858–1867.

4 Thin-Film Optics

Thin-film coatings are ubiquitous on all optical components, such as lenses, mirrors, cameras, windows, etc. Antireflection coatings are used on nearly all flat glasses, such as on windows, and architectural glasses. Antireflection is also used on nearly all refractory optical components, including common eyeglasses. High-reflection coatings are used on mirrors. These can be simple metal coatings or several layers of dielectric films. Long-pass, short-pass, and bandpass filters are used in spectral filtering applications. In all of these applications, the design principle is based on manipulating the reflection and transmission amplitudes and phase shift of each thin film so that the superposition of all the waves produces the desired output spectrum. The basic building block for most optical filters is a quarter-wave-thick film, although that is not a fundamental requirement. While the design principles provide the strategy and starting point for these filters, real-life filters often require subsequent numerical refinement.

In this chapter, we will discuss the fundamental principles of designing thin films for optical coatings, specifically antireflection coatings, high-reflection coatings, and metal film optics. Obviously, the subject of optical thin films extends well beyond these applications and also includes optical filters such as dichroic, bandpass, and notch filters. The references at the end of this chapter should provide the interested reader with more details on this subject [1–4]. Some prior knowledge in optics is beneficial, but it is not essential since the concepts are developed from basic principles.

4.1 ANTIREFLECTION COATINGS

4.1.1 Fresnel Reflection

Consider a substrate with a refractive index of n_s and an outside medium (air) with refractive index n_a as shown in Figure 4.1. The incident field from the left can be represented as a traveling plane wave as $e^{-jk_0 n_a z}$ and the reflected field as $re^{+jk_0 n_a z}$. The transmitted field on the substrate side can be represented as $te^{-jk_0 n_s z}$. We can derive the reflection coefficient at normal incidence by ensuring the continuity of the electric field amplitude and its derivative at the interface.

From Figure 4.1, the field continuity can be written by taking the field values at $z = a$ on both sides of the interface and making them equal:

$$e^{-jk_0 n_a a} + re^{+jk_0 n_a a} = te^{-jk_0 n_s a}. \tag{4.1}$$

The continuity of the field derivative is done by taking the derivative of each component on either side of $z = a$ and making them equal:

$$-jk_0 n_a e^{-jk_0 n_a a} + rjk_0 n_a e^{+jk_0 n_a a} = -jk_0 n_s te^{-jk_0 n_s a}. \tag{4.2}$$

We can simplify this by setting the interface coordinate to zero, $a = 0$, which results in

$$1 + r = t \tag{4.3}$$

$$-n_a + rn_a = -n_s t. \tag{4.4}$$

Eliminating t from equations (4.3) and (4.4) results in

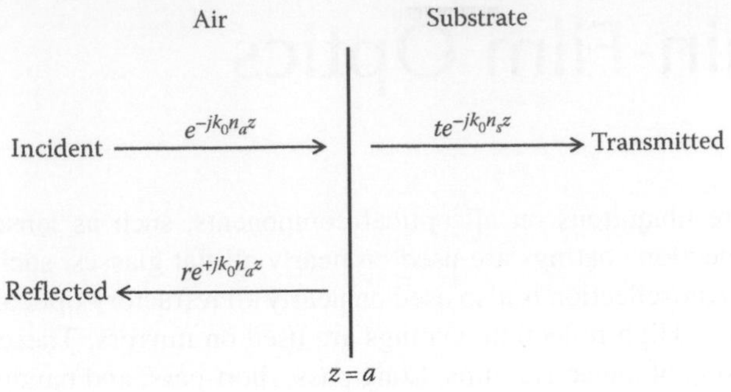

FIGURE 4.1 Electromagnetic fields at a substrate/air interface.

$$r = \frac{n_a - n_s}{n_a + n_s} \tag{4.5}$$

where r is the field reflection coefficient. The power reflection coefficient will be $R = |r|^2$. Therefore,

$$R = \left| \frac{n_a - n_s}{n_a + n_s} \right|^2. \tag{4.6}$$

This is the well-known equation for reflection at normal incidence. For silica glass with a refractive index of 1.48, the air/glass interface produces a reflection of 3.7%.

4.1.2 Single-Layer Antireflection Coating

The concept of antireflection is based on introducing a thin dielectric film with a refractive index n_f on the substrate that will produce an equal reflection amplitude but opposite in phase to cancel out the substrate reflection. The electromagnetic fields at the two interfaces are shown in Figure 4.2. As will be seen from the brief derivation, the film has to be a quarter-wave thick ($\pi/2$ phase), such that the round trip phase will be π to produce a negative sign that subtracts from the reflection.

The reflection coefficient can be derived similar to the earlier example. Setting $b = 0$,

$$1 + r = A + B \tag{4.7}$$

$$-n_a + rn_a = -n_f A + n_f B \tag{4.8}$$

and

$$Ae^{-jk_0 n_f a} + Be^{+jk_0 n_f a} = te^{-jk_0 n_s a} \tag{4.9}$$

$$-jk_0 n_f Ae^{-jk_0 n_f a} + jk_0 n_f Be^{+jk_0 n_f a} = -jk_0 n_s te^{-jk_0 n_s a} \tag{4.10}$$

The four equations (4.7), (4.8), (4.9) and (4.10) can be solved for A, B, r, and t. It involves some messy (but straightforward) algebra, but the solution for r works out to be

$$r = \frac{(n_a - n_{eff})}{(n_a + n_{eff})} \tag{4.11}$$

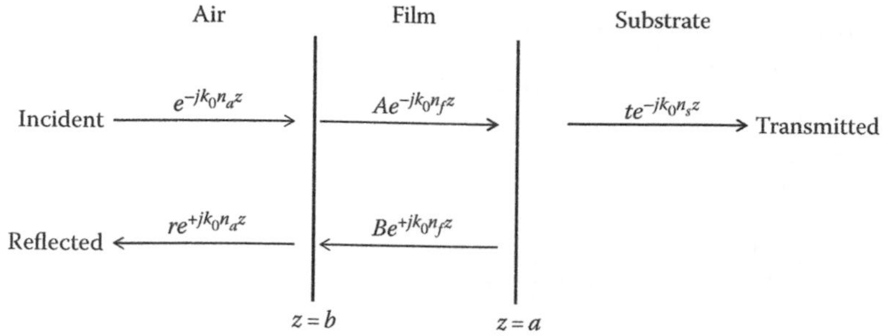

FIGURE 4.2 Electromagnetic fields at two interfaces due to a single dielectric thin film on a substrate.

where

$$n_{eff} = n_f \frac{(n_s + n_f) + (n_s - n_f)e^{-2jk_0 n_f a}}{(n_s + n_f) - (n_s - n_f)e^{-2jk_0 n_f a}}. \tag{4.12}$$

Comparing this to the previous expression for r without the film (Equation 4.5), we can see that the substrate index has changed from n_s to n_{eff}. Therefore, we can consider n_{eff} as the effective index of the substrate due to the presence of the thin film.

We can obtain a perfect antireflection condition if n_{eff} can be made equal to n_a. Since n_a is real, n_{eff} also has to be real, and the complex exponential factor has to satisfy $e^{-2jk_0 n_f a} = 1$. In other words, the exponent has to be multiples of π such as $2k_0 n_f a = N\pi$, where N is an odd integer. This can be recast as $a = N(\lambda_0/4n_f)$. In other words, the film has to be an odd integer multiple of a quarter-wave thick film. For $N = 1$ (one quarter-wave film), Equation 4.12 collapses to

$$n_{eff} = \frac{n_f^2}{n_s}. \tag{4.13}$$

Since we need $n_{eff} = n_a$ for perfect antireflection condition, Equation 4.13 can be written as

$$n_f = \sqrt{n_a n_s}. \tag{4.14}$$

From this, we can conclude that the required refractive index of the film for perfect antireflection is $n_f = \sqrt{n_a n_s}$ and the required film thickness is $\lambda_0/4n_f$. Note that this only works when the film thickness is a quarter-wave thick. This means that perfect antireflection can be achieved only at one specific wavelength λ_0.

As an example, consider a silica glass substrate, with a refractive index of $n_s = 1.48$ at $\lambda_0 = 550\,\text{nm}$. If we need an antireflection condition at $\lambda_0 = 550\,\text{nm}$, the required film must have a refractive index of $n_f = \sqrt{1.48} = 1.217$, and the film thickness has to be $\lambda_0/4n_f = 112.9\,\text{nm}$.

Since n_{eff} in Equation 4.12 is a complex number in general, we can plot its contour on a complex plane as the film thickness increases from $a = 0$ to $a = \lambda_0/4n_f$. This is shown in Figure 4.3. The curve describes an arc in the complex plane. When there is no film on the substrate, the value of n_{eff} is the same as the substrate index ($a = 0$, $n_{eff} = n_s = 1.48$). As a increases, n_{eff} becomes complex, and eventually approaches $n_{eff} = 1.0$ (antireflection condition) when $a = \pi/(2k_0 n_f)$ or equivalently $a = \lambda_0/4n_f$.

We can also examine what happens when a extends beyond the quarter-wave thickness of $\lambda_0/4n_f$. Graphically, n_{eff} would describe a complete circle starting from $n_{eff} = n_s = 1.48$ to $n_{eff} = 1.0$ for quarter-wave-thick film and then back to $n_{eff} = 1.48$ for half-wave-thick film. This is shown in Figure 4.4.

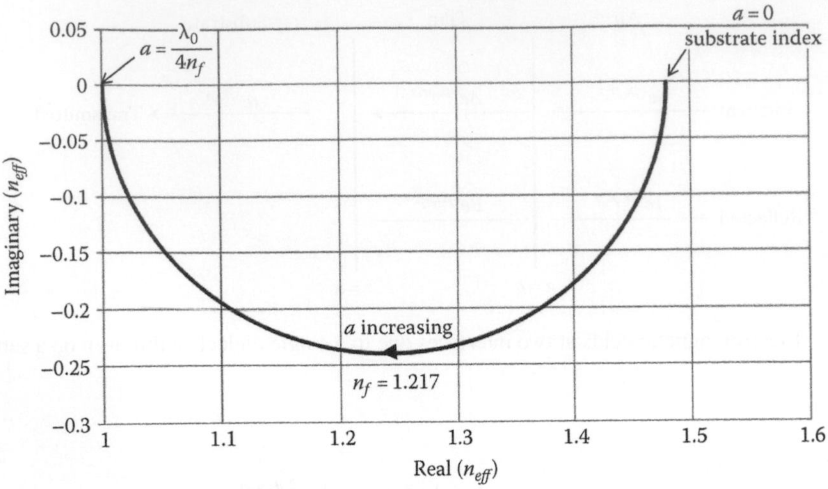

FIGURE 4.3 Contour of n_{eff} as a function of the film thickness for a single-layer antireflection film ($n_f = 1.217$) on a substrate with $n_s = 1.48$.

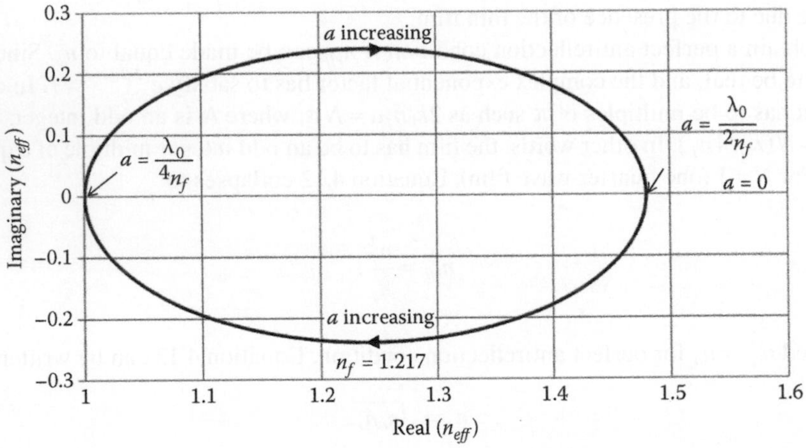

FIGURE 4.4 Contour of n_{eff} as film thickness goes from zero to half-wave thickness. $n_f = 1.217$ and $n_s = 1.48$.

This means that a half-wave film has no effect on the substrate—the film behaves as if it does not exist at the design wavelength of λ_0. For this reason, half-wave-thick films are also known as absentee films.

The calculated reflection spectrum using a transfer matrix method (described later in this chapter) for $n_f = 1.217$ and $t_f = 113\,\text{nm}$ is shown in Figure 4.5. As expected, we can see that the reflection drops to zero at the design wavelength of $\lambda_0 = 550\,\text{nm}$.

There is, however, a small problem with implementing this design in practice. The required refractive index of 1.217 is much smaller than any of the dielectric materials used in thin-film depositions. The lowest refractive index in thin solid films is MgF_2 that has a refractive index of 1.38. Polymers do exist with lower refractive indices, but not as low as 1.217. The only way to solve this problem is by increasing the variables in the design space. One way to do that is by increasing the number of layers. If we can use two films instead of one, it can provide more degrees of freedom to allow us to select more realistic materials.

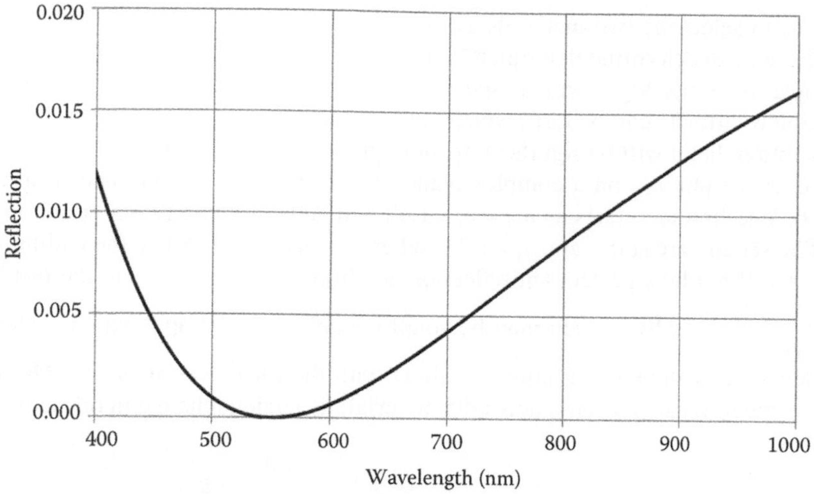

FIGURE 4.5 Calculated reflection spectrum of a single-film antireflection coating ($n_f = 1.217$ and $n_s = 1.48$) using the transfer matrix method.

4.1.3 Two-Layer Quarter-Wave Film Designs

The reflection from several quarter-wave films can be calculated by simply extending the single quarter-wave film reflection calculations. With one quarter-wave film, we found that the effective index of the substrate was $n_{eff} = n_f^2/n_s$. We will assign a subscript to each film to make a distinction. The first film on the substrate will be represented as $n_{eff1} = n_{f1}^2/n_s$. When a second quarter-wave film of index n_{f2} is deposited on top of the first film, the effective index of the system becomes $n_{eff2} = n_{f2}^2/n_{eff1} = \left(n_{f2}^2/n_{f1}^2\right)n_s$.

In general, the effective index of the substrate due to N quarter-wave layers can be written as

$$n_{effN} = \frac{n_{fN}^2}{n_{fN-1}^2}\frac{n_{fN-2}^2}{n_{N-3}}\ldots n_s \quad \text{if } N \text{ is an even number} \tag{4.15}$$

and

$$n_{effN} = \frac{n_{fN}^2}{n_{fN-1}^2}\frac{n_{fN-2}^2}{n_{N-3}}\ldots\frac{1}{n_s} \quad \text{if } N \text{ is an odd number.} \tag{4.16}$$

For a two-layer structure, the overall reflection can be written as

$$r = \frac{n_a - n_{eff2}}{n_a + n_{eff2}} = \frac{n_a - \left(n_{f2}^2/n_{f1}^2\right)n_s}{n_a + \left(n_{f2}^2/n_{f1}^2\right)n_s}. \tag{4.17}$$

Using Equation 4.17, the required condition to eliminate reflection is

$$\frac{n_{f2}^2}{n_{f1}^2}n_s = n_a \tag{4.18}$$

or equivalently,

$$\frac{n_{f2}^2}{n_{f1}^2} = \frac{n_a}{n_s}. \tag{4.19}$$

This allows us to select the two materials as long as they can satisfy $n_{f2}^2/n_{f1}^2 = 1/1.48$. Looking at Equation 4.19, we can determine that film #2 must have a lower index than film #1.

For example, if we use MgF_2 with an index of 1.38 as film #2, then film #1 must have an index of 1.68. Therefore, film #1 can be LaF_3 which has a refractive index close to 1.60. This may not be a perfect condition, but it will be significantly more practical than the single film of index 1.217.

As before, if we plot n_{eff} on a complex plane, it will describe two connected arcs as shown in Figure 4.6. The first arc starts as $n_{eff} = n_s = 1.48$ and ends at $n_{eff1} = 1.73$ at a quarter-wave film thickness. The second arc starts at $n_{eff1} = 1.73$ and ends at $n_{eff2} = 1.1$. Since the ending value is not exactly 1.0, it will not be a perfect antireflection condition. We can estimate the resulting reflection as $R = \left|\frac{1-1.1}{1+1.1}\right|^2 = 0.22\%$, which may be small enough for most applications. The calculated spectral reflectance is shown in Figure 4.7, along with the ideal case of $n_{f1} = 1.68$. It should be noted that the spectral calculation ignores the material dispersion (change in refractive index with wavelength).

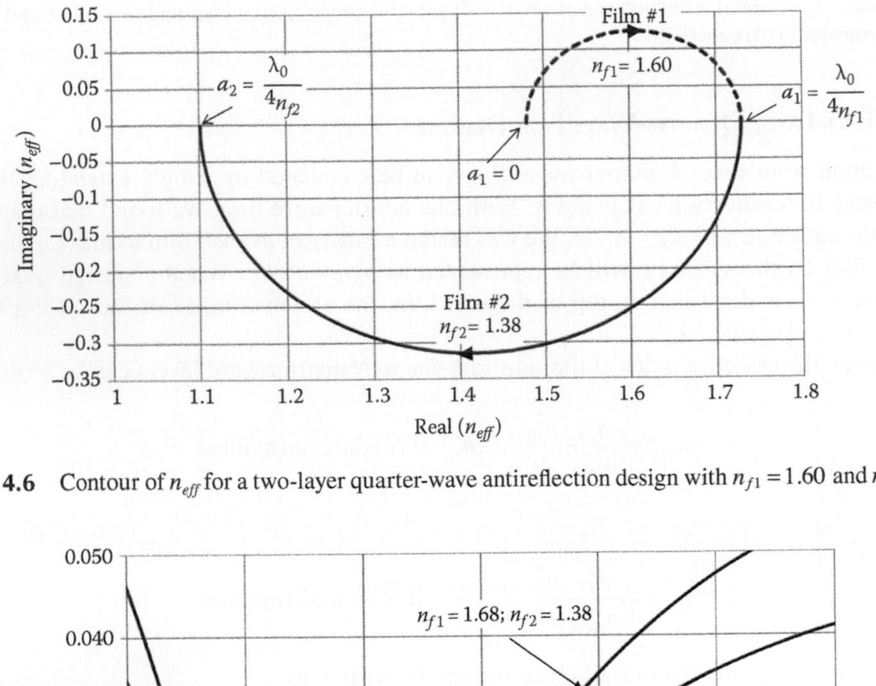

FIGURE 4.6 Contour of n_{eff} for a two-layer quarter-wave antireflection design with $n_{f1} = 1.60$ and $n_{f2} = 1.38$.

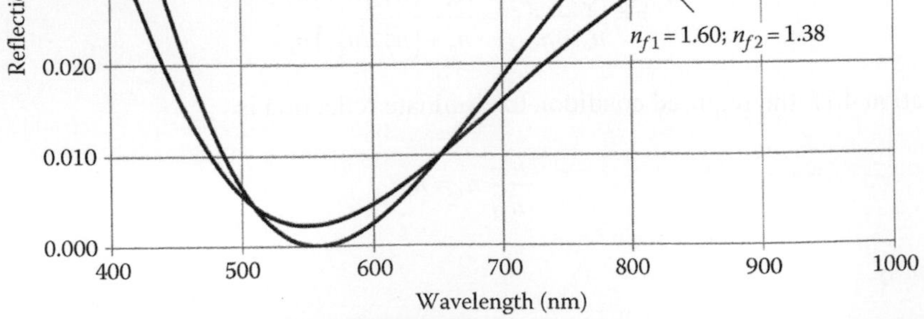

FIGURE 4.7 Spectral reflectance plots of the two-layer antireflection design with $n_{f1} = 1.60$ and $n_{f2} = 1.38$, and $n_{f1} = 1.68$ and $n_{f2} = 1.38$.

Obviously, one can easily extend this method to many more layers. As the number of layers increase, a simplified notation becomes necessary to represent the layer structure. In the aforementioned example with LaF_3 and MgF_2, the notation that will be used is silica | *HL* | air, where *H* is the quarter-wave high-index material (LaF_3) and *L* is the quarter-wave low-index material (MgF_2).

We can examine the complex effective index plot in Figure 4.6 to understand the direction traced by the arcs. Since the first layer's index of 1.60 is higher than the substrate index (1.48), the arc moves toward higher refractive index values, from 1.48 to 1.73. The geometrical mean of the starting and the ending values of the arc is the layer's refractive index of 1.60. Since the second layer's index of 1.38 is lower than the starting index of 1.73, the arc continues its trace toward smaller values, ending at 1.1. We can verify that the geometrical mean of 1.73 and 1.1 is 1.38. This is the general behavior of the complex effective index—each film moves the n_{eff} toward higher or lower values depending on if the film index is higher or lower than the previous effective index.

4.1.4 Two-Layer Non-Quarter-Wave Film Designs

In the previous example, we chose LaF_3 with an index of 1.60 even though the required refractive index of the film was 1.68. Although this produced an acceptably small reflection at the reference wavelength, there is an even better method, where we can use these same two materials to achieve a perfect antireflection condition. This is done by relaxing the quarter-wave requirement, but still insisting that the final n_{eff2} becomes 1.0.

For a single layer, we showed that the substrate index became modified to

$$n_{eff1} = n_{f1}\frac{\left(n_s + n_{f1}\right) + \left(n_s - n_{f1}\right)e^{-2jk_0 n_{f1} a_1}}{\left(n_s + n_{f1}\right) - \left(n_s - n_{f1}\right)e^{-2jk_0 n_{f1} a_1}}. \tag{4.20}$$

After the second film, the effective index becomes

$$n_{eff2} = n_{f2}\frac{\left(n_{eff1} + n_{f2}\right) + \left(n_{eff1} - n_{f2}\right)e^{-2jk_0 n_{f2} a_2}}{\left(n_{eff1} + n_{f2}\right) - \left(n_{eff1} - n_{f2}\right)e^{-2jk_0 n_{f2} a_2}} \tag{4.21}$$

where we have used n_{eff1} in place of n_s. In order to achieve antireflection this time, only n_{eff2} has to be equal to n_a. The value of n_{eff1} does not have to be constrained to any specific value. The requirement for a quarter-wave-thick film came primarily from forcing the n_{eff} values to be real. In this case, film #1 and film #2 do not have to be quarter-wave thick as long as their net effect makes the final n_{eff2} equal to n_a. Though it may seem trivial, this is actually a very useful result because now we can start with actually available materials and then find the required film thicknesses to achieve the antireflection condition rather than insisting on all films being quarter-wave thick.

For example, let us consider Al_2O_3 with an index of 1.77 as film #1 and MgF_2 with an index of 1.38 as film #2. In the complex plane, we can draw two complete circles corresponding to half-wave-thick films as shown in Figure 4.8. The first arc starts from $n_{eff} = n_s = 1.48$ and crosses the real axis again at $n_{eff1} = 2.12$ and then back to 1.48 at a half-wave film thickness. Instead of starting the second arc from $n_{eff1} = 2.12$ as we did in the previous example, we will start the arc from its desired end point, which would be $n_{eff2} = 1$. This circle will cross the real axis at 1.92. The intercept between the two circles (there are two) are the solutions for achieving perfect antireflection with these two materials.

Considering the solution at the first intersection (on the upper half of the plot in Figure 4.8), the first film has to be 0.134-wave thick and the second film has to be 0.297-wave thick. The resulting complex contour plot is shown in Figure 4.9. Notice that neither of the arcs are a quarter-wave thick. This technique allows us to use arbitrary materials to meet the perfect antireflection condition.

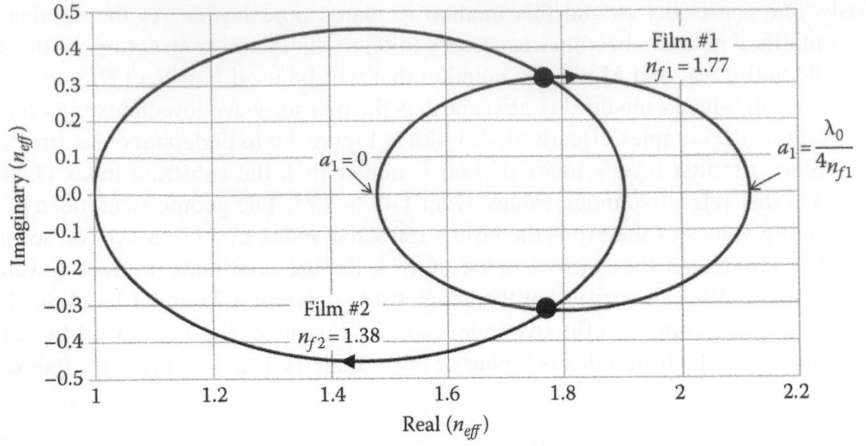

FIGURE 4.8 Intersection of the two contours of n_{eff} for a two-film antireflection design ($n_{f1} = 1.77$ and $n_{f2} = 1.38$).

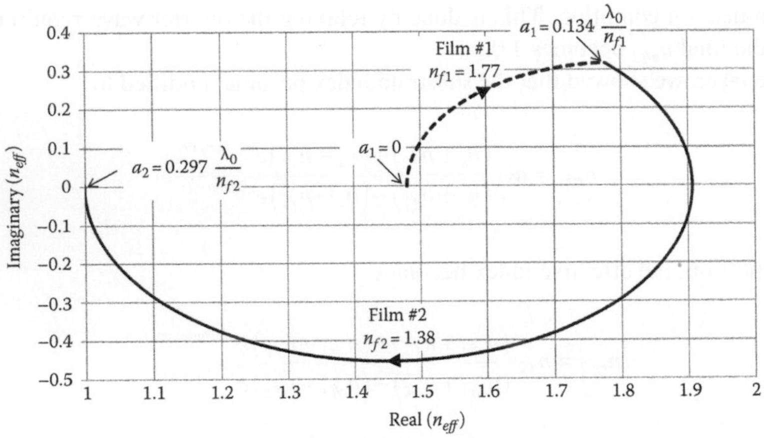

FIGURE 4.9 Contour of n_{eff} for a two non-quarter-wave antireflection stack with 0.134-wave and 0.297-wave films ($n_{f1} = 1.77$ and $n_{f2} = 1.38$).

An equally valid solution could have been obtained by selecting the second intersection (on the lower half of the plot in Figure 4.8). This occurs when the first film is 0.365-wave thick and the second film is 0.205-wave thick, as shown in Figure 4.10.

The spectral reflection for both designs shown in Figures 4.9 and 4.10 using the transfer matrix method is shown in Figure 4.11. We can see that the antireflection property at the reference wavelength $\lambda_0 = 550\,\text{nm}$ is the same for both cases, although they vary in their spectral behavior at all other wavelengths. This is to be expected because the analysis only holds for the reference wavelength λ_0.

However, not all combinations of materials will yield a solution. For example, consider our first example, with LaF_3 with an index of 1.60 and MgF_2 with an index of 1.38. We found that the quarter-wave combinations resulted in an effective index of 1.1. This, unfortunately, cannot be corrected by relaxing the quarter-wave requirement. We can understand why by examining the contour of the complex effective index plot shown in Figure 4.12, where both films are varied from 0- to 0.5-wave thickness starting from the index values of 1.48 and 1.0, respectively. We can see that the two circles do not intersect, representing a case where there is no solution for these two films.

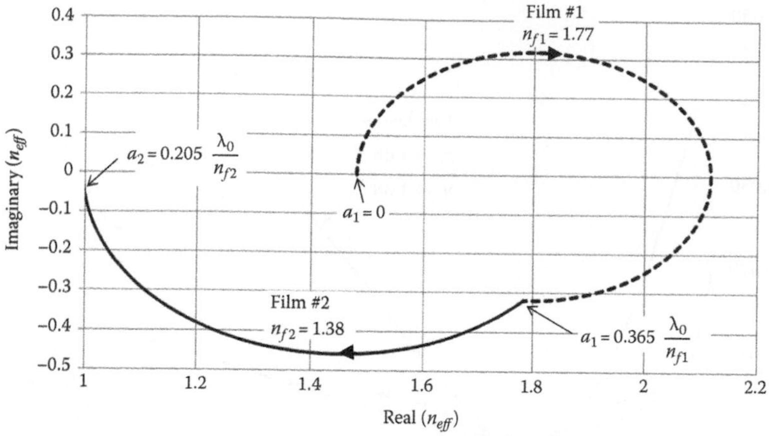

FIGURE 4.10 Contour of n_{eff} for a two non-quarter-wave antireflection stack with 0.365-wave and 0.205-wave films ($n_{f1} = 1.77$ and $n_{f2} = 1.38$).

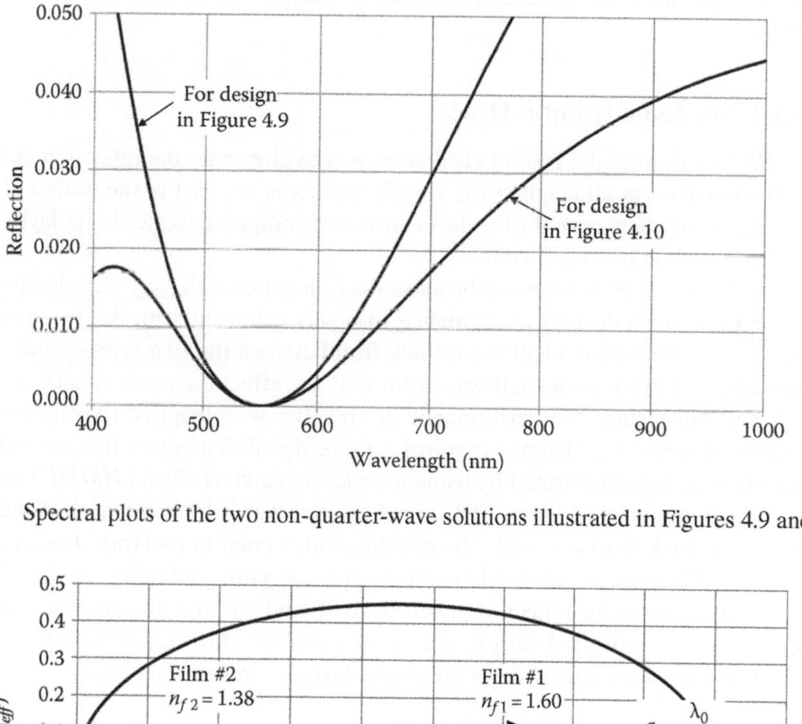

FIGURE 4.11 Spectral plots of the two non-quarter-wave solutions illustrated in Figures 4.9 and 4.10.

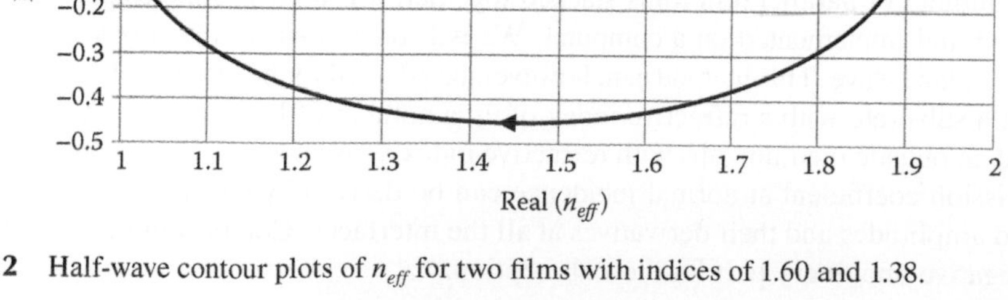

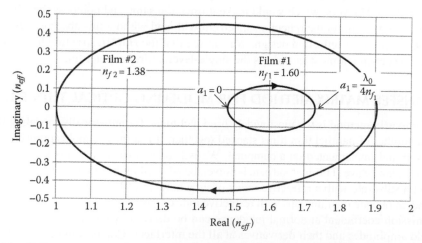

FIGURE 4.12 Half-wave contour plots of n_{eff} for two films with indices of 1.60 and 1.38.

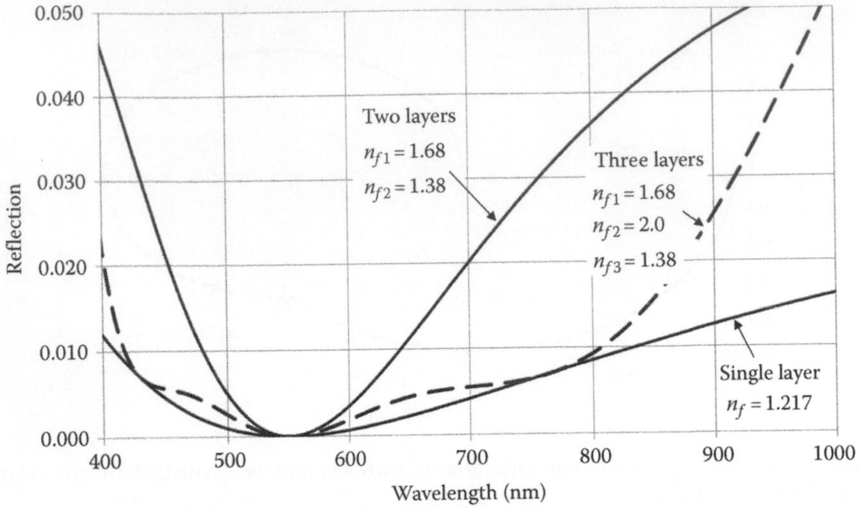

FIGURE 4.13 Comparison of the three-layer antireflection design with absentee layer with the single- and two-layer designs.

4.1.5 Three-Layer Antireflection Design

In the silica | *HL* | air design, the goal of each layer was to eliminate the reflection at the reference wavelength. The behavior at all neighboring wavelengths was not part of the design objective. One of the undesirable consequences of the two layer structure compared to the single-layer design is the narrowing of the antireflection bandwidth.

A common technique used to improve the antireflection bandwidth is by introducing a third film, without affecting the antireflection performance that was achieved with the first two films. One way to do this is by introducing a half-wave-thick film between the two layers. Since this behaves as an absentee film, insertion of the half-wave film will not affect the performance at the reference wavelength, but it will change the performance at all other wavelengths. Exactly how it changes the reflectance at all other wavelengths requires a more detailed analysis that we will not pursue here, but it can be easily demonstrated by using a structure such as silica | *HMML* | air, where *H* is a quarter-wave-thick material of index 1.68, *L* is a quarter-wave-thick material of index 1.38, and *M* is a quarter-wave-thick third material. The complex index contour plot (not shown) would trace a full circle for the *MM* layers, which would make it trace the same end points as the silica | *HL* | air structure. Figure 4.13 shows the reflection spectrum with and without this absentee film, as well as the ideal single-layer antireflection design. We can see that the *HMML* structure has a much wider antireflection spectrum, almost as wide as the single-layer antireflection film.

4.2 TRANSFER MATRIX METHOD FOR MODELING OPTICAL THIN FILMS

The transfer matrix method is the most widely used technique for numerically modeling the optical effects of a number of parallel thin films stacked together. It is a fairly simple approach that can be easily derived and implemented on a computer. We will only consider the simplest case of normal incidence of a plane wave. This method can, however, be adapted for beams and off-normal incidence.

Consider a substrate with a refractive index of n_s; a number of layers with refractive indices n_{f1}, n_{f2}, etc.; and an outside medium (air) with refractive index n_a as shown in Figure 4.14. The reflection and transmission coefficient at normal incidence can be derived by ensuring the continuity of the electric field amplitudes and their derivatives at all the interfaces. Complex indices will be assumed to have a negative imaginary part for their absorption.

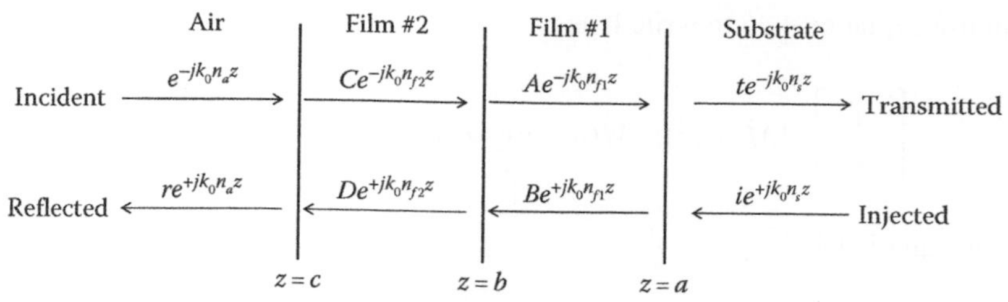

FIGURE 4.14 Multilayer film structure for deriving the transfer matrix method equations.

Considering the interface at $z=c$ (air–film #2 interface), the field continuity can be written as

$$e^{-jk_0n_ac}+r\,e^{+jk_0n_ac}=Ce^{-jk_0n_{f2}c}+De^{-jk_0n_{f2}c}. \tag{4.22}$$

The derivatives of the fields can be written as

$$-jk_0n_ae^{-jk_0n_ac}+rjk_0n_ae^{+jk_0n_ac}=-jk_0n_{f2}Ce^{-jk_0n_{f2}c}+jk_0n_{f2}De^{-jk_0n_{f2}c}. \tag{4.23}$$

This can be cast into a 2×2 matrix form as

$$\begin{bmatrix} e^{-jk_0n_ac} & e^{-jk_0n_ac} \\ -jk_0n_ae^{-jk_0n_ac} & jk_0n_ae^{+jk_0n_ac} \end{bmatrix}\begin{bmatrix} 1 \\ r \end{bmatrix}=\begin{bmatrix} e^{-jk_0n_{f2}c} & e^{-jk_0n_ac} \\ -jk_0n_{f2}e^{-jk_0n_{f2}c} & jk_0n_{f2}e^{+jk_0n_{f2}c} \end{bmatrix}\begin{bmatrix} C \\ D \end{bmatrix} \tag{4.24}$$

or written more compactly as

$$M(n_a,c)\begin{bmatrix} 1 \\ r \end{bmatrix}=M(n_{f2},c)\begin{bmatrix} C \\ D \end{bmatrix}. \tag{4.25}$$

Now, moving to the $z=b$ interface, the equation becomes

$$M(n_{f2},b)\begin{bmatrix} C \\ D \end{bmatrix}=M(n_{f1},b)\begin{bmatrix} A \\ B \end{bmatrix}. \tag{4.26}$$

and at $z=a$

$$M(n_{f1},a)\begin{bmatrix} A \\ B \end{bmatrix}=M(n_s,a)\begin{bmatrix} t \\ i \end{bmatrix}. \tag{4.27}$$

We can solve for r and t by rewriting these equations until the whole system of equations is only in terms of $\begin{bmatrix} 1 \\ r \end{bmatrix}$ and $\begin{bmatrix} t \\ i \end{bmatrix}$ by eliminating all the other field amplitudes A, B, C, D, etc.

Going back to Equation 4.25, we can write it as

$$\begin{bmatrix} 1 \\ r \end{bmatrix}=[M(n_a,c)]^{-1}M(n_{f2},c)\begin{bmatrix} C \\ D \end{bmatrix}. \tag{4.28}$$

Then, we can use Equation 4.26 to write it as

$$\begin{bmatrix} 1 \\ r \end{bmatrix} = [M(n_a,c)]^{-1} M(n_{f2},c)[M(n_{f2},b)]^{-1} M(n_{f1},b) \begin{bmatrix} A \\ B \end{bmatrix} \tag{4.29}$$

and then using Equation 4.27

$$\begin{bmatrix} 1 \\ r \end{bmatrix} = [M(n_a,c)]^{-1} M(n_{f2},c)[M(n_{f2},b)]^{-1} M(n_{f1},b)[M(n_{f2},b)]^{-1} M(n_s,a) \begin{bmatrix} t \\ i \end{bmatrix}. \tag{4.30}$$

Since all of the intermediate matrices are 2×2 matrices, the product of all these matrices will also be a 2×2 matrix. Therefore, we can compactly express it as

$$\begin{bmatrix} 1 \\ r \end{bmatrix} = M \begin{bmatrix} t \\ i \end{bmatrix} \tag{4.31}$$

where M is the 2×2 matrix that contains the product of all of the matrices in Equation 4.30. Furthermore, i can be considered to be zero because normally there will be no field incident from the substrate direction (unless there is an external feedback such as the reflection from the backside of the substrate). Therefore, Equation 4.31 becomes

$$\begin{bmatrix} 1 \\ r \end{bmatrix} = M \begin{bmatrix} t \\ 0 \end{bmatrix}. \tag{4.32}$$

From this, we can get

$$t = \frac{1}{M_{11}} \tag{4.33}$$

and

$$r = M_{21} t = \frac{M_{21}}{M_{11}}. \tag{4.34}$$

Therefore, we can compute the transmission and reflection amplitudes t and r from the elements of the transfer matrix M. This process has to be repeated for each wavelength one at a time to produce a spectrum. However, since the calculation is simply a product of many 2×2 matrices, it is a fast process, so repeating it for a large number of wavelengths to generate a spectrum is usually not computationally intensive. This is the process that is known as the transfer matrix method (TMM).

However, the reverse problem is not that straightforward. The TMM equation (4.32) cannot be inverted to produce the thin-film structure when a desired spectral performance is specified as the input. This can only be done by numerical iterations and optimizations. There are a number of commercial thin-film design software that take different optimization methods to solve this problem, and they also come with a large database of thin-film material properties. For our purposes, we draw our attention to the open-source software OpenFilters. It contains most of the essential features for design and optimization for basic research as well as academic users, lacking only an extensive database of materials [5].

4.3 HIGH-REFLECTION DIELECTRIC COATINGS

As derived in the previous section, the effective index of the substrate due to N quarter-wave layers was given by Equations 4.15 and 4.16, which are repeated as follows:

$$n_{effN} = \frac{n_{fN}^2}{n_{fN-1}^2}\frac{n_{fN-2}^2}{n_{N-3}}\ldots n_s \quad \text{if } N \text{ is an even number} \tag{4.15}$$

$$n_{effN} = \frac{n_{fN}^2}{n_{fN-1}^2}\frac{n_{fN-2}^2}{n_{N-3}}\ldots\frac{1}{n_s} \quad \text{if } N \text{ is an odd number} \tag{4.16}$$

The reflection coefficient was

$$r = \frac{n_a - n_{effN}}{n_a + n_{effN}}. \tag{4.11}$$

Examining Equation 4.11, there are two ways in which this expression can yield a very high reflection coefficient: (1) $|n_{effN}| \gg n_a$ or (2) $|n_{effN}| \ll n_a$. Both of these conditions will produce the desired result of $r \approx 1$. From the expressions for n_{effN} in Equations 4.15 and 4.16, the conditions for high reflection can be achieved by making the ratio n_{fN}/n_{fN-1} either very large or very small, and by selecting a large N. In other words, we need a large number of alternating quarter-wave layers with a large refractive index ratio. This is the basic requirement for creating a highly reflective coating.

For symmetry reasons, we choose a unit cell such as $\left(\frac{H}{2}L\frac{H}{2}\right)$ or $\left(\frac{L}{2}H\frac{L}{2}\right)$ as the repeating structure where $H/2$ is a one-eighth-wave high-index layer and $L/2$ is a one-eighth-wave low-index layer. However, unlike the case with antireflection designs, in this case n_{effN} does not have to be a real. As long as n_{effN} is large, Equation 4.11 will produce the desired high reflection.

Let us consider two fictitious dispersion-free materials with refractive indices of 2.5 and 1.4, representing them as H and L with quarter-wave thicknesses of 55 nm and 98.2 nm, respectively at $\lambda_0 = 550$ nm. For simplicity, we will consider the thin films without any substrate to isolate the effects of the films from the substrate, though in practice this is clearly not possible. We can examine the effects of adding $\left(\frac{H}{2}L\frac{H}{2}\right)$ one unit cell at a time, up to 10 cells (21 total layers). The results are summarized in Table 4.1, and the corresponding spectral plots are shown in Figure 4.15. We can clearly see that the reflection at the reference wavelength increases as the number of unit cells is increased.

The complex effective index plot provides valuable insights into how this high reflection emerges. Let us consider the two unit-cell structure air$\left|\left(\frac{H}{2}L\frac{H}{2}\right)^2\right|$air with H index = 2.5 and L index = 1.4. The complex effective index contour looks as shown in Figure 4.16. It traces a continuously expanding spiral. Starting from an index of 1.0 (since we assumed the substrate to be air), it progresses through $n_{eff} = 4.3$, $n_{eff} = 0.5$, $n_{eff} = 11.7$, and $n_{eff} = 0.168$, ending at a complex value of $n_{eff} = 0.3 + j2.14$. We could also see that the termination value is not necessarily the best point, and we can improve the reflectivity slightly by moving the termination point to $n_{eff} = 0.168$ (along the real axis) by reducing the thickness of the last film to 0.01-wave thick instead of one-eighth-wave thick. This example illustrates that, though we started the design with the symmetric unit cell $\left(\frac{H}{2}L\frac{H}{2}\right)$, the performance could be improved even further by making minor adjustments to the film thicknesses. The complex index contour plot gives some directions and insights on such

TABLE 4.1
Effects of Increasing the Number of Alternating Layers on a High-Reflection Design

Structure	Reflection (%) at $\lambda_0 = 550$ nm
$\text{Air}\left\|\left(\frac{H}{2}L\frac{H}{2}\right)\right\|\text{air}$	44.1
$\text{Air}\left\|\left(\frac{H}{2}L\frac{H}{2}\right)^2\right\|\text{air}$	81.3
$\text{Air}\left\|\left(\frac{H}{2}L\frac{H}{2}\right)^3\right\|\text{air}$	93.3
$\text{Air}\left\|\left(\frac{H}{2}L\frac{H}{2}\right)^4\right\|\text{air}$	98.2
$\text{Air}\left\|\left(\frac{H}{2}L\frac{H}{2}\right)^5\right\|\text{air}$	99.4
$\text{Air}\left\|\left(\frac{H}{2}L\frac{H}{2}\right)^{10}\right\|\text{air}$	99.99

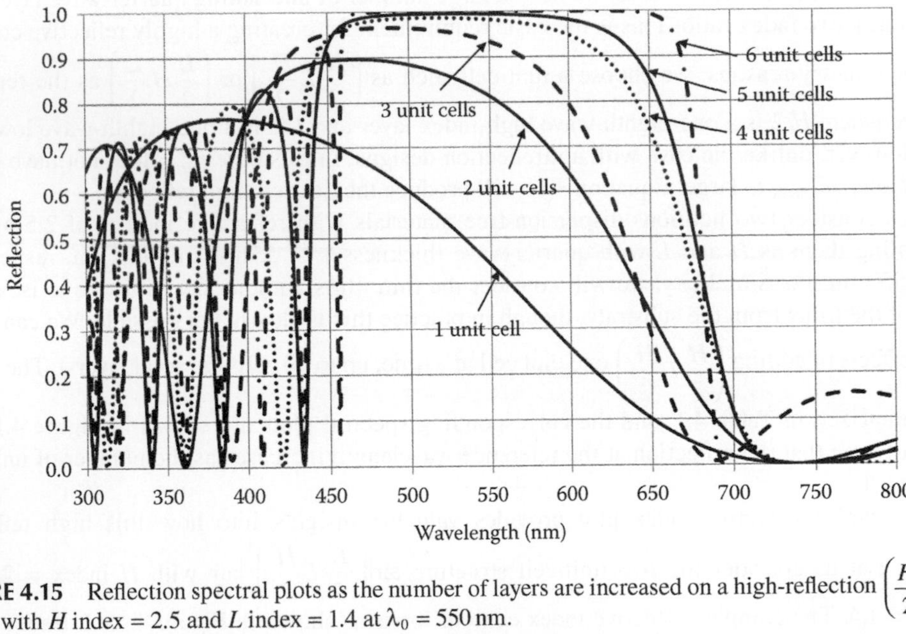

FIGURE 4.15 Reflection spectral plots as the number of layers are increased on a high-reflection $\left(\frac{H}{2}L\frac{H}{2}\right)^n$ design with H index = 2.5 and L index = 1.4 at $\lambda_0 = 550$ nm.

adjustments. However, it does not tell us how the film structure performs at wavelengths other than the reference wavelength λ_0. Therefore, practical designs of this type often require numerical optimizations using the transfer matrix method.

4.4 METAL FILM OPTICS

4.4.1 Reflectance Properties of Metals

Metals exhibit much higher optical reflectance than any dielectrics. This arises primarily due to the high values of the imaginary part of the refractive index, represented by the symbol κ. The real part,

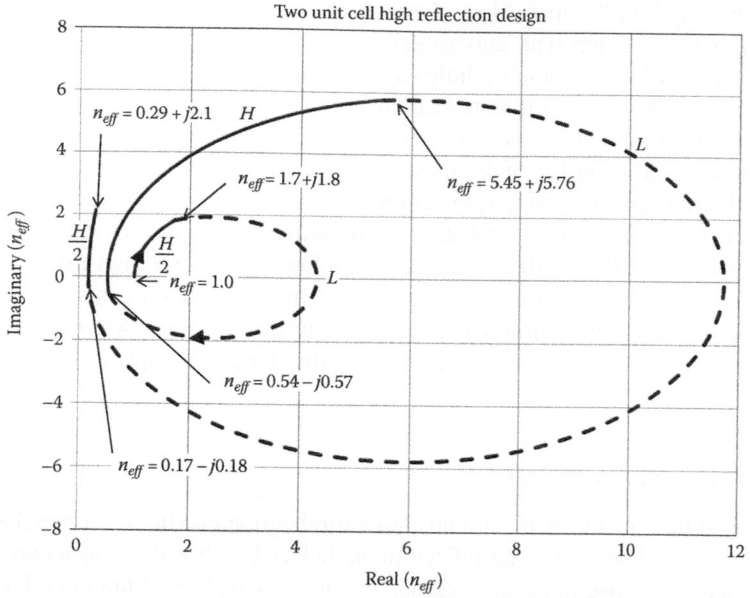

FIGURE 4.16 Contour of n_{eff} for the two-unit-cell high-reflection design air$\left|\left(\frac{H}{2}L\frac{H}{2}\right)^2\right|$air with H index = 2.5 and L index = 1.4.

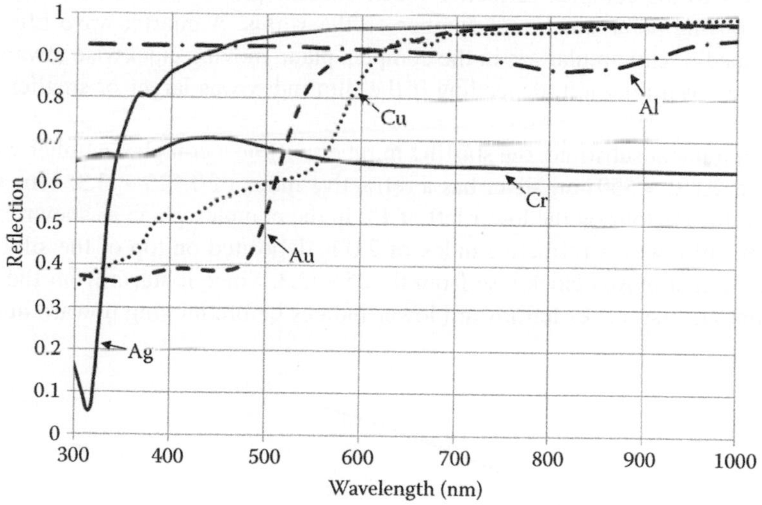

FIGURE 4.17 Optical reflectance of several metals.

represented by n, can be close to 1.0 or even lower depending on the wavelength and type of metal. Assuming the metal has a low n and a high κ, we can apply the normal incidence reflection formula to understand why metals reflect light:

$$r = \frac{n_a - n_s}{n_a + n_s} = \frac{n_a - (n - j\kappa)}{n_a + (n - j\kappa)} \approx \frac{+j\kappa}{-j\kappa} \rightarrow -1. \tag{4.35}$$

Metal also exhibit a very broad reflection spectrum. The reflection spectra of several metals are shown in Figure 4.17. In the visible spectrum, silver has the highest reflection of nearly 98%, but it quickly falls close to the blue and UV part of the spectrum. Silver is actually very transparent near the UV wavelength of 325 nm. Also, silver is very soft and often unsuitable as a robust coating. It tarnishes due to reaction with atmospheric sulfur. Aluminum has a broader spectral reflection from UV to IR, but its reflection is a little smaller. It dips below 90% near 800 nm due to a drop in κ. An advantage of aluminum is that its oxide makes a very tough and robust natural passivation. Both copper and gold are unsuitable as reflectors in the visible spectrum, but gold makes an excellent infrared reflector, and is also inert to environmental factors. Chromium has a low-visible reflection, but it has excellent adhesion to many surfaces, therefore it is often used as an under-layer for gold and silver.

The design of metal film optics revolves around creating antireflection and high transmission coatings. However, unlike the all-dielectric designs, a low reflective design using metals does not necessarily mean it will producc a high transmission. This is because metals are inherently lossy. Antireflection designs on metals simply mean that the reflection is low without any consideration of transmission. A high transmission design, on the other hand, attempts to reduce reflection and absorption simultaneously.

4.4.2 Antireflection for Metals

As odd as it may sound, it is possible to make an antireflection coating for a metal substrate. This works on the same principles as the antireflection on dielectric substrates. Layers are designed such that the accumulated effective index approaches that of the outside medium (air). This allows all of the incident energy to be transmitted into the metal substrate. However, unlike with dielectric substrates, the transmitted energy will be entirely absorbed and nothing will emerge from the back side of the metal substrate. The reflection will be zero, but the transmission will also be zero.

Referring back to the complex refractive index contour plots, when the substrate was a pure dielectric, the starting point was a real number on the *x*-axis. A quarter-wave film deposited on the substrate traced a semicircular arc in the complex plane moving clockwise toward the right, or counterclockwise toward the left, depending if the film index was larger or smaller than the substrate index.

In the case of a metal substrate, the starting number will be a complex number with a negative imaginary value. At $\lambda_0 = 550$ nm, silver has a refractive index of $0.125 - j3.3$. This will place the starting point of the contour on the lower left of 1.0 in the complex plane as shown in Figure 4.18. If a quarter-wave film with a refractive index of 2.0 is deposited on top of the silver substrate, it will create an arc that moves clockwise from $0.125 - j3.3$. Since it starts from the lower half of the plot, the contour first moves left toward lower indices before moving upward and to the right.

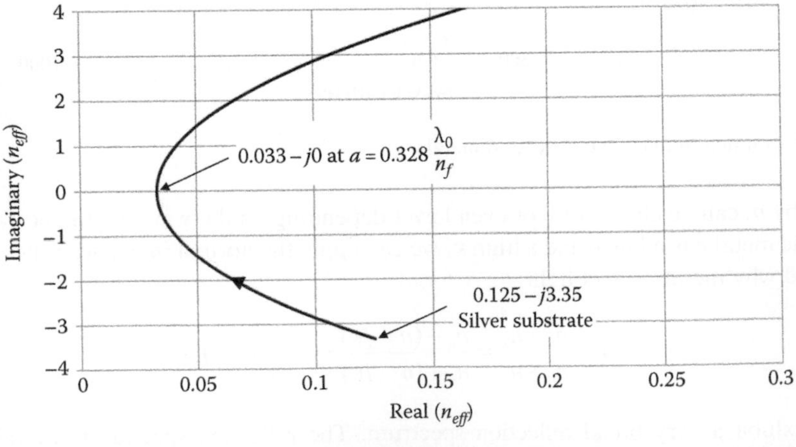

FIGURE 4.18 Contour of the complex n_{eff} close to the starting point for a dielectric thin film of index 2.0 on a silver substrate at $\lambda_0 = 550$ nm.

It crosses the real axis at an index of 0.033 when the film thickness is 0.328-wave thick. From that point on, the silver substrate can be thought of as having an equivalent index of 0.033, even though a physical material cannot have such a refractive index. Since the center of the circle is the geometrical mean of the end points, and the first point has an extremely small value of 0.033, the right side intercept will be a very large value—in this case, 121. A half-wave-thick film will trace an entire circle back to $0.125-j3.3$, as shown in Figure 4.19.

To create antireflection, the contour has to intercept the x-axis at 1.0. Unfortunately, this contour does not intercept anywhere close to 1.0, irrespective of the dielectric index or its thickness. Physically, this problem arises due to the significantly lower reflection from the film–air interface compared to the metal–dielectric interface. Even if a π phase shift could be achieved within the dielectric film, it would still only result in an incomplete destructive interference. One way to resolve this problem is by adding another higher reflective film on top of the dielectric film. The reflection of the silver substrate at $\lambda_0 = 550\,\text{nm}$ is about 96%. We could create a multilayer high-reflective stack on the front side such as Ag | dielectric film | reflective stack | air. This technique will work, but an even simpler approach is to use a single metal film as the top reflector. Since it has to allow light to pass through, it also has to be ultra-thin. A possible structure is Ag | dielectric | ultrathin Ag | air. Before studying this structure, it is instructive to examine what kind of contour is traced by a film such as silver with a complex refractive index.

We know that a dielectric film traces a full circle when its thickness reaches a half-wave. For larger thicknesses, it keeps repeatedly tracing over the same circle. A film thickness of $2H+h$ has the same effect as h at the reference wavelength. The location and trajectory of this repeating circle greatly depends on other layers that are beneath that layer. But films with complex indices have a very different behavior. We know that any material with a large imaginary part, such as silver, when sufficiently thick will prevent light from penetrating past that layer and reaching the underlying films. Therefore, regardless of what is underneath that thick metal film, the effective index of the structure will be that of the metal alone when the film becomes thick.

For example, consider a silica glass substrate with an effective index is 1.48. As a thin layer of silver is deposited on the glass, the effective index will trace a contour that moves from 1.48 almost directly toward the silver index of $0.125-j3.3$, as shown in Figure 4.20. After that, regardless of the thickness of the silver film, the effective index will remain pinned at $0.125-j3.3$ and would not make the circuitous rounds like dielectric films. This is an interesting but somewhat obvious conclusion that can be put to use in our antireflection designs.

Now, returning to the Ag | dielectric | ultrathin Ag | air structure, we can select the thickness of the dielectric (index of 2.0) such that the ultrathin silver's contour will intersect the real axis

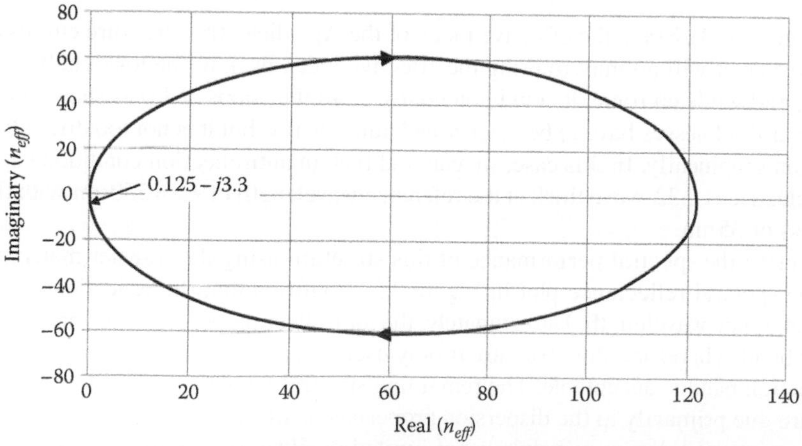

FIGURE 4.19 Continuation of Figure 4.18 for a half-wave film of index 2.0 on a silver substrate at $\lambda_0 = 550\,\text{nm}$.

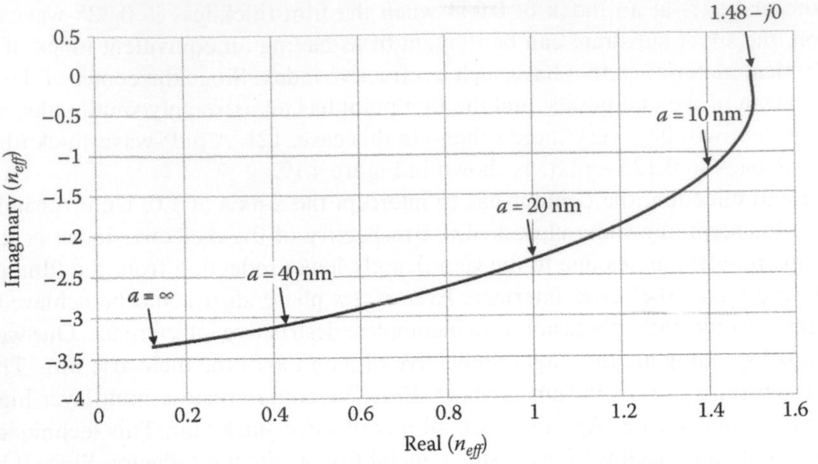

FIGURE 4.20 Contour of n_{eff} for a thick silver film on a silica glass substrate at $\lambda_0 = 550\,\text{nm}$.

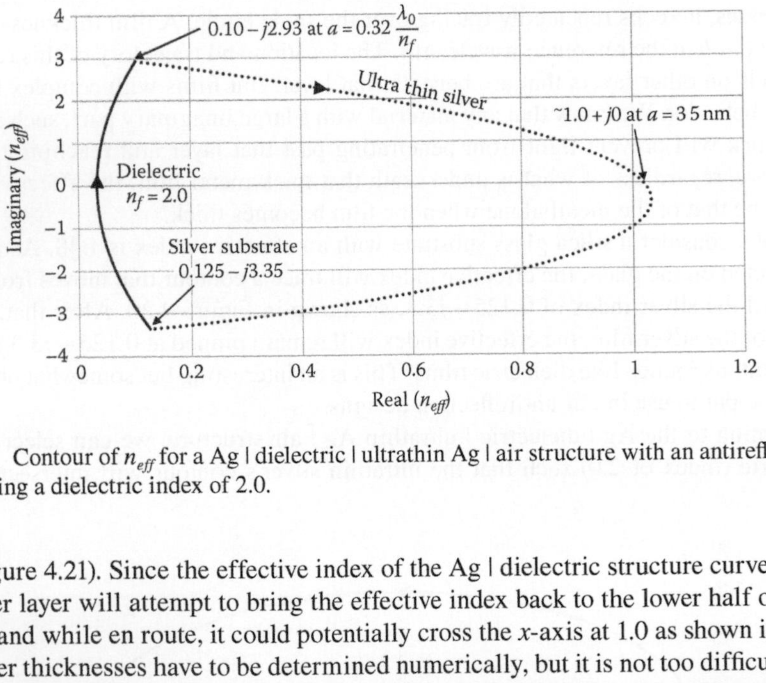

FIGURE 4.21 Contour of n_{eff} for a Ag | dielectric | ultrathin Ag | air structure with an antireflection effect at $\lambda_0 = 550\,\text{nm}$ using a dielectric index of 2.0.

at 1.0 (see Figure 4.21). Since the effective index of the Ag | dielectric structure curves upward, the ultrathin silver layer will attempt to bring the effective index back to the lower half of the plane to $0.125 - j3.3$, and while en route, it could potentially cross the x-axis at 1.0 as shown in Figure 4.21. The exact layer thicknesses have to be determined numerically, but it is not too difficult to converge on the solution graphically. In this case, we can find that an antireflection condition can be reached when the dielectric is 0.32-wave thick at the reference wavelength of $\lambda_0 = 550\,\text{nm}$ with the top silver film thickness of 35 nm.

We can verify the spectral performance of this structure using the transfer matrix method calculation. The spectral reflectance plot in Figure 4.22 confirms that the reflection indeed drops to zero at the reference wavelength. Unfortunately, the antireflection performance only exists within a very narrow band. Therefore, this structure is only useful in a small number of applications where a narrow reflection band is acceptable. The remaining spectral features on the plot, including the dip at 320 nm, are due primarily to the dispersion properties of silver.

The aforementioned design principle is not limited to silver, or to having the same top and bottom metals. We could design an antireflection for a silver substrate using, for example, chromium as

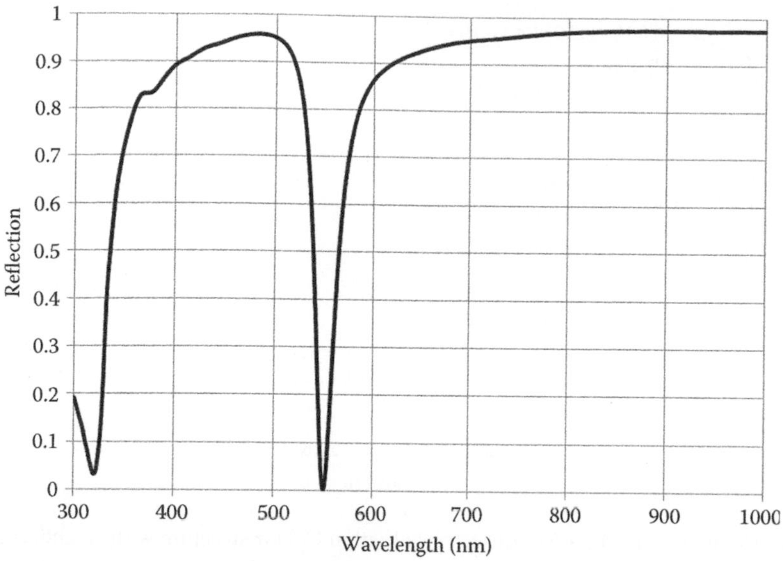

FIGURE 4.22 Spectral reflectance plot of the Ag | dielectric | ultrathin Ag | air structure with an antireflection effect at $\lambda_0 = 550$ nm using a dielectric index of 2.0.

the top metal. At 550 nm, chromium has a refractive index of $3.1 - j4.4$. The main difference here will be the end point of the trace. After a thick chromium layer, the effective index will converge toward $3.1 - j4.4$ instead of $0.125 - j3.3$ for silver. However, the same strategy as before can be used to make the trace cross the x-axis at 1.0 to achieve the antireflection condition. The required chromium thickness is 3 nm, and the dielectric is 0.195-wave thick. The corresponding plots for this example are shown in Figures 4.23 and 4.24. The antireflection spectral width in this case is larger, which arises due to the lower reflectivity of chromium compared to silver.

4.4.3 High Optical Transmission through Metals

In the previous examples, even though the reflection from the metal substrate was reduced, the transmission through the back of the metal substrate was zero. All of the incident power was absorbed in the metal substrate. Therefore, the structure behaves as a perfect absorber at the reference wavelength. However, thin metal films could also be used to make transmission filters, such as linepass and bandpass filters, as long as they are constructed on a transparent substrate such as glass. Compared to all-dielectric filters, metal filters require far fewer layers and exhibit a very broad reflection spectrum outside the transmission band. While they are not suitable for every optical filtering application, they can be very useful in specific cases. An example is the metallic filter used on energy-saving architectural windows known as low-emissive glass.

A metal-based linepass filter consists of a dielectric cavity enclosed between two ultrathin metallic reflectors. The ultrathin requirement is so that they will be partially transparent at the reference wavelength. For example, let us consider a cavity structure such as silica | ultrathin Ag_1 | dielectric ultrathin Ag_2 | air, with the substrate having an index of 1.48, and the dielectric having an index of 2.0. We start by selecting a reasonably small thickness for the first Ag film, say 15 nm, which exhibits a reflection of 51% and a transmission of 42% at $\lambda_0 = 550$ nm. The effective index of the substrate | ultrathin Ag_1 structure traces a clockwise contour starting from 1.48 and terminating at $1.19 - j1.9$. Using this as the starting point, it is relatively easy to estimate the thicknesses of the dielectric cavity and the second silver layer. First, consider a very thick silver layer. This will have a contour that ends at $0.125 - j3.3$ regardless of where it starts from. Then, we progressively change

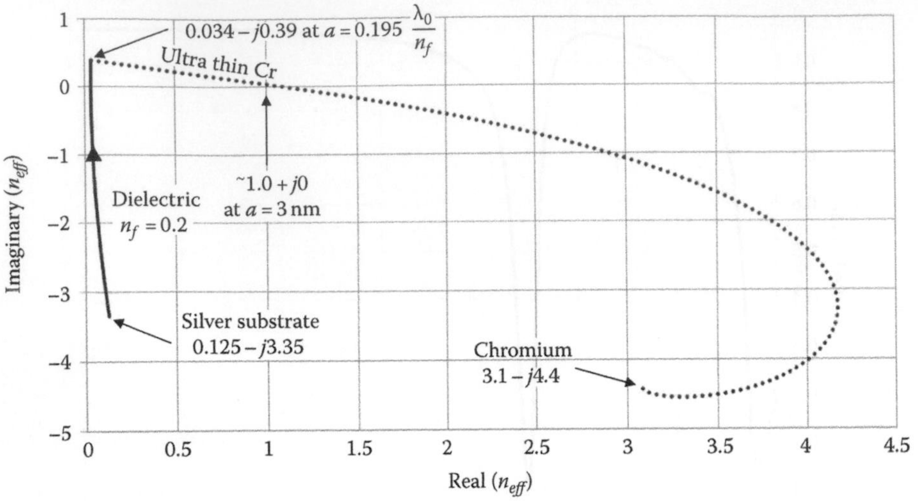

FIGURE 4.23 Contour of n_{eff} for a Ag | dielectric | ultrathin Cr | air structure with an antireflection effect at $\lambda_0 = 550\,\text{nm}$ using a dielectric index of 2.0.

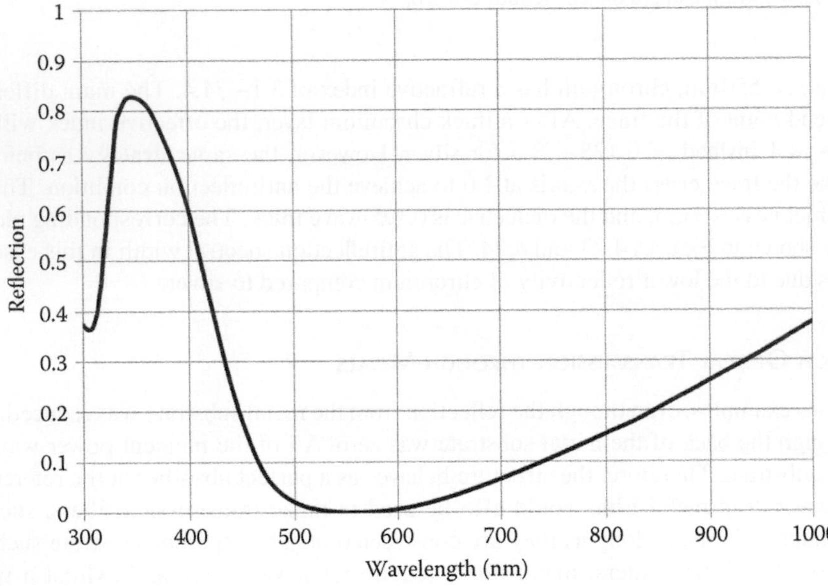

FIGURE 4.24 Spectral reflectance plot of Ag | dielectric | ultrathin Cr | air structure with an antireflection effect at $\lambda_0 = 550\,\text{nm}$ using a dielectric index of 2.0.

the dielectric layer thickness until the contour intersects the real axis at 1.0. In this case the crossing occurs when the silver thickness is 10.8 nm, as shown in Figure 4.25.

Figure 4.26 shows the reflection and transmission spectra of this structure. Note that the transmission and reflection do not add up to 1.0 due to the absorption loss. As expected, the transmission has a peak value of 80% at $\lambda_0 = 550\,\text{nm}$. Compared to dielectric bandpass filters, a notable feature of this filter is the rapidly increasing reflection at all infrared wavelengths that extends all the way into RF frequencies. This structure is also significantly easier to make, compared to an all-dielectric structure with a similar performance, because it contains just three layers. Because they transmit a portion of the visible spectrum but reflect all infrared, these filters are also known

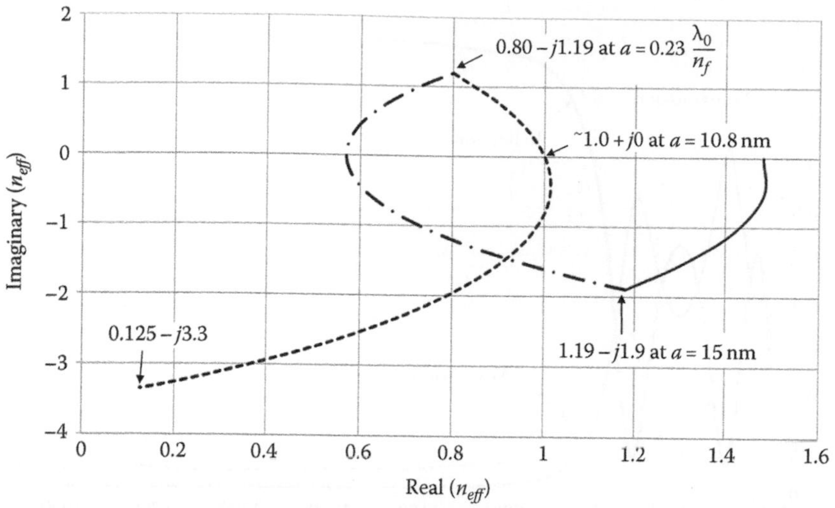

FIGURE 4.25 Contour of n_{eff} for a silica|ultrathin Ag_1 | dielectric | ultrathin Ag_2 | air structure for high transmission at $\lambda_0 = 550\,\text{nm}$.

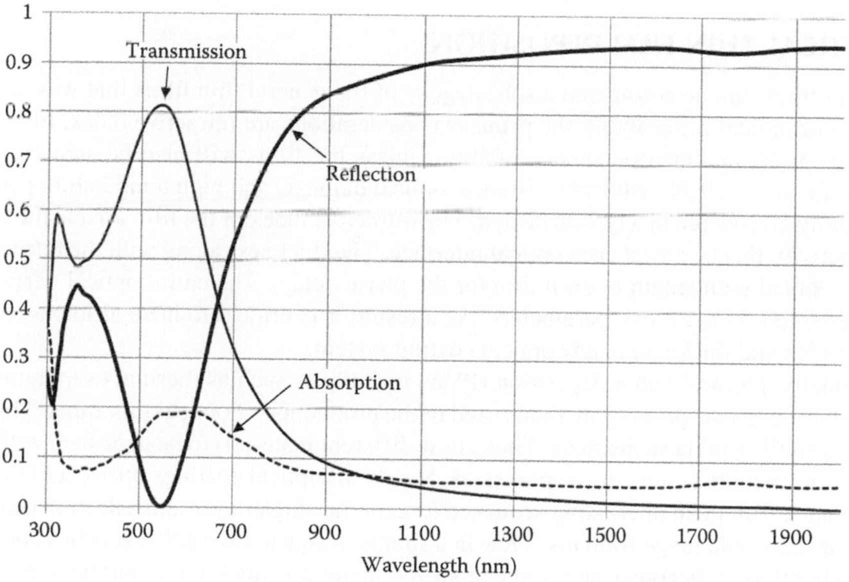

FIGURE 4.26 Spectral plots of the silica | ultrathin Ag_1 | dielectric | ultrathin Ag_2 | air structure at $\lambda_0 = 550\,\text{nm}$.

as low-emissive (low-E) filters and are extensively utilized as energy-efficient coatings in architectural windows [6]. The blue tint that we often see in high-rise glass buildings is due to the low-E coatings.

Expanding further on the silica | ultrathin Ag_1 | dielectric ultrathin Ag_2 | air structure, the transmission bandwidth can be broadened by creating a coupled cavity structure, such as silica | ultrathin Ag_1 | dielectric | ultrathin Ag_2 | dielectric | ultrathin Ag_3 | air. This contains two cavities back to back that produces two adjacent transmission peaks and results in an overall enhancement of the transmission bandwidth. This result is shown in Figure 4.27 where Ag_1 and Ag_3 are 15 nm thick, and the center Ag_2 is 10.8 nm with a dielectric thickness of 0.23-wave.

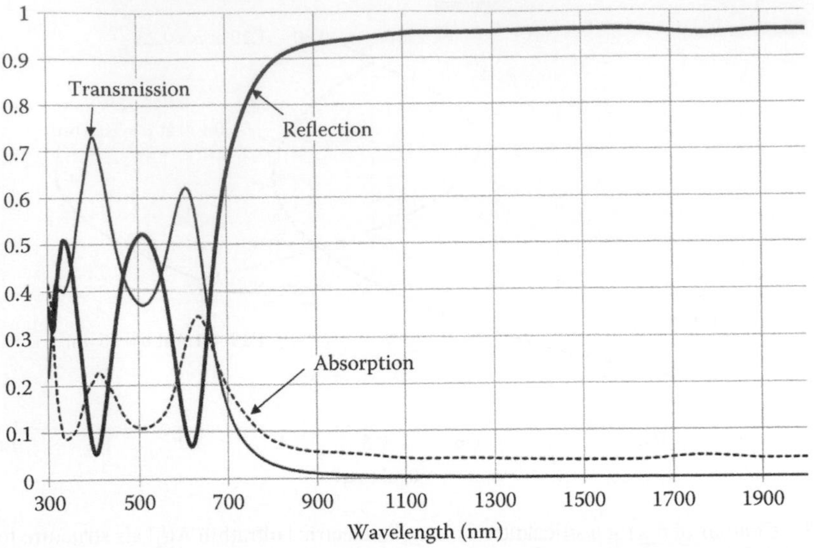

FIGURE 4.27 Spectral plot of the coupled cavity structure with three metal layers and two dielectric spacers using the silica | ultrathin Ag_1 | dielectric | ultrathin Ag_2 | dielectric | ultrathin Ag_3 | air structure.

4.5 OPTICAL THIN-FILM DEPOSITION

Optical thin films can be considered a subcategory of the general thin films that was discussed in Chapter 3. For optical applications, the primary considerations are refractive index, film thickness, optical absorption, and environmental stability. Unless the films will also be subjected to harsh conditions, factors such as chemical resistance, optical damage, and high temperature performance are not usually considered in a typical design. The refractive index of the film affects the magnitude of the reflectivity that occurs at each optical interface. The thickness along with the refractive index affects the optical path length of each film (or the phase delay). The entire optical performance is therefore dictated by these two parameters. As a result, it is critical to have absolute control over refractive index and thickness in any optical coating system.

Traditionally, physical vapor deposition (PVD) techniques such as thermal evaporation and ion sputtering have been the primary methods used in the production of optical thin films. This is partly due to the versatility of these methods. They allow different materials to be deposited without major changes to the vacuum system or the processes. Nearly all optical coatings utilize a high and a low refractive index film in an alternating sequence (except the single-layer antireflection coating). The number of layers could range from just a few in a simple design to over 100 layers in a complex optical filter. Therefore, it becomes necessary to switch materials quickly without having to open the chamber and reload a different process after each layer. In thermal evaporation, a number of different source materials can be preloaded into separate crucibles, and rotated during use. Some systems also allow replenishing the source materials in situ without having to interrupt the process. This becomes necessary for designs that use a large number of layers or thicker films such that a single filling of the crucible will not be adequate to complete the deposition. Since the film thicknesses are usually some fraction of the optical wavelength (quarter-wave being the most common), the total thickness of the design linearly scales with the intended wavelength of operation. An optical filter designed for the long-wave infrared spectrum (8–12 μm) will require significantly thicker films than the same filter designed for the visible spectrum.

Due to the low energies of the deposition species, simple thermal evaporation tends to produce loosely packed structures with pores and columnar structures. Although porous structures have

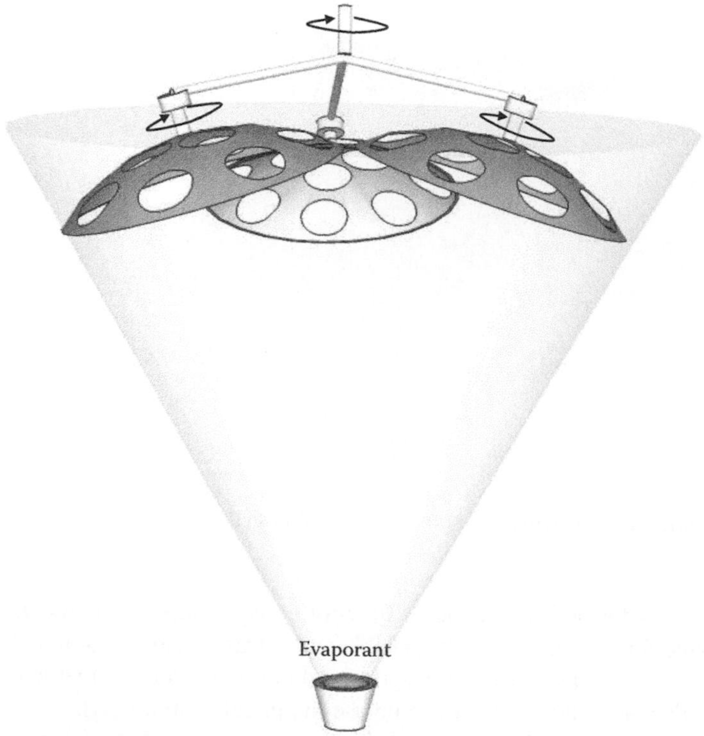

FIGURE 4.28 Planetary rotation system used in thermal evaporation systems for improving uniformity with multiple substrates. Individual substrates are mounted in their slots, and each hemispheric assembly is rotated in its own axis, and the entire assembly is rotated around the central axis.

been intentionally used in some optical applications because they lend themselves to fine tuning of the refractive index and dispersion [7], robust coatings that withstand environmental moisture and temperature changes generally require a denser film. Bombarding the substrate with energetic ions during deposition is a commonly used method to densify thermally evaporated films. Known as ion-assisted deposition (IAD), this has become the de facto standard in optical coating systems. IAD also enables one to increase the surface energy without raising the substrate temperature, which is important in plastic substrates and ophthalmic coatings [8,9]. Film uniformity also requires careful attention to the placement of the substrates. To improve uniformity with multiple substrates, a planetary rotational system is often utilized, as shown in Figure 4.28. Individual assemblies loaded with substrates are rotated along their own axes, and the entire system is rotated along a different axis to improve uniformity.

Because the substrates have to be mounted upside down, thermal evaporation does not lend itself for continuous coating of large substrates such as windows. RF magnetron sputtering can be an attractive option when the substrates are large or irregularly shaped because sputter targets and cathodes can be manufactured in many of different geometries to accommodate the substrates. A planar rectangular or cylindrical target can be used to coat large windows by constantly moving the glass panes under the cathode as shown in Figure 4.29. Additionally, cylindrical magnetron cathodes have a uniform erosion rate, providing greater utilization of the target material. The low sputter yield of dielectrics is one of the man drawbacks of sputtering, but it can be overcome to some extent by using reactive sputtering from metal targets.

CVD techniques are less popular for optical coatings due to the limited material inventory, requirement for precursors, and high process temperatures. However, PECVD has been shown to

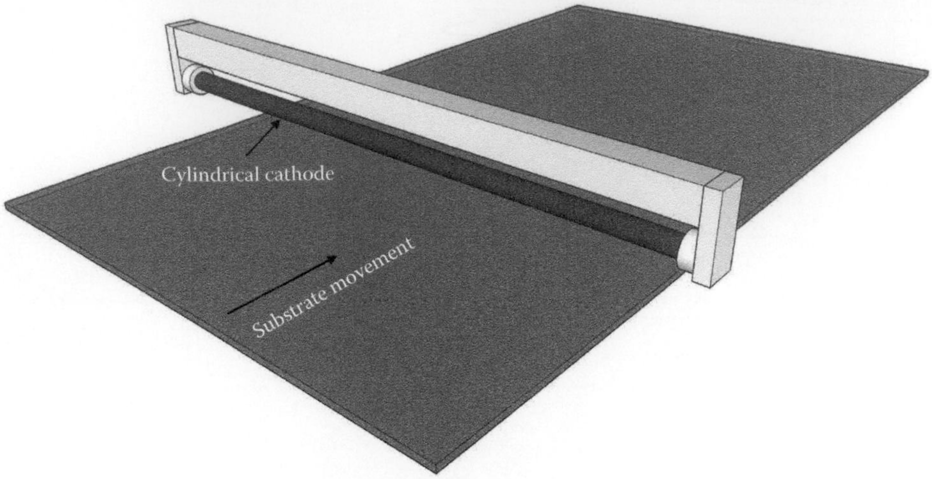

FIGURE 4.29 Sputtering large windows with a cylindrical target.

have some advantages for optical coatings [10]. Atomic layer deposition (ALD) is very slow and generally unsuitable for optical films in the visible and infrared region, but may become an attractive option for extreme ultraviolet (EUV) optics. Additionally, sol–gel methods based on liquid chemistries have also been shown to have some use in optical coatings [11].

For thickness measurement, the traditional method of using quart crystal microbalance (QCM) has many deficiencies for producing optical thin films with consistent quality and reproducibility. As discussed in Chapter 3, QCM works by measuring the shift in acoustic resonance, which is then translated into an equivalent physical thickness by assuming a certain acoustic impedance and film density. Neither of these parameters are accurately known and are most often treated as fitting parameters. The tooling factor used to account for the geometric position of the QCM also rarely works out to be a constant during actual depositions. Additionally, QCM readings are also susceptible to changes in temperature and pressure, as well as the accumulated thicknesses from prior deposition cycles. Therefore, there will be significant run-to-run variations that will make it difficult to achieve consistent films. Instead of QCM, a more commonly used approach is optical thickness monitoring. This can be realized by installing a light source and measuring its reflection (or transmission) from the substrate in real time. Optical viewports within easy reach of the substrate will be necessary, and they have to be adequately shielded from the deposition flux to prevent the viewports from getting coated. This provides a true measure of the performance being sought rather than translating the desired information from indirect measurements. A single wavelength monitoring using a laser source is the simplest to deploy because the minimal divergence of the laser beam makes it easy to direct the beam toward the substrate and also extract the reflection, as shown in Figure 4.30. The transmitted beam can also be extracted if there is adequate visibility from the opposite side of the coating system. Since there will be significant amounts of extraneous light generated from the deposition process itself, especially with thermal evaporation or sputtering, some means to mitigate this noise is necessary. One could chop the incident light with a lock-in amplifier or use narrow band filters in the path of the reflected beam, or both, to reduce this noise. Multiple lasers at different wavelengths could also be used to expand the measurement space, as well as a broadband incoherent light source with a spectrometer. It is also possible to install an ellipsometry setup to extract more accurate optical properties of the film, although extracting the data in real time would require significant computational effort for the mathematical fitting of the data [12].

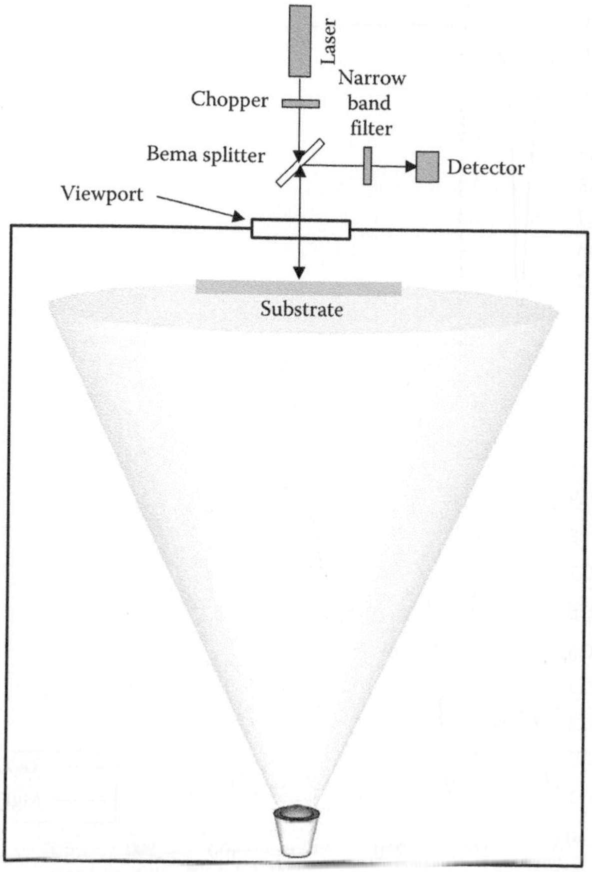

FIGURE 4.30 Real-time optical monitoring of the deposition process by using a single laser source through a viewport to measure the reflected signal.

Considering silica$\left|\left(\frac{H}{2}L\frac{H}{2}\right)^4\right|$air at $\lambda_0 = 550\,\text{nm}$, where H is TiO_2 and L is MgF_2, Figure 4.31 shows the expected reflection spectrum (Figure 4.31a) and the expected progression of the reflection at 532 nm (which is a common laser line) as a function of the total film thickness (Figure 4.31b). By monitoring the trajectory of this reflection, it becomes possible to make adjustments in real time to account for small deviations due to thickness or refractive index variations.

It is worthwhile pointing out that while an optical monitoring system is critical in a production environment, it is less attractive in a research environment because of the limited spectral range that can be accessed with such a system. For instance, UV coatings will require a UV source and UV detectors as well as UV transparent viewports. For the spectral range used in optical telecommunications (1.55 μm), a different set of lasers and detectors will be required. For mid- and long-wave infrared, it becomes exceedingly difficult to set up a system that can be operated compactly and conveniently next to a deposition chamber. For this reason, in a research environment where a wide range of materials are being deposited, despite its many shortcomings, the quartz crystal microbalance still remains the next best choice. The disadvantages of the QCM can be traded off with greater labor effort. The substrate can be retrieved after one-half (or some other fraction) of a layer has been deposited to make detailed measurements. Based on these measurements, adjustments can be made to the subsequent fractions of the film to compensate for any deviations.

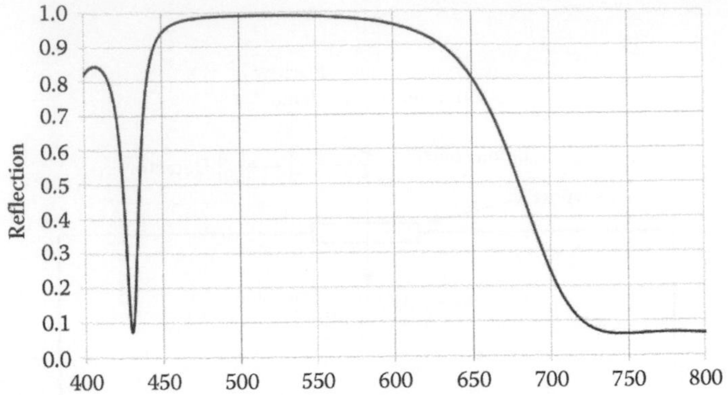

(a) Wavelength (nm)

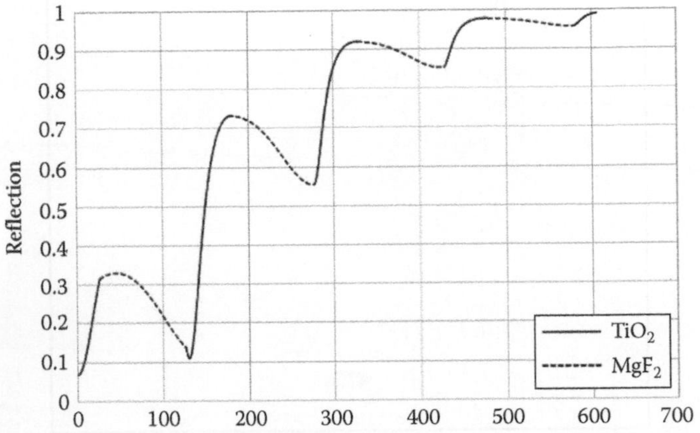

(b) Total film thickness (nm)

FIGURE 4.31 (a) The plot shows the expected reflection spectrum from silica$\left|\left(\frac{H}{2}L\frac{H}{2}\right)^4\right|$air using $\lambda_0 = 550\,\text{nm}$, where H is TiO_2 and L is MgF_2. (b) Shows the evolution of the reflection as a function of deposition thickness at an interrogation wavelength of 532 nm.

PROBLEMS

4.1 For ZnSe substrate at $\lambda_0 = 5\,\mu\text{m}$ ($n = 2.43$) and SiO_2 as the antireflection thin film ($n = 1.4 - j0.00$), plot the complex n_{eff} as a function of film thickness in the complex plane. Find the film thickness that results in the smallest reflection.

4.2 Using TiO_2 and MgF_2 as the layers at $\lambda_0 = 550\,\text{nm}$ ($n = 2.5$ and 1.38, respectively), design a highly reflecting structure on a silica substrate ($n = 1.48$) to achieve greater than 98% reflection. Using the OpenFilters software, plot the reflection spectrum and calculate the reflection bandwidth using the minima on either side of the reflection peak as the width of the reflection band.

4.3 For the aforementioned problem, design an additional antireflection layer on the backside of the silica substrate to improve the transmission of the long-wavelength side of the reflection band minima.

4.4 Considering aluminum as the metal substrate, design a two-layer structure with SiO_2 and an ultrathin aluminum as the top layer to produce antireflection at the laser wavelength of 532 nm. The refractive indices of Al and SiO_2 at $\lambda_0 = 532\,\text{nm}$ are $0.9 - j6.2$ and 1.48, respectively.

LABORATORY EXERCISES

4.1 Deposit a thin film of SiO_2 by three different methods, (1) thermal evaporation, (2) RF sputtering, and (3) ion-assisted evaporation, and examine the differences in their optical constants.

4.2 Design a two-layer antireflection coating for sapphire (Al_2O_3) using SiO_2 and Si_3N_4 for a wavelength of 1.0 μm. The refractive indices for Al_2O_3, SiO_2, and Si_3N_4 at 1.0 μm are 1.62, 1.48, and 2.0, respectively. Perform this deposition on a sapphire substrate and measure the performance.

4.3 Silver has a high percolation thickness and does not generally form a continuous film until about 10 nm film thickness is reached. For this exercise, deposit a 5 nm thick silver film on a clean silica substrate and compare the reflection measurements against the transfer matrix model. At least one calibration run may be necessary prior to the actual run. The measurement will most likely reveal a significant discrepancy due to the non-planar silver film. Next, deposit a 1 nm of germanium followed by the same 5 nm silver. The measurements from this system should be closer to the model because germanium acts as a good wetting agent for silver [13].

4.3 Using MgF_2 and TiO_2 as the low- and high-index dielectric layers on a fused silica substrate, design a thin film structure to produce > 98% reflection at a wavelength of 532 nm. Deposit this film structure and compare the measurement against the predictions.

REFERENCES

1. H. A. Macleod, *Thin-Film Optical Filters*, 4th edn., CRC Press, Boca Raton, FL, 2010.
2. H. K. Pulker, Characterization of optical thin films, *Appl. Opt.*, 18(12) (June 15, 1979) 1969–1977.
3. R. R. Willey, *Practical Design and Production of Optical Thin Films*, 2nd edn., Marcel Dekker, Inc., New York, 2002.
4. J. D. Rancourt, *Optical Thin Films: User Handbook*, SPIE Optical Engineering Press, Bellingham, WA, 1996.
5. S. Larouche and L. Martinu, OpenFilters: Open-source software for the design, optimization, and synthesis of optical filters, *Appl. Opt.*, 47(13) (May 1, 2008) C219–C230.
6. R. J. Martín-Palma, L. Vázquez, J. M. Martínez-Duart, and Malats-Riera, Silver-based low-emissivity coatings for architectural windows: Optical and structural properties, *Sol. Energy Mater. Sol. Cells*, 53 (1998) 55–66.
7. A. Barranco, A. Borras, A. R. Gonzalez-Elipe, and A. Palmero, Perspectives on oblique angle deposition of thin films: From fundamentals to devices, *Prog. Mater. Sci.*, 76 (March 2016) 59–153.
8. U. Schulz, U. B. Schallenberg, and N. Kaiser, Antireflection coating design for plastic optics, *Appl. Opt.*, 41(16) (June 1, 2002) 3107–3110.
9. U. Schulz and N. Kaiser, Vacuum coating of plastic optics, *Prog. Surf. Sci.*, 81 (2006) 387–401.
10. L. Martinu and D. Poitras, Plasma deposition of optical films and coatings: A review, *J. Vac. Sci. Technol. A*, 18 (2000) 2619.
11. D. Chen, Anti-reflection (AR) coatings made by sol–gel processes: A review, *Sol. Energy Mater. Sol. Cells*, 68(3–4) (June 2001) 313–336.
12. W. M. Duncan and S. A. Henck, Insitu spectral ellipsometry for real-time measurement and control, *Appl. Surf. Sci.*, 63(1–4) (January 1993) 9–16.
13. W. Chen, M. D. Thoreson, S. Ishii, A. V. Kildishev, and V. M. Shalaev, Ultra-thin ultra-smooth and low-loss silver films on a germanium wetting layer, *Opt. Express*, 18(5) (2010) 5124–5134.

5 Substrate Materials

Silicon is easily the most common substrate material used in any cleanroom. It is the preferred substrate for microelectronics, MEMS, and solar cells. As a result, it is also one of the most mature and well-understood materials. However, one of the major limitations of silicon is its inability to emit light. The emission of light from semiconductors depends on the alignment in k-space of the lowest point of the conduction band and the highest point of the valence band [1,2]. In silicon, these points are offset from each other as shown in Figure 5.1a. This makes it an indirect bandgap material with an extremely low probability of photon emission. Despite this limitation, silicon is used in many photonic devices including photodetectors, waveguides, and modulators. However, for applications that require light emission, such as LEDs and lasers, other materials with a direct bandgap had to be developed. GaAs is arguably the second most popular semiconductor substrate after silicon. It has a direct bandgap, as shown in Figure 5.1b, so it can emit photons. This property and its fixed lattice constant over a wide range of alloy compositions with aluminum are some of the reasons for the popularity of GaAs in optoelectronics. In addition to LEDs and lasers, it is also used in some high-speed electronics due to its favorable carrier mobilities. InP is also a direct-bandgap material that is used for making optoelectronic devices. The bandgap of InP and its alloys (InGaAsP) covers the 1.3 and 1.6 μm optical emission bands, which are two of the most important bands in optical fiber communications. Sapphire is also a substrate that has been gaining popularity, driven by the LED lighting market. It is one of the few substrates on which crystalline GaN films can be grown. GaN has a large direct bandgap, and is used to produce ultraviolet photons, which are then turned into white light using phosphor materials. Other substrates used in device fabrication include SiC, SiO_2, Ge, GaP, GaSb, InAs, InSb, and ZnO.

Substrates serve multiple functions during a fabrication process: (1) they provide mechanical rigidity, flatness, and temperature stability during processes such as photolithography, deposition and etching; (2) they provide a semiconducting electrical material whose conductivity can be modified by selectively introducing impurities to create diodes and transistors; and (3) they provide a defined crystal orientation on which subsequent layers can be grown as lattice-matched crystalline thin films.

Most substrates for device fabrication come in thin circular discs known as wafers, with thickness around 250–500 μm. The nominal thickness is determined by the material hardness. Softer materials like InP and GaAs tend to come in thicker substrates (> 500 μm). Silicon substrates tend to be thinner (< 400 μm). Ultrathin flexible silicon wafers (< 25 μm) are also available for specialty applications in MEMS.

Substrate diameters vary from 1 in. all the way up to 18 in. [3,4]. Figure 5.2 shows a photograph of a 6-in silicon wafer. Larger diameters require advanced manufacturing processes with a greater level of uniformity, but also bring greater manufacturing efficiencies and yield. Technology has been developed to manufacture silicon wafers of all sizes up to 18 in. to meet the demands of CMOS microelectronics. In research laboratories, where tooling and material costs are bigger concerns than manufacturing yield, the substrate sizes tend to be limited to 4 in. or smaller. On the other hand, GaAs and InP substrates are typically no larger than around 4 in. This is limited not only by the crystal growth maturity, but also by their smaller application space and smaller market demand. Less common substrates like GaP and InAs will only be available in even smaller sizes, such as 3 in., with many other less common exotic substrates manufactured only on demand.

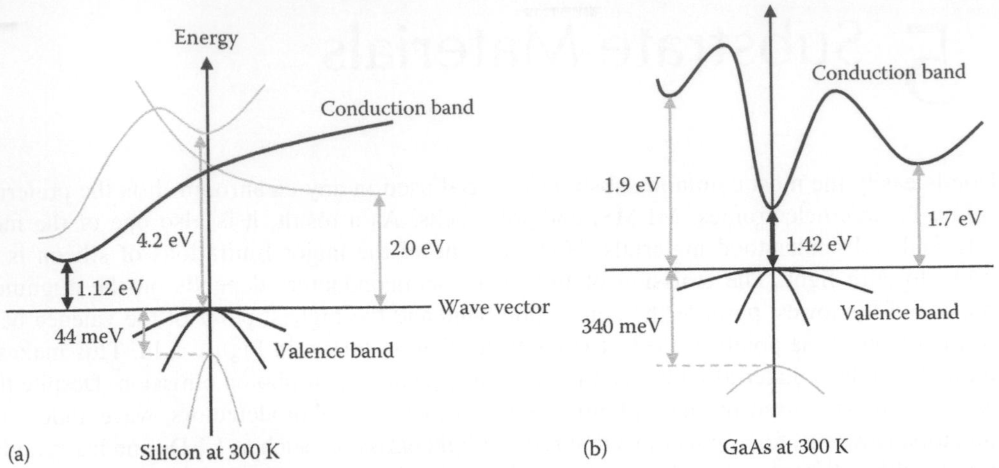

FIGURE 5.1 Electronic band structure of (a) Si; (b) GaAs. The bold lines show the relevant conduction and valence bands.

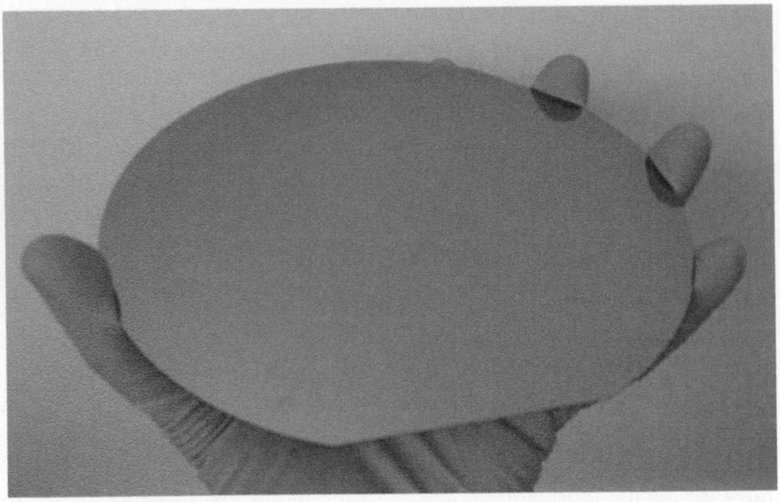

FIGURE 5.2 A 6 in. (150 mm) silicon wafer.

5.1 SILICON

5.1.1 Silicon Wafer Manufacture

5.1.1.1 Raw Material

The raw material for silicon wafers is silica sand mined from quarries. This is not all that different from the common beach sand. It consists mostly of silica (SiO_2). This silica is chemically purified and reduced to polycrystalline silicon (Si), which is a grey rock-like material. The chemical purification is extremely critical because it affects the electrical properties of the silicon wafers. The purity level required is of the order of 1 impurity atom for every 10^9 silicon atoms [5]. The lowest grade of silicon, with the highest concentration of impurities, is known as metallurgical grade silicon, and is intended for general industrial applications. Electronic grade contains the least concentration of impurities, and is also the most expensive. Solar grade silicon is inferior to electronic grade, but was developed to meet the needs of the photovoltaic industry where cost was a greater concern than purity [6–8].

5.1.1.2 Crystal Growth

The most common growth method is to melt the polycrystalline silicon in a large slowly rotating chamber made of graphite or quartz, and allow it to condense over a seed crystal. The chamber is evacuated and filled with argon to prevent oxidation of the polysilicon during growth. A controlled amount of impurities such as phosphorous, boron, or arsenic is also added to the melt to provide a known doping species and concentration. This determines the polarity (n-type or p-type) and the resistivity of the substrate. The seed crystal is lowered into the melt and pulled slowly out of the melt. The rate of extraction affects the crystallization rate and the diameter of the resulting crystal. The extraction is a very slow process, extending over several days depending on the desired diameter. The result is a long column of crystal that is referred to as the ingot. This growth process is known as the Czochralski growth, abbreviated as CZ [8]. Though it is the most widely used method, one of the disadvantages of the CZ method include axial inhomogeneity of the dopant species and unintentional impurities from the crucible. This makes it difficult to obtain high-resistivity (low-doped) substrates. An alternative growth method is the float zone (FZ) process [9]. In this process, the starting material is a solid polysilicon rod. Starting from a seed crystal attached to one end of this rod, an induction heating coil is moved along the length of this rod to melt and recrystallize the polysilicon rod into a single crystal. Since there is no crucible, high purities and high resistivities can be obtained with this method.

5.1.1.3 Ingot Processing

The ingot is sliced off at each end and ground to a circular rod to the required diameter. X-ray diffraction analysis is performed on the ingot to determine its crystal orientation. We need two directions to indicate the precise crystal orientation. Using the end-face of the ingot as one reference, a primary flat is ground along the length of the ingot to orient this reference. A secondary flat may also be used to indicate whether the substrate is n-type of p-type.

5.1.1.4 Wafer Saw

Crystalline silicon is extremely hard and difficult to slice. Diamond-coated blades are required to slice the ingot into wafers. Even then, it is impractical to slice the wafers one at a time since each slice would take several hours to complete. Instead, the slicing process is done using an array of closely spaced diamond-coated wire saws [10]. The ingot gets sliced into wafers simultaneously by all the wires at once. Tens of thousands of wafers can be sliced simultaneously this way. The process has to be done slowly with plenty of coolant to prevent cracking and chipping. The resulting sliced wafers will have a rough texture on both faces. Further polishing and refining is necessary to turn these into usable wafers. The process is illustrated in Figure 5.3.

5.1.1.5 Etching, Lapping, and Polishing

The sawed wafers are chemically etched to remove most of the mechanical damage from the sawing process. Then, they are lapped down to a more precise thickness and subsequently polished by chemical and mechanical means to improve the surface roughness, planarity, and thickness variations. In most applications, only one side of the wafer is used for making devices so the wafers are only polished on one side. These are referred to as single-side-polished (SSP) wafers. In some optical and MEMS applications, both sides are used and will require double-side-polished (DSP) wafers. Since polishing is one of the most time-consuming portions of the wafer manufacturing process, DSP wafers are generally twice as expensive as SSP wafers. For CMOS circuit applications, an ultra high-quality epitaxial thin film of silicon is grown on top of the polished wafer to provide a high-grade silicon layer.

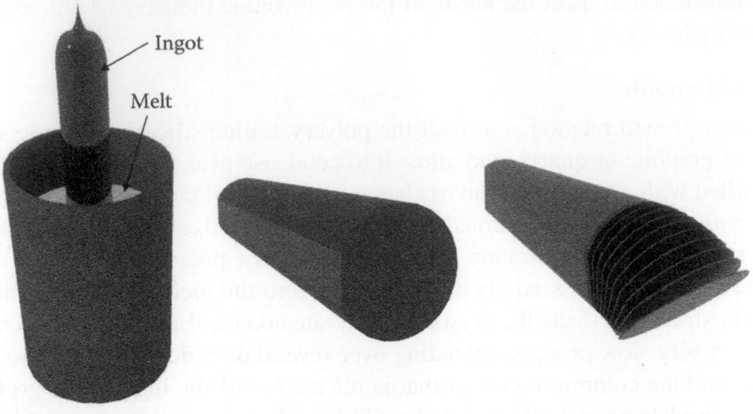

FIGURE 5.3 Crystal growth and ingot processing.

5.1.1.6 Finished Silicon Wafers

After the manufacturing process, each wafer is inspected for defect density, resistivity, warp, and thickness variations. Based on these numbers, they are sorted into prime grade, test grade, and mechanical grade.

Silicon wafers are mass produced for the electronics and solar industry, and as a result, their cost is relatively low even for ultra high-purity wafers with extremely low defect densities. Other substrate materials with similar levels of purity and defect density would cost 10–100 times more, mostly due to the economy of scale rather than technological complexity.

Commercially produced wafers will contain several basic specifications. These include (1) diameter, (2) thickness, (3) dopant species, (4) resistivity (which is related to dopant concentration), (5) surface crystal orientation, (6) SSP or DSP, (7) growth method (CZ or FZ), (8) grade (prime or test), (9) total thickness variation (TTV), (10) warp, etc.

5.1.2 Silicon Crystal Orientations

Silicon has a cubic crystal structure. That means it is symmetric along the *x*, *y*, and *z* Cartesian coordinates. The unit cell of a silicon crystal consists of two face-centered-cubic (FCC) structures interpenetrating each other at ¼ distance along the body diagonal. This structure is known as the diamond structure because it is identical to the real diamond structure (carbon). Germanium also has the same diamond crystal structure.

III–V semiconductors such as GaAs, InP, AlAs, etc., also have the same crystal structure, except, of the two interpenetrating FCC structures, one FCC lattice contains the first element (Ga) and the other FCC lattice contains the second element (As). In all other respects, the structure is identical to silicon. This structure is known as the zinc-blende structure.

Each unit cell of silicon contains eight atoms. This includes the atoms of one FCC and part of the second FCC that penetrates into the first FCC. The lattice constant, measured along one of its cubic faces is 5.43 Å. Therefore, there are eight atoms within each 5.43 Å cubic volume.

Even though silicon has a cubic symmetry, it can be cut and polished along any direction. It can be cut parallel to one of its cubic faces, or it can be cut diagonally from one edge to another edge of the same cube, or it can be cut diagonally through three corners of the cube. These are considered “standard cuts” because the starting and ending points of the cuts fall within the corners or edges of the same unit cell. Standard notations have been developed to designate the crystal orientations of the cut planes. These are described in Fourier coefficients (k-vectors) of the 3D

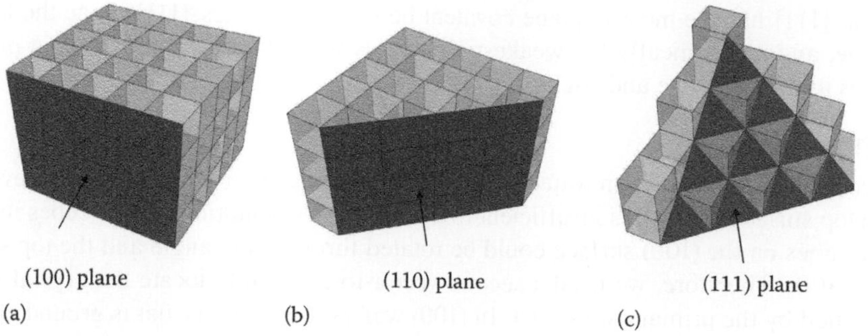

FIGURE 5.4 The standard planes of a silicon crystal structure: (a) the (100) plane is any plane that is parallel to one of the faces of the cubes that make up the lattice; (b) the (110) plane is one that cuts across two edges of the same cube; (c) the (111) plane goes through three diagonal points of the same cube.

periodic lattice. Figure 5.4a through c shows the (100), (110), and (111) crystal planes. When the starting and ending points fall in different unit cells, these can produce an infinite combination of "nonstandard cuts."

5.1.2.1 {100} Planes

This is shown in Figure 5.4a. When the cut is parallel to one of the cubic faces, the k vector normal to that face can be described as $(k_x,k_y,k_z)=(1/a,0,0)$ where a is the lattice constant. It is more common to normalize this k-vector by $1/a$ and write it as (1,0,0) or simply [100] where the square brackets indicate the vector whose direction is normal to the crystal plane. The crystal plane itself is written with the notation (100), with a regular parenthesis. However, this could have been (010) or (001) depending on which axes are x, y, or z. Instead of distinguishing between the x, y, or z axes, all of which are identical, the entire family of these equivalent planes is simply written as {100}. In other words, all faces of the cubes are treated as being equivalent.

5.1.2.2 {110} Planes

When the cube is sliced diagonally from one edge toward the opposite edge, the k vector normal to that plane will be [110], [101], or [011] depending on how the x, y, or z axes are chosen. The entire family of these equivalent planes is written as {110} and is shown in Figure 5.4b.

5.1.2.3 {111} Planes

When the cube is sliced through all three axes equally, the k vector normal to that face becomes [111], and the planes are represented as {111}. This is shown in Figure 5.4c.

5.1.2.4 Other Crystal Planes

It can be seen that the crystal can be cut in more ways than just {100}, {110}, or {111}. For example, {210}, {311}, etc., are all valid cut faces. However, in silicon wafers, these are considered "nonstandard cuts" and can only be acquired by specifying the orientation before the ingot is manufactured. Most silicon surfaces are manufactured with the standard surfaces of {100}, {110}, or {111}.

5.1.2.5 Crystal Orientations and Their Properties

The orientation of each face has different chemical, mechanical, and optical properties. Most optical crystals are birefringent along different crystal directions. Chemical etching properties of the crystals are also different along different axes. The mechanical strength and thermal expansion coefficients can also be different. Therefore, the crystal orientation of the wafer surface is an important parameter that governs the properties of devices that are fabricated on these wafer surfaces.

In silicon, {111} has the most in-plane covalent bonds. This makes {111} plane the least chemically reactive, and mechanically the weakest. {110} has the next most number of in-plane bonds, which makes it more reactive and slightly stronger.

5.1.2.6 (100) Wafer

The most commonly found silicon wafers have a (100) plane as their top surface. However, specifying just the top surface as (100) is insufficient to describe the orientations of the cubes in the crystal lattice. The cubes on the (100) surface could be rotated through any angle and the top surface will still remain (100). Therefore, we need a second vector to accurately locate the crystal orientation. This is specified by the primary wafer flat. In (100) wafers, the primary flat is ground along a (110) plane. This allows us to accurately place the orientation of each cube on the wafer surface as illustrated in Figure 5.5a and b.

The {111} planes on a (100) wafer surface are tilted from the top face by exactly 54.7°. When a defect is scribed on the primary flat and pressure is applied, the wafer will cleave perpendicular to the flat along a (111) plane or a (110) plane. Both the (111) and (110) planes intersect the (100) surface at right angles, so the cleaved pieces will generally be rectangular. If the cleaved side face is perpendicular to the top surface, then the cleaved plane is (110); if it appears tilted, then it is (111). Both types of cleave planes are possible, although (110) tend to be the dominant cleave plane in thin wafers. This is because the applied tensile stress during a cleaving process will be a perpendicular (110) plane, which more than compensates for the preference for (111) planes to cleave [11].

As illustrated in Figure 5.6, it is easy to show from geometry that the angle between the {111} planes and the (100) top wafer surface is 54.7°. If the cube has a side of length a, the diagonal along any face will be $\sqrt{2}a$. Half of this diagonal will be $a/\sqrt{2}$. Therefore, the indicated angle is $\tan^{-1}\sqrt{2} = 54.7°$. This angle has important implications in some chemical etching processes as we will see in Chapter 7. For example, potassium hydroxide (KOH) will etch silicon along {100} and {110} planes, but not along {111} planes. When the etchant encounters the {111} planes, the etch rate will come to a stop. The end result is that the etched holes will have an exact 54.7° taper away from the top (100) surface.

5.1.2.7 (110) Wafer

These wafers have (110) as their top surface and are less common than (100) wafers. The primary wafer flat is a (100) plane as shown in Figure 5.7a and b, but it can also be another plane from the {110} family by rotating the cube 90° on the surface. The cubes on the surface are therefore lying on one of their edges, rather than flat on their faces as in the case of (100) wafers.

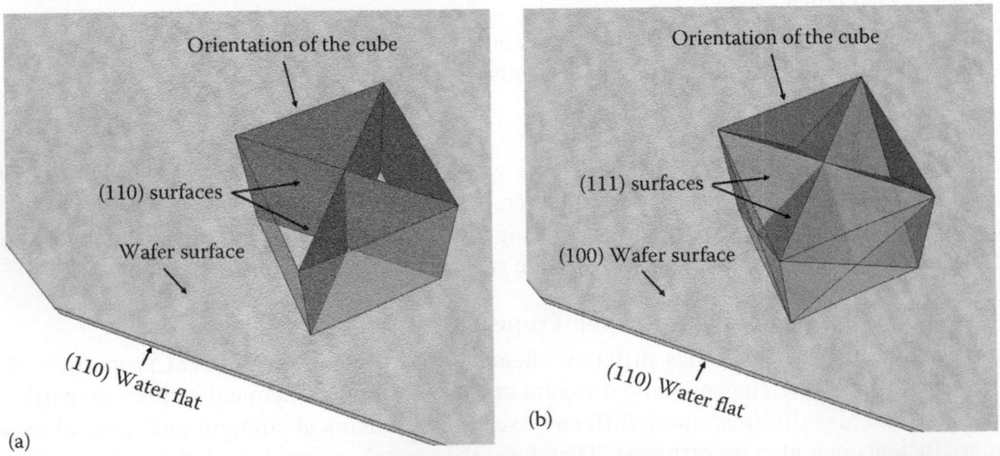

FIGURE 5.5 (a) Orientation of the (110) planes on a wafer that has a (100) surface and a (110) flat; (b) orientation of the (111) planes on a wafer that has a (100) surface and a (110) flat.

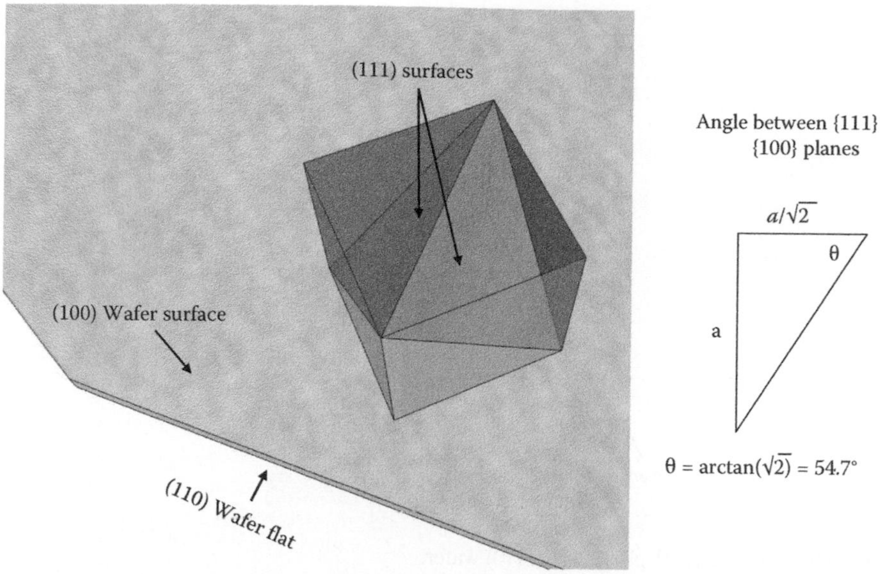

FIGURE 5.6 Angle of the {111} planes on a (100) surface wafer.

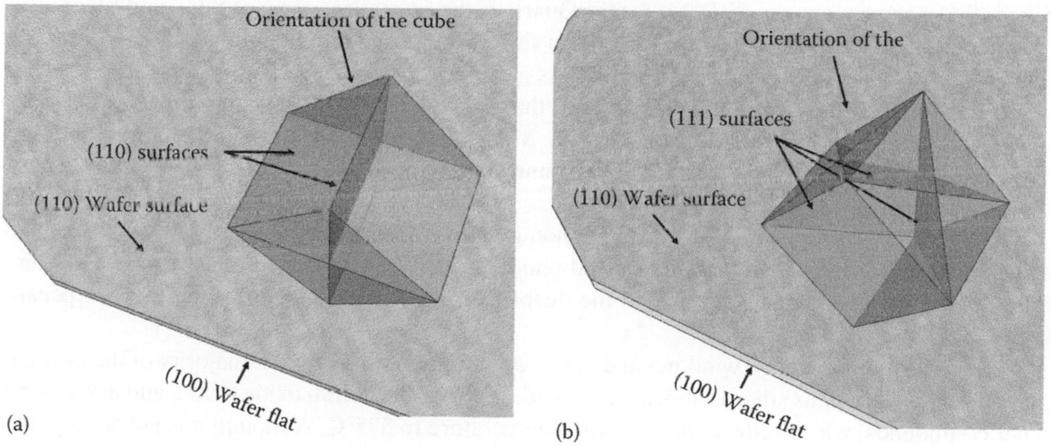

FIGURE 5.7 (a) Orientation of the (110) planes on a wafer that has a (110) surface and a (100) flat; (b) orientation of the (111) planes on a wafer that has a (110) surface and a (100) flat.

An interesting aspect of the (110) wafers is that some of the {111} planes (the two that are shown in Figure 5.8) intersect the wafer surface at right angles. Therefore, if a (110) wafer cleaves along {111} planes it will produce parallelograms with 70.5° and 109.5° corners. The cleaved faces will be at right angles to the wafer surface. If the wafer cleaves along {110} planes, it will produce rectangles with right-angled cleaved faces.

A more useful property of (110) wafers is that they can produce vertical sidewall profiles when etched with potassium hydroxide, whereas (100) wafers produce a 54.7° taper. However, the etched hole shape cannot be arbitrary. It has to match the {111} parallelogram orientations at 70.5° and 109.5°. As illustrated in Figure 5.8, it follows from geometry that the intersection between the {111} planes and the (110) top wafer surface forms a parallelogram with 109.5° and 70.5°. Another sloped {111} plane also intersects the surface at $\tan^{-1}(1/\sqrt{2}) = 35.3°$. The etch result will therefore contain four {111} vertical sidewalls and two {111} sidewalls sloped at 35.3°.

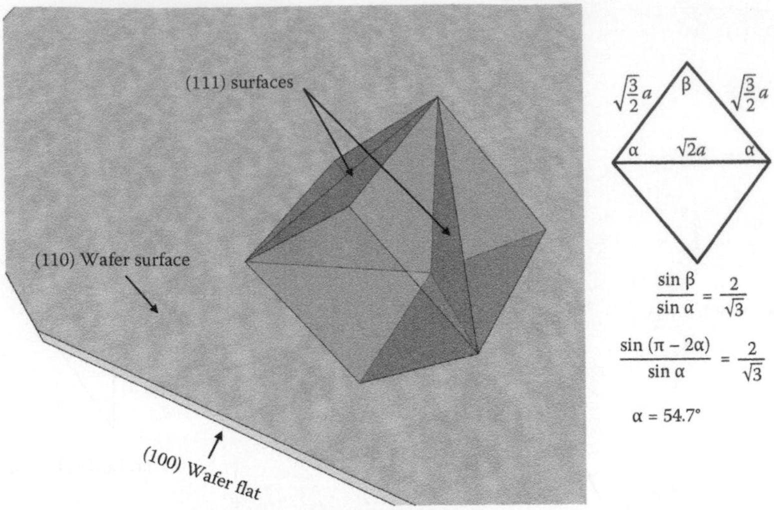

FIGURE 5.8 Angles of the {111} planes on a (110) wafer.

5.2 SILICA

Silica (SiO_2) comes in many different forms. Quartz is the crystalline form of SiO_2, and fused silica is the amorphous form of SiO_2. Fused silica can sustain very high temperatures and does not soften until 1600°C. What we refer to as ordinary glass is a mixture of SiO_2 and other compounds such as calcium oxide, boron oxide, sodium oxide, etc., that are added to reduce the melting temperature of the glass so that it becomes easier to work with. Many different types of glasses are used in device fabrication and are commonly used in many cleanroom, so it is useful to have a basic understanding of these different types [12].

Quartz and fused silica are the most common glass wafers used in device fabrication. They are also used as substrates for optical filters and photomasks. Compared to ordinary glass, they have high optical transparency, especially in the deep-UV range but are significantly more expensive than other types of glasses.

Soda lime glass is used in windows and glasswares, and accounts for the majority of the ordinary glass market. It contains silica (75%), sodium oxide (13%) and calcium oxide (10%), and a number of other compounds, which reduce the softening temperature to 575°C. Although it is not widely used in device fabrication, it is the least expensive glass and is used in some photomasks where deep-UV transmission is not required.

Borosilicate glass contains silica and boron trioxide, and is known for its low thermal expansion properties, and is sold in many consumer products under the trade name Pyrex. It is also used in high temperature components like light bulbs. It softens around 800°C. Borosilicate wafers are used in cleanrooms when a glass wafer needs to be irreversibly bonded to a silicon wafer. This is done through a process known as anodic bonding, and the glass needs to have a significant ionic component as well as a thermal expansion coefficient that is closely matched to that of silicon. Borosilicate glass meets both of these requirements. Its thermal expansion coefficient is $3 \times 10^{-6}\ K^{-1}$ and for silicon it is $2.6 \times 10^{-6}\ K^{-1}$ [13,14].

5.3 SAPPHIRE

Sapphire is the crystalline form of aluminum oxide (Al_2O_3) and has a very high melting temperature of 2050°C. Amorphous aluminmum oxide is a white ceramic, while sapphire is an optically clear crystal. Its crystal structure is hexagonal. It has a wide optical transparency from the deep UV to the mid-infrared, which makes it an attractive material for a number of optical applications. It is also one of the hardest material, second only to diamond. Sapphire is widely used in the growth of GaN epitaxial thin films, which is a key enabling material in the production of white LEDs. The ingots are prepared in much the same way as with silicon by heating and melting raw aluminum oxide powder and using a seed crystal to produce crystallization [15,16].

5.4 COMPOUND SEMICONDUCTORS

Compound semiconductors contain two or more elements. If one element is from group III and the other elements is from group V of the periodic table, the resulting compound semiconductor is referred to as III–V semiconductors. If they are from group II and group VI, they are referred to as II–VI semiconductors. A number of compound semiconductors including their bandgaps and lattice constants are shown in Figure 5.9.

GaAs is probably the most widely used III–V semiconductor substrate. It has a higher electron and hole mobility than silicon and is therefore used in high-speed electronic devices in microwave communications and radar systems. It has a larger bandgap than silicon (1.42 eV compared to 1.12 eV), but still narrow enough to behave as a semiconductor at room temperature. The larger bandgap results in a lower temperature sensitivity, which makes GaAs attractive in high-power electronic devices. The larger bandgap also makes it less sensitive to radiation, which is important in satellite and space electronic components [17,18]. GaAs is significantly denser than silicon, but it is also softer and weaker.

One of the unique properties of GaAs is its ability to form alloys with aluminum in the form of $Al_xGa_{1-x}As$, with nearly the same lattice constant as GaAs. Hence, $Al_xGa_{1-x}As$ of any composition can be grown on a GaAs substrate without significant lattice mismatch or strain. This is indicated by the nearly vertical line between GaAs and AlAs in Figure 5.9. The bandgap of $Al_xGa_{1-x}As$ ranges from 1.42 eV (bandgap of pure GaAs) to 2.12 eV (bandgap of pure AlAs). At a composition of $Al_{0.45}Ga_{0.55}As$ (bandgap of 1.98 eV), it switches from a direct to an indirect bandgap [19,20]. Therefore, the range of $x = 0–0.45$ is of great interest in optical emission sources such as LED and laser diodes. Additionally, the ability to alter the bandgap and still maintain the same lattice

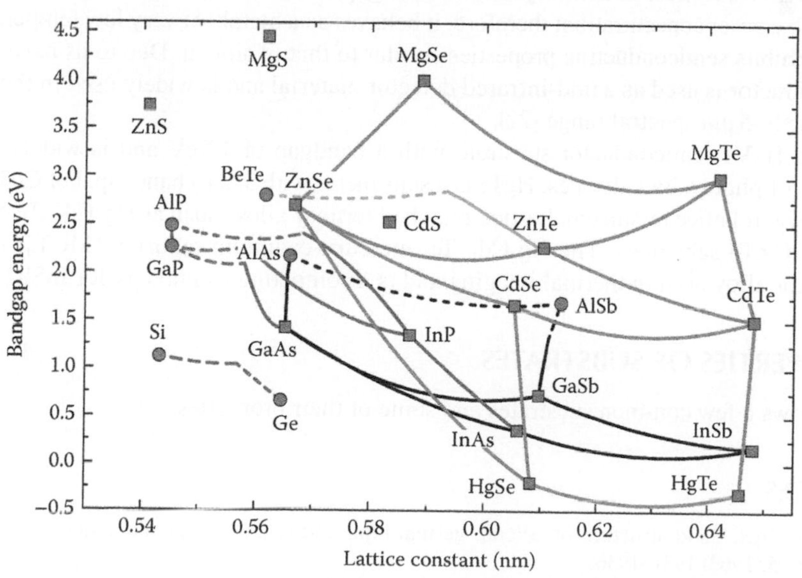

FIGURE 5.9 Compound semiconductors and their alloys, lattice constants and bandgaps. (Figure reproduced with permission from Arizona State University MBE Optoelectronics Group; Courtesy of Ding Ding, Michael DiNezza, Yong-Hang Zhang.)

TABLE 5.1
Properties of a Few Common Semiconductors

	Silicon	Quartz	Sapphire	GaAs	InSb
Crystal structure	Diamond	Quartz	Hexagonal	Zinc-blende	Zinc-blende
Lattice constant (Å)	5.43	$a = 4.194$ $c = 5.40$	$a = 4.78$ $c = 12.99$	5.65	6.48
Bandgap (eV at 25°C)	1.1 indirect	9.3 indirect	9.9 indirect	1.42 direct	0.17 direct
Melting temperature (°C)	1412	1700	2050	1238	527
Optical transparency	1.1–10 μm	250 nm to 2.5 μm	300 nm to 5 μm	900 nm to 15 μm	
Density (g/cm^3)	2.33	2.65	3.98	5.32	5.75
Intrinsic carrier density (cm^{-3})	1×10^{10}			2×10^6	2×10^{16}

constant as the underlying substrate has allowed the GaAs/$Al_xGa_{1-x}As$ family of semiconductors to be used in a large number of quantum engineered electronic devices. Examples of this include quantum-well infrared photodetectors, quantum cascade lasers, and quantum well lasers. The red laser pointer that everyone is familiar with is made from $Al_xGa_{1-x}As$ on a GaAs substrate.

InP is another III–V substrate that is widely used in semiconductor lasers and detectors for optical communication equipment. There is only one composition, $In_{0.53}Ga_{0.47}As$, that is lattice matched to InP, with a bandgap of 0.74 eV (wavelength of 1.67 μm). However, InP can be alloyed with $In_{0.53}Ga_{0.47}As$ to produce a quaternary semiconductor $In_xGa_{1-x}As_yP_{1-y}$ that can be tailored from the bandgap of $In_{0.53}Ga_{0.47}As$ (0.74 eV) to the bandgap of InP (1.34 eV or a wavelength of 0.92 μm). Therefore, this material system can access nearly the entire near infra-red (NIR) spectrum. The two most important spectral bands in fiber optic telecommunications are 1.3 and 1.55 μm, both of which fall within this spectral range [21].

InSb is a III–V substrate with a very narrow bandgap of 0.17 eV. At room temperature this results in a very high carrier concentration; therefore, it behaves as a metal. At very low temperatures, such as 77 K, it exhibits semiconducting properties similar to that of silicon. Due to its narrow bandgap, this semiconductor is used as a mid-infrared detector material and is widely used in thermal image sensors in the 3–5 μm spectral range [22].

CdTe is a II–VI semiconductor substrate with a bandgap of 1.5 eV and is widely used in the manufacture of photovoltaic devices. HgTe is a semimetal with a zero bandgap, but CdTe and HgTe have only a small lattice mismatch. Lattice-matched ternary alloys such as $Hg_xCd_{1-x}Te$ can be made on CdZnTe (CZT) substrates. The $Hg_xCd_{1-x}Te$ (mercury-cadmium telluride [MCT]) is a popular semiconductor alloy used in thermal imaging and is a competing technology for InSb [23,24].

5.5 PROPERTIES OF SUBSTRATES

Table 5.1 shows a few common substrates and some of their properties.

REFERENCES

1. J. C. Phillips, Band structure of silicon, germanium, and related semiconductors, *Phys. Rev.*, 125(6) (March 15, 1962) 1931–1936.
2. F. Herman, The electronic energy band structure of silicon and germanium, *Proc. IRE*, 43(12) (1955) 1703–1732.
3. P. O. Hahn, The 300 mm silicon wafer—A cost and technology challenge, *Microelectron. Eng.*, 56 (2001) 3–13.

4. Z. Lu and S. Kimbel, Growth of 450 mm diameter semiconductor grade silicon crystals, *J. Cryst. Growth*, 318(1) (March 1, 2011) 193–195.
5. W. Zulehner, Historical overview of silicon crystal pulling development, *Mater. Sci. Eng. B*, 73 (2000) 7–15.
6. B. R. Bathey and M. C. Cretella, Solar grade silicon, *J. Mater. Sci.*, 17 (1982) 3077–3096.
7. S. Pizzini, M. Acciarri, and S. Binetti, From electronic grade to solar grade silicon: Chances and challenges in photovoltaics, *Phys. Status Solidi (a)*, 202(15) (2005) 2928–2942.
8. W. Zulehner, Czochralski growth of silicon, *J. Cryst. Growth*, 65 (1983) 189–213.
9. P. H. Keck, W. Van Horn, J. Soled, and A. MacDonald, Floating zone recrystallization of silicon, *Rev. Sci. Instrum.*, 25 (1954) 331.
10. H. J. Möller, Chapter two—Wafering of silicon. In: *Advances in Photovoltaics: Part 4. Semiconductors and Semimetals*, Vol. 92, Amsterdam, the Netherlands, 2015, pp. 63–109.
11. K. Wasmer, C. Ballif, R. Gassilloud, C. Pouvreau, R. Rabe, J. Michler, J.-M. Breguet, J.-M. Solletti, A. Karimi, and D. Schulz, Cleavage fracture of brittle semiconductors from the nanometer to the centimeter scale, *Adv. Eng. Mater.*, 7(5) (2005) 309–317.
12. N. P. Bansal and R. H. Doremus, *Handbook of Glass Properties*, Elsevier, Amsterdam, the Netherlands, 2013.
13. A. Cozma and B. Puers, Characterization of the electrostatic bonding of silicon and Pyrex glass, *J. Micromech. Microeng.*, 5 (1995) 98–102.
14. A. Berthold, L. Nicola, P. M. Sarro, and M. J. Vellekoop, Glass-to-glass anodic bonding with standard IC technology thin films as intermediate layers, *Sens. Actuators*, 82 (2000) 224–228.
15. H. Tang, H. Li, and J. Xu, Chapter 10—Growth and development of sapphire crystal for LED applications. In: *Advanced Topics on Crystal Growth*, Sukarno, O. F. (ed.), Intech, http://dx.doi.org/10.5772/54249. Published: February 20, 2013.
16. D. C. Harris, A century of sapphire crystal growth, *Proceedings of the 10th DoD Electromagnetic Windows Symposium*, Norfolk, Virginia, May 2004.
17. F. Schwierz and J. J. Liou, Semiconductor devices for RF applications: Evolution and current status, *Microelectron. Reliab.*, 41(2) (February 2001) 145–168.
18. C. Y. Chang and F. Kai, *GaAs High-Speed Devices: Physics, Technology, and Circuit Applications*, John Wiley & Sons, New York, 1994.
19. S. Adachi, GaAs, AlAs, and $Al_xGa_{1-x}As$: Material parameters for use in research and device applications, *J. Appl. Phys.*, 58 (1985) R1.
20. I. Vurgaftman, J. R. Meyer, and L. R. Ram-Mohan, Band parameters for III–V compound semiconductors and their alloys, *J. Appl. Phys.*, 89 (2001) 5815.
21. S. Adachi, Material parameters of $In_{1-x}Ga_xAs_yP_{1-y}$ and related binaries, *J. Appl. Phys.*, 53 (1982) 8775.
22. R. A. Stradling, InSb-based materials for detectors, *Semicond. Sci. Technol.*, 6 (1991) C52–C58.
23. A. Rogalski, HgCdTe infrared detector material: History, status and outlook, *Rep. Prog. Phys.*, 68 (2005) 2267–2336.
24. W. Lei, J. Antoszewski, and L. Faraone, Progress, challenges, and opportunities for HgCdTe infrared materials and detectors, *Appl. Phys. Rev.*, 2 (2015) 041303.

6 Lithography

Lithography is a technique used for creating structures in the lateral direction, similar to thin-film deposition that was used for creating precise structures in the vertical direction. Although it is relatively easy to control the film thicknesses down to a fraction of a nanometer, the same statement cannot be said about the lateral dimension. Structures smaller than 100 nm are extremely difficult to make and require very elaborate and expensive methods. Lithography is arguably the most expensive step in a device fabrication process, requiring equipment that range from a few million dollars in a small R&D laboratory to hundreds of millions in a manufacturing environment.

Photolithography is a subset of lithography that uses photons to define the lateral structures. Additionally, there are also nonoptical lithography techniques that use charged beams such as electrons or ions to define the structures, but photolithography is the most widely used system today. In photolithography, a photomask is used to project an optical image onto a photosensitive film and permanently alter its chemical or mechanical properties. It is most commonly used as a sacrificial film and discarded after using it as stencil to pattern another underlying or overlying film. In a small number of cases, the photoresist is left as a structural film, and not discarded, such as the SU-8 photoresist.

There are several critical defining steps in the photolithography process: (1) substrate cleaning and preparation, (2) application of a photoresist film on a substrate, (3) designing and creating a photomask, (4) aligning the photomask image to existing features on the substrate, (5) projecting and exposing the image of the photomask onto the photoresist, (6) chemically removing the exposed (or unexposed) areas of the photoresist, and (7) patterning the underlying (or overlying) film.

This chapter covers only the basic information for working with lithography tools in the laboratory. For a detailed discussion of its history, photochemistry, and the optical systems used in photolithography, the reader is directed to a few excellent textbooks on this subject [1–4].

6.1 SUBSTRATE CLEANING AND PREPARATION

Prior to performing any lithographic process, the substrate has to be thoroughly cleaned. The cleaning of silicon wafers is a vast but often overlooked subject, so only the basic aspects are discussed in this chapter. For more details, the reader is referred to an entire textbook devoted to this subject [5].

6.1.1 Acetone–Methanol–Isopropyl Alcohol (AMI) Cleaning

The simplest and most widely used cleaning process in research laboratories is the acetone–methanol–isopropyl alcohol cleaning, abbreviated as the AMI cleaning process. It is performed primarily for removing organic contaminations from the substrate, including photoresists. Acetone can soften or dissolve a large number of organic residues, but it can leave an organic film behind. Therefore, methanol is used to dissolve the acetone before it dries, and then isopropyl alcohol (IPA) is used to rinse off the methanol. IPA is very hygroscopic, so it will leave the surface dehydrated. In many cases, an IPA-cleaned surface is equivalent to a dehydration-baked surface.

In most cleaning processes, the wafer is never allowed to dry by letting the liquid evaporate away from the surface. Even the cleanest liquid contains some level of contaminants, and if the liquid is allowed to dry by evaporation, all of these contaminants will be deposited on the wafer. Instead, the wafer should be dried by forcing the liquid meniscus off the wafer with clean, dry, high-velocity nitrogen, starting from the center and working radially outward. The wafer can also be spin-dried, which has the same effect as blow-drying with nitrogen.

6.1.2 Piranha (Sulfuric Peroxide Mixture) Cleaning

For more aggressive organic removal, a sulfuric acid + hydrogen peroxide mixture can be used. This is an extremely exothermic and aggressive mixture and hence known as the piranha clean (named after the flesh-eating fish) or sulfuric peroxide mixture (SPM). The solution is mixed immediately prior to use and then discarded. The reaction is self-heating, so external heating is usually not necessary, but it can also be placed on a hot plate to maintain an elevated temperature during prolonged use. In addition to organics, piranha also removes a number of metals such as aluminum, titanium, and nickel. This is an excellent cleaning recipe, but many laboratories ban its use due to its safety hazards. A heavily contaminated wafer can produce a violent reaction that can make the solution boil over, and because the solution will continue to give off gases well after it has cooled off, it can even cause an explosion if the waste is left in a sealed container.

6.1.3 RCA Cleaning

The RCA clean, named after the Radio Corporation of America, consists of two separate steps for removing organic contaminations and particles and for removing ionic contaminations [6,7]. The first step, termed standard clean 1 (SC-1), is performed with a solution of ammonium hydroxide, hydrogen peroxide, and water in a volumetric ratio of 1:1:5, although a number of different dilution ratios are also used. SC-1 is also sometimes referred to as the ammonia peroxide mixture (APM). Organic residues are removed by dissolution in NH_4OH and by the strong oxidizing action of hydrogen peroxide. SC-1 also etches the native SiO_2 film slowly from a silicon surface and simultaneously forms a new oxide at the same rate. Hence, it consumes silicon at a very slow rate. As a result, very small particles that are attached to the surface by van der Waals or electrostatic forces can be removed. Such particles are almost impossible to remove by rinsing or drying, but etching the underlying surface is an effective way to remove these particles. The second step in the RCA clean, termed SC-2, is a mixture of hydrochloric acid, hydrogen peroxide, and water in a volumetric ratio of 1:1:5. It is also known as the hydrochloric peroxide mixture (HPM). This can remove traces of metals and metal hydroxides, some of which could originate from the SC-1 step. Subsequently, several alternative processes have been developed to replace the original RCA process [8,9], but it still continues to play a dominant role in many laboratories and manufacturing facilities.

6.1.4 Buffered Oxide Etch (BOE) Clean

Buffered oxide etch (BOE) clean is a very popular method used for removing SiO_2 from silicon surfaces, including native oxides and grown oxide films. It is a mixture of hydrofluoric acid and ammonium fluoride [10]. This is a self-terminating reaction, that is, the reaction comes to a stop when all of the oxides are consumed, leaving the bare unoxidized silicon surface. The termination point is identified by the characteristic change in the surface property from hydrophilic to hydrophobic. Even with a very thin oxide layer, the silicon surface will be hydrophilic. It will turn to hydrophobic when all of the oxides have been etched away, a condition easily identified by the liquid simply rolling off the substrate without wetting it. Buffering is the process of mixing an acid (HF, in this case) with its conjugate base (NH_4F) so that it maintains a constant pH value throughout its use. The buffered HF provides a stable etch rate and a longer shelf life compared to HF alone. When used in conjunction with photoresists, BOE will not penetrate the photoresist through microscopic pinholes and cracks as much as HF. Nevertheless, BOE and HF should be used with extreme caution due to their health hazard. HF is a very small molecule, so it is able to easily penetrate skin and tissue and cause internal organ damage.

6.1.5 Plasma Cleaning

Gas plasmas can be used to remove surface contaminations from a substrate. The most common gas used for cleaning organic contaminants is oxygen [11]. The oxygen molecules in the plasma are converted into a variety of high-energy metastable species, including ions and neutral radicals such as O^{2-}, O^{2+}, O_3, O^-, and O^+. These metastable species readily combine with organic species to produce CO_2 and H_2O, as well as other organic species with a lower molecular weight. As long as the reaction products are volatile, they will be pumped out of the vacuum chamber. In addition, ultraviolet emission from the plasma discharge also helps to break down the organic species. One drawback of oxygen plasma cleaning is the oxidation of the underlying substrate. In the case of bare silicon wafers, this is not usually a problem because the oxidation is a self-limiting reaction that only forms a very thin oxide layer. For other materials, if oxidation is a concern, an inert gas plasma such as argon can be used. In this case, cleaning takes place by physical sputtering rather than chemical action. This will remove organic and inorganic contaminations, but sputter damage and surface roughening could become a concern. In general, it is useful to run a combination of the two gases, so both chemical and physical cleaning take place simultaneously.

6.1.6 Megasonic Cleaning

Sonic cleaning is a technique used for dislodging small particles adhered to surfaces. When these particles are a few microns in diameter or smaller, the van der Walls forces can be so significant that they cannot be removed easily, either by rinsing or by blowing it with compressed nitrogen.

Megasonic cleaning is a variant of the more common ultrasonic cleaning. The latter is widely used for cleaning industrial parts, jewelry and tools. The item being cleaned is immersed in a fluid (water or another appropriate solvent) and a high frequency acoustic wave is generated in the fluid around the item being cleaned. The rapidly oscillating pressure waves create microscopic areas of compression and partial vacuum bubbles (cavitations). The collapse and appearance of the liquid/gas interface of the cavitations can exert tremendous forces on small particles, and release them from surfaces. Ultrasonic cleaning operates at around 200 kHz. At this frequency, the cavitations can be large, so it is not very effective in removing particles smaller than 1 μm in size. Furthermore, the longer duration between cycles at 200 kHz can cause significant energy to be build up in the cavitations that can adversely affect the surfaces. For example, soft semiconductors such as InSb or GaAs will simply fracture and break apart in an ultrasonic cleaner. Megasonic cleaning utilizes the same principles as ultrasonic cleaning, except the acoustic frequency is much higher, on the order of 1–2 MHz. At this frequency, very small particles on the order or 100 nm can be dislodged, and the energy build up is also much smaller, so it can be used with soft semiconductors and fragile items. Megasonic cleaning is used in many semiconductor processes, including wafer cleaning and photomask cleaning [12].

6.1.7 Evaluation of Surface Quality

The water droplet contact angle measurement is a simple and straightforward technique to examine the presence of even one monolayer of contamination. Electrical or optical measurements will not usually detect the presence of monolayer contaminations without more elaborate techniques. A small volume of water is dispensed on the substrate surface and the shape of the droplet is captured with a horizontal microscope camera. The angle of the droplet will be characteristic of the interface where all three phases—the substrate, liquid and atmosphere—come together [13]. Assuming the properties of the latter two quantities are known, the contact angle (measured through the liquid) can be used to evaluate the nature of the substrate. Many plastic surfaces such as polyethylene and PTFE have high contact angles. Metals and glass have smaller contact angles. Plasma cleaning of inorganic surfaces will dramatically increase the surface energy, which will make the water contact angle nearly zero. Figure 6.1 shows an example of two droplet images on a silicon substrate. The left figure is typical when some hydrocarbon film is present on the surface, and the right figure corresponds to a more hydrophilic case, such as with a native oxide on silicon.

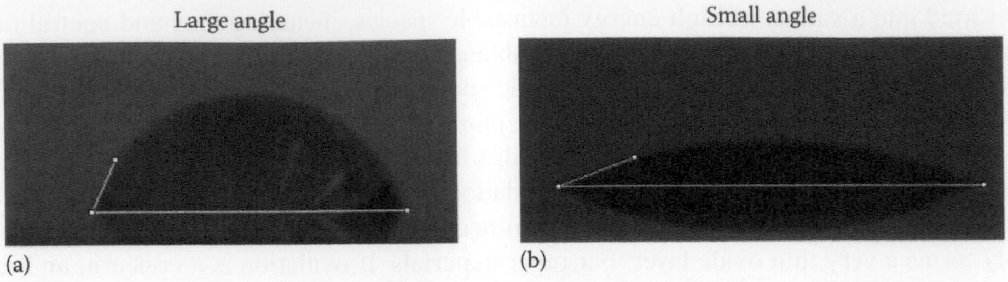

FIGURE 6.1 Water contact angle measurement with a horizontal microscope. (a) A fairly large angle, when some hydrophobic layer is present on the silicon surface, and (b) a smaller contact angle, such as when there is a native SiO_2 layer on silicon.

6.2 SPIN COATING

Photoresists are ultraviolet (UV)-sensitive organic polymers. Unlike inorganic materials used in sputtering and evaporation, polymers are generally incompatible with low pressures, high temperatures, and radiation, which make them unsuited for vacuum deposition methods. Instead, the polymers are typically dissolved in a liquid solvent and applied by spin coating or spray coating. Spin coating is by far the most common method. Virtually, any flat substrate can be coated by spinning, and the thickness can be as uniform as ±2% over large areas. Spray coating is used when the substrates are not flat or if there are large topographical features, but the uniformity will be worse than with spin coating [14,15].

In spin coating, the substrate is held by a vacuum chuck on a motorized rotating stage, and the solvent-laden photoresist is dispensed on the substrate and spun at high speeds to create a thin uniform film. This is a very simple and straightforward process, but it is still widely used in industry for producing photoresist films. There are four stages in spin coating: (1) dispense, (2) spread, (3) thin-out, and (4) evaporation.

6.2.1 Stage 1: Dispense Stage

In the first stage, a sufficient quantity of photoresist is dispensed in a puddle with a pipette or syringe, as illustrated in Figure 6.2. If too little photoresist is dispensed, it can result in an incomplete coverage; too much will result in the excess photoresist being thrown off during the spin, resulting in waste. However, excess photoresist will not change the film thickness—film thickness is determined only by the final spin speed and the viscosity. Normally, the photoresist is dispensed while the substrate is stationary, but it is also possible to slowly rotate the substrate during dispensing to reduce the time delay between the dispense stage and the spread stage. Too much delay can cause the dispensed photoresist to start evaporating, and lead to nonuniformities due to changes in the viscosity.

6.2.2 Stage 2: Spread Stage

During the spread stage, the substrate is slowly spun to allow the fluid front to advance toward the edges of the substrate and wet the entire surface. The angular acceleration will produce a tangential force in the fluid, and the angular velocity will produce a centrifugal force in the fluid along the radial direction. The combination of the two will cause the fluid to flow in a spiraling path toward the edge, against the direction of rotation, as shown in Figure 6.3. The spiraling flow is beneficial to compensate for inconsistencies in the dispense volume and location. If the angular acceleration is too slow, the tangential force will be too small, and the fluid will move more radially. This could result in parts of the substrate not getting covered, especially if the dispensed volume was not perfectly symmetrical about the center of rotation. At the other extreme, if the angular acceleration is

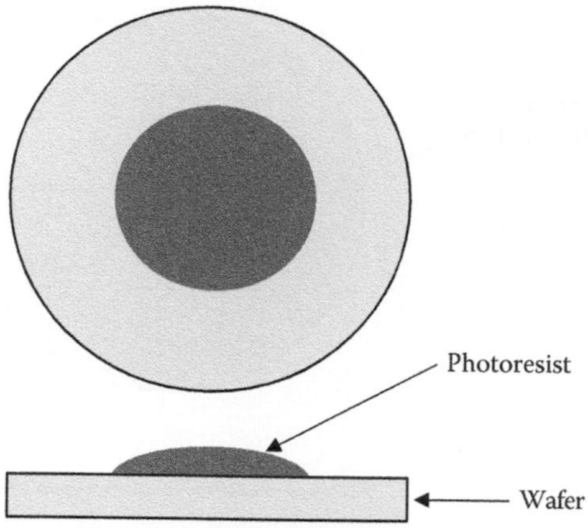

FIGURE 6.2 Stage 1: photoresist dispensed on a substrate.

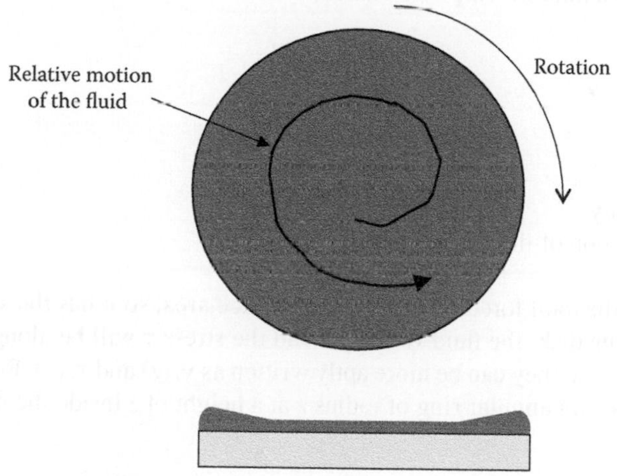

FIGURE 6.3 Stage 2: spreading stage.

too rapid, the tangential force will be too high, and the fluid may separate from the substrate and be thrown off. Some experimentation is necessary to minimize the dispense volume while achieving full coverage. In R&D laboratories, it is common practice to use more than the minimum required dispense amount to ensure full coverage and accept the excess waste.

6.2.3 Stage 3: Thin-Out Stage

The thin-out stage starts when the rotation has reached full speed and the angular acceleration has fallen to zero. In this stage, there are no tangential forces, and the fluid flows only radially outward, as shown in Figure 6.4. The centrifugal force will be exactly balanced by the shear force in the boundary layer of the fluid. As the flow continues, the film will thin out and asymptotically reach a steady-state thickness.

Since spin coating is so widely used in photolithography, it is useful to examine the equations that govern the film thickness. The following analysis is based on the theory originally presented in [16–19].

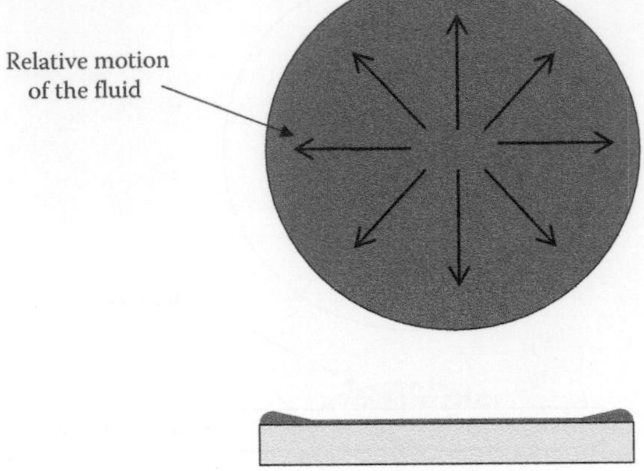

FIGURE 6.4 Stage 3: thin-out stage.

The shear stress in a fluid flowing on a stationary surface is

$$\tau = \eta \frac{\partial \nu}{\partial z} \tag{6.1}$$

where

η is the viscosity
ν is the fluid velocity
z is along the direction of fluid thickness

The shear stress τ is the total force divided by the surface area, so it has the same unit as pressure. In the case of a rotating disk, the fluid velocity ν and the stress τ will be along the radial direction, and also a function of z, so they can be more aptly written as $\nu_r(z)$ and $\tau_r(z)$. Referring to Figure 6.5, the centrifugal force on an annular ring of radius r at a height of z inside the fluid will be

$$F_c = mr\omega^2 \tag{6.2}$$

where

m is the mass of the fluid in the annular ring
ω is the angular frequency of rotation

The mass of the fluid in the annular ring can be written as

$$m = 2\pi r \Delta r z \rho \tag{6.3}$$

where ρ is the fluid density.

Equating the centrifugal force F_c from Equation 6.2 to the shearing force τA, where the surface area of the annular ring is $A = 2\pi r \Delta r$, we can obtain the following equation:

$$2\pi r \Delta r z \rho r \omega^2 = 2\pi r \Delta r \tau, \tag{6.4}$$

which can then be simplified to

$$\tau_r(z) = \rho z r \omega^2. \tag{6.5}$$

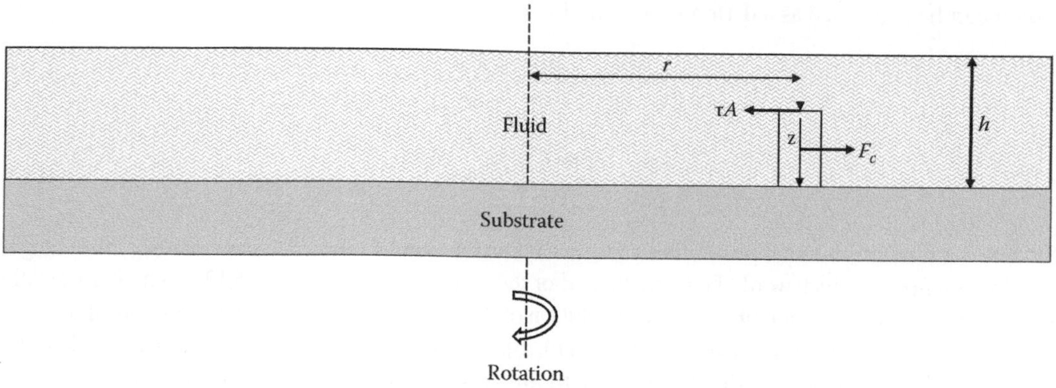

FIGURE 6.5 Shearing force and centrifugal force in a fluid on a rotating substrate.

Furthermore, since the shearing force disappears to zero at the top of the fluid surface, we can apply this boundary condition and arrive at

$$\tau_r(z) = \rho r\omega^2(h - z) \tag{6.6}$$

where h is the film thickness at the radial position r. Finally, substituting Equation 6.1 into 6.6, and taking a derivative, we can get

$$-\eta\frac{\partial^2 v_r(z)}{\partial z^2} = \rho r\omega^2. \tag{6.7}$$

Equation 6.7 can be solved by assuming the boundary condition that $v_r(0) = 0$ (fluid velocity at the substrate/fluid interface is zero) and $\frac{\partial v_r(h)}{\partial z} = 0$ (shearing stress at the air/fluid interface is zero). This results in the following solution:

$$v_r(z) = \frac{1}{\eta}\rho r\omega^2\left(hz - \frac{1}{2}Z^2\right). \tag{6.8}$$

The net radial flow can be found by integrating the radial velocity over the film height:

$$Q = \int_0^h v_r(z)\,dz = \frac{\rho r\omega^2 h^3}{3\eta}. \tag{6.9}$$

The net loss (or gain) of the fluid flow in a certain volume will result in a decline (or increase) in the film thickness at that radial position, that is,

$$\frac{\partial h}{\partial t} = -\nabla \cdot Q. \tag{6.10}$$

Since the flow is limited to radial flow only, the divergence will only contain the radial derivative term as follows:

$$\frac{\partial h}{\partial t} = -\frac{1}{r}\frac{\partial (rQ)}{\partial r} \tag{6.11}$$

which can be expressed as a differential equation of the form

$$\frac{\partial h}{\partial t} = -\frac{\rho\omega^2}{3\eta r}\frac{\partial(r^2h^3)}{\partial r} \tag{6.12}$$

$$\frac{\partial h}{\partial t} = -\frac{\rho\omega^2 h^2}{\eta}\left[\frac{2h}{3} + r\frac{\partial h}{\partial r}\right]. \tag{6.13}$$

Since h is a function of radial position r and time t, it can be written as $h(r, t)$. Given an initial film height distribution $h(r, 0)$ (which would be from the end of the spread stage), Equation 6.13 can be numerically solved for $h(r, t)$ as a function of t and r. The solution of this equation for different starting profiles $h(r, 0)$ indicates that $\partial h / \partial r$ quickly converges to zero [14,16]. In other words, the film flattens out quickly and becomes independent of the radial position. With this assumption, we can write Equation 6.13 as

$$\frac{dh}{dt} = -\frac{2\rho\omega^2 h^3}{3\eta}. \tag{6.14}$$

If the initial film thickness $h(r,0) = h_0$ (i.e., assuming the film thickness is uniform after the spread cycle), the thickness as a function of time can be obtained by solving Equation 6.14 as follows:

$$\int_{h_0}^{h}\frac{dh}{h^3} = -\frac{2}{3}\int_0^t \frac{\rho\omega^2}{\eta}dt \tag{6.15}$$

which results in an analytical solution

$$h = \frac{h_0}{\left(1 + \frac{4\rho\omega^2 h_0^2 t}{3\eta}\right)^{\frac{1}{2}}}. \tag{6.16}$$

Figure 6.6 shows Equation 6.16 plotted for different spin speeds and different viscosity values representative of typical photoresists, with a starting uniform film thickness of $h_0 = 100\ \mu m$. This initial

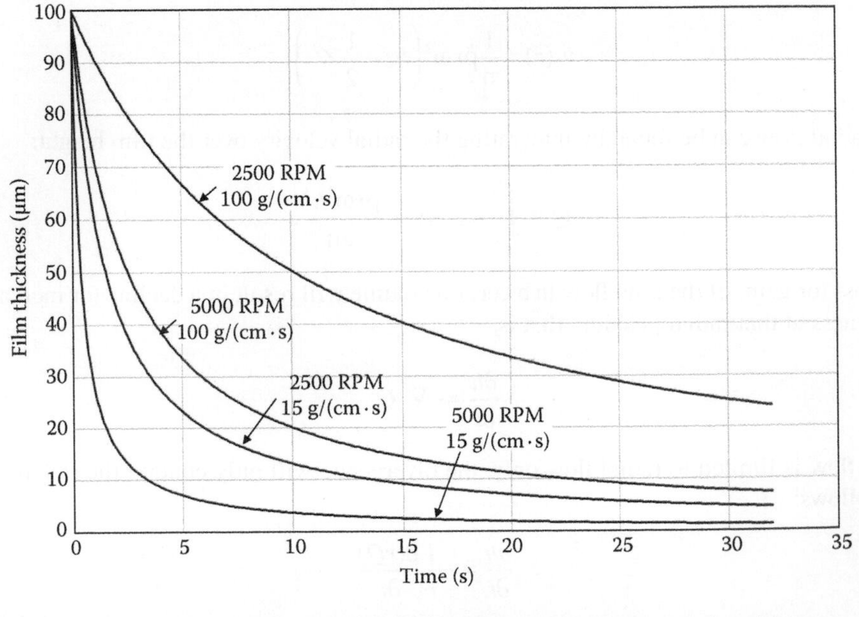

FIGURE 6.6 Temporal evolution of film thickness for different spin speeds and viscosities.

film thickness will be determined by the outcome of the spread stage. As expected, we can verify that the higher spin speeds and lower viscosities result in thinner films. Furthermore, we can notice that the square-root time dependence results in the film thicknesses approximately leveling off as time increases. The final film thicknesses for this example are 1.2 μm and 4.5 μm for the viscosity value of 15 g/(cm·s) and 7.3 μm and 24 μm for the viscosity value of 100 g/(cm·s). In addition, the initial film thickness h_0 also plays a significant role in the decay rate of h. Films with larger thicknesses decay faster and reach a stable thickness a lot quicker.

Figure 6.7 shows the film thickness after a fixed 30 s spin for two viscosity values. As mentioned earlier, the starting thickness was assumed to be $h_0 = 100\,\mu m$. This is the most commonly shown curve in many photoresist data sheets. It shows a decline in film thickness with increasing spin speed and an even greater change in film thickness with viscosity. In fact, most photoresist manufacturers will supply the same photoresist in multiple viscosity values, simply by changing the solvent content, and a family of spin speed curves will be shown in the data sheet corresponding to their viscosities. An example of the spin speed curves for a commercially available photoresist is shown in Figure 6.8.

6.2.4 Stage 4: Evaporation Stage

Although Equation 6.16 indicates a square-root time dependence, in practice, the film thickness levels off with time even quicker than suggested by the equation. This arises due to the evaporation of the solvents from the film as illustrated in Figure 6.9. As solvents evaporate, the viscosity of the film will increase, which will reduce the rate at which the film thins out. Evaporation occurs throughout all stages of the spin process but becomes dominant when the surface-to-volume ratio becomes large (i.e., after the film has become sufficiently thin). During this stage, the evaporation rate of the solvents will become a much larger component than the radial flow. As a result, the film will continue to become thinner not only due to the radial flow considered in Equation 6.16 but also due to simple evaporation of the solvents.

As mentioned earlier, the general solution of Equation 6.13 predicts that the film thickness will become uniform after the first few seconds of a spin process. However, in the presence of significant

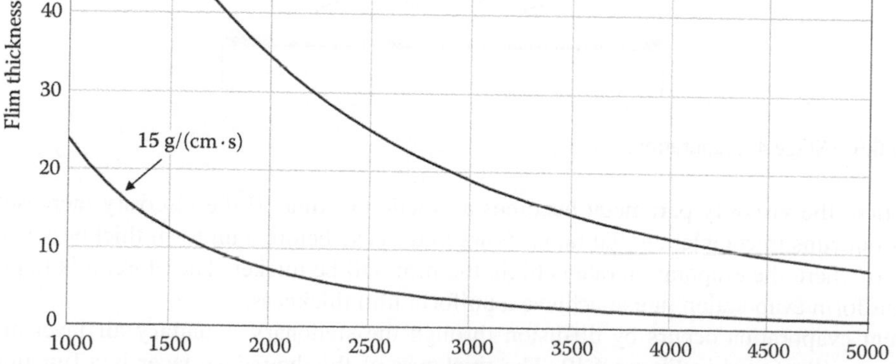

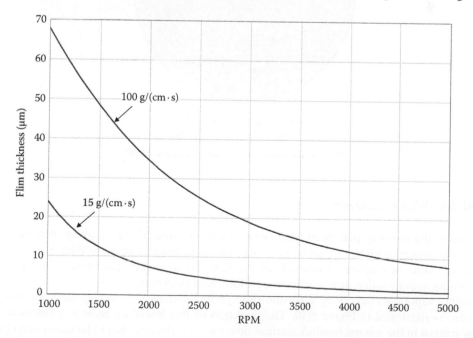

FIGURE 6.7 Film thickness as a function of spin speed for two viscosity values.

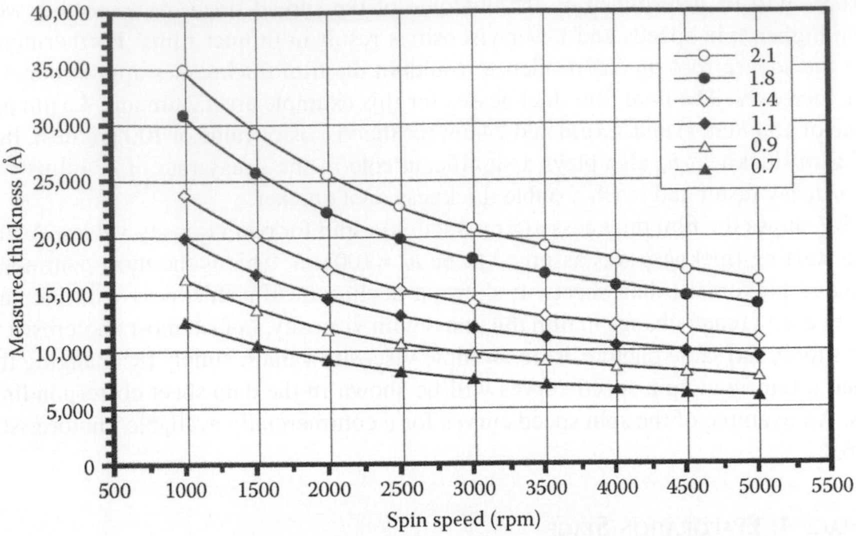

FIGURE 6.8 Film thickness as a function of spin speed for the Rohm and Haas SPR-955 family of photoresists. Each curve corresponds to a different dilution of the same photoresist. (Reproduced with permission from Dow Electronic Materials.)

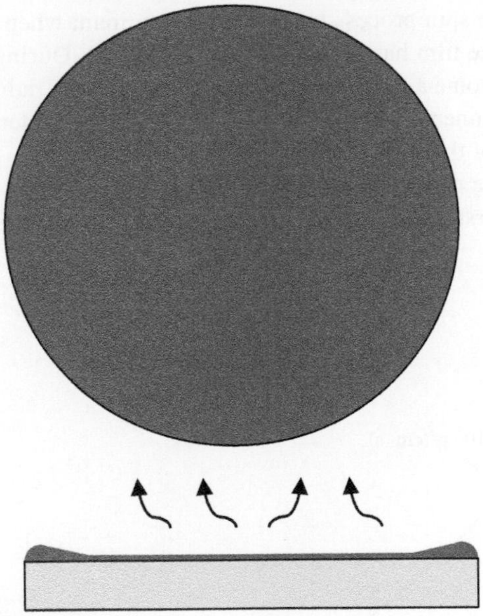

FIGURE 6.9 Stage 4: evaporation.

evaporation, the viscosity parameter becomes a function of time. If the viscosity increases before the thin-out runs to completion, the radial flows may cease before a uniform thickness is reached. In regions where the evaporation rate is high, the film will be thicker. Therefore, it is important to have a uniform evaporation rate to achieve a uniform film thickness.

Solvent evaporation occurs by diffusion through the stationary boundary air layer above the substrate, as illustrated in Figure 6.10. The thickness of this boundary layer is a function of the airflow pattern in the spinner bowl. A laminar flow free of turbulence has to be maintained to have a predictable boundary layer thickness and evaporation rate. If not, the evaporation rate and the

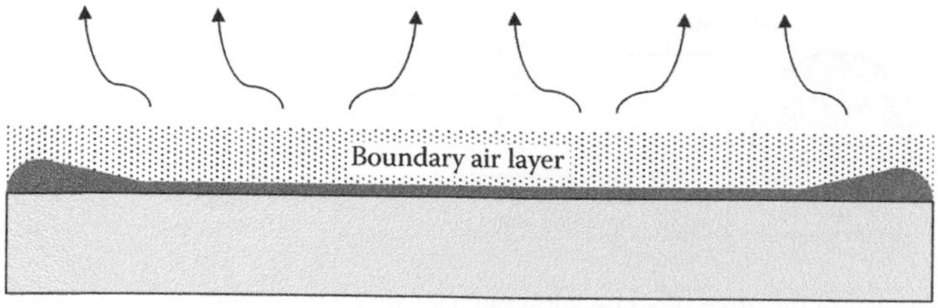

FIGURE 6.10 Evaporation of solvents through a boundary layer.

film thicknesses could vary across the substrate surface. Therefore, spin coating is not just about the spinning substrate but also about controlling the airflow around the substrate in the spin bowl. Careful attention to all of these parameters is important to achieve a uniform film thickness.

After significant evaporation of the solvents, the viscosity of the film will have increased to a high enough value to make it impossible to make further changes to the film thickness. This is the reason for why spinning the wafer again at a higher spin speed will not reduce the film thickness that was achieved after the initial spin.

Mathematical modeling of spin coating that includes the radial flow and the evaporation rate can be done, but rarely necessary [20]. Viscosity is not always precisely known, and each spin coater could have variations in airflow pattern that make each laboratory installation unique. In most cases, one starts off with the spin speed curves provided by the photoresist data sheets and refine these through empirical data from process trials.

6.2.5 Edge Bead

Another important characteristic of spin coating is the appearance of the edge bead. This is a narrow region near the edge of the substrate where the film will be thicker. As the fluid approaches the edge of the substrate, the surface tension from the outer edge will produce an additional force against the fluid flow. This will add to the shear stress in the film and will appear as if the fluid has a greater viscosity. As a result, the film will be slightly thicker along the edges than the remainder of the substrate. The edge bead areas will generally be unusable for making devices, because they can compromise the contact between the photoresist film and the photomask. In such cases, the edge bead can be removed with a solvent. This can be done by injecting the solvent via a nozzle directed at the edge bead while spinning the substrate, as illustrated in Figure 6.11. However, this does not entirely eliminate the edge bead. The overspray from the solvent will soften parts of the film immediately interior to the original edge bead and will cause it to flow radially and accumulate, creating a new smaller edge bead. An alternative technique is to mask the substrate and expose the edge bead with UV light and then develop and dissolve away the edge bead. One drawback of this method is that the first develop step will slightly alter the chemical composition of the photoresist film for the subsequent exposures.

6.2.6 Common Problems Encountered in Spin Coating

If large areas of the substrate near the edges remain uncoated after spin coating, as shown in Figure 6.12, it is most likely due to an insufficient dispense quantity. The bare spots typically begin at a defect site or particle on the substrate that deflects the flow of fluid. If sufficient fluid remains elsewhere, the deflected areas will get covered due to the tangential flow during the spread cycle. In the absence of excess fluid, these areas will remain dry. Larger dispense volumes and an adequate spread stage are necessary to prevent such bare spots from appearing on the substrate.

Particles or defects similar in size to the film thickness will deflect the fluid flow in the radial direction during the thin-out stage. As the fluid flows around this obstacle, it will create a downstream

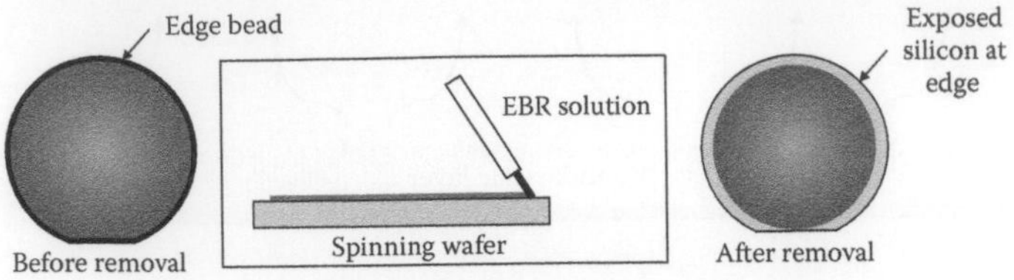

FIGURE 6.11 Removal of the edge bead by spraying a solvent while spinning.

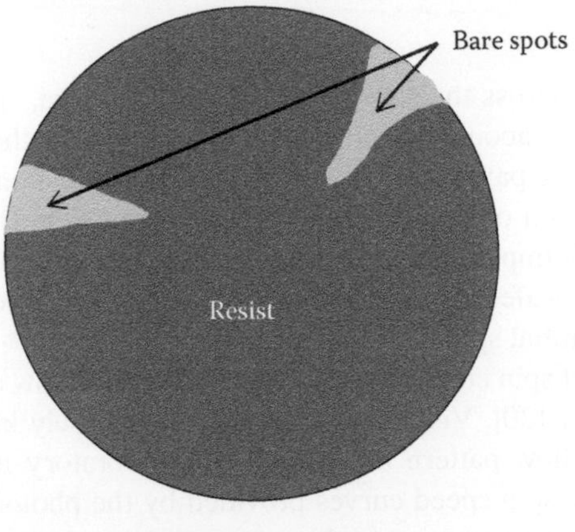

FIGURE 6.12 Common problem: bare spots on the substrate.

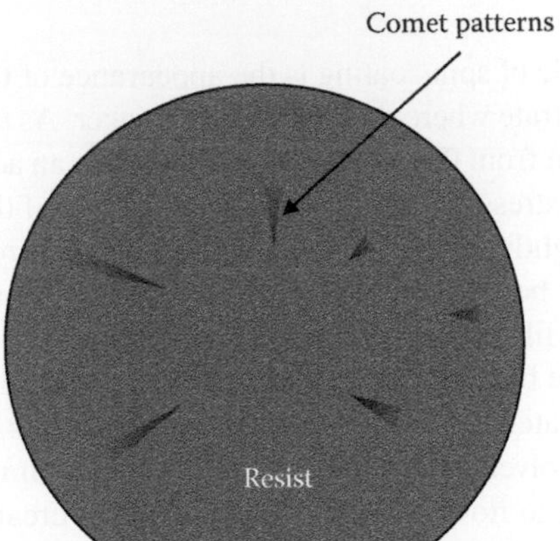

FIGURE 6.13 Common problem: particles on the substrate or in the photoresist.

disturbance (similar to the wake of a ship) and the particles will become permanently frozen after the solvents evaporate. This produces characteristic comet-like patterns all pointed toward the rotational center of the substrate, as shown in Figure 6.13. The comets are usually visible to the eye, but the particles that produced the disturbances may be too small to see. This is one reason why spin coating has to be performed in a cleanroom—even particles that are too small to be visible will produce an effect much larger than their physical size. Similar effects are also seen from substrates that contain etched features. If the wafer contains very deep topographical features, spin coating may even become unsuitable. Spray coating is more effective in such cases, although the uniformity will be worse [14].

If the spread stage is terminated too early, the initial thickness profile h_0 may not have reached a uniform profile; and, if the evaporation rate is also high, the radial flow may stop prematurely before the film planarizes into a uniform thickness profile. This will result in a thickness variation along the radial direction and will appear as slowly varying concentric rings, as shown in Figure 6.14. These are fluid wave fronts that became frozen in time due to the premature evaporation of the solvents. This can be avoided by increasing the spread stage and reducing the evaporation rate by controlling the airflow pattern.

On circular substrates, the width of the edge bead will be constant everywhere along the edge. With other shapes, the merging of two edges near the corners can cause the edge bead to become more pronounced and cover a much wider area. With rectangular or square substrates, a large portion of the substrate can be rendered unusable due to the edge beads, as shown in Figure 6.15. This can become an important consideration when working with very small substrate sizes.

6.2.7 Solvent Bake (Soft Bake)

The main purpose of the solvents in a photoresist is to allow liquid dispensing and spin coating. The solvent content also controls the viscosity, which has a dramatic effect on the final film thickness. After spin coating has been completed, the solvents should be removed as much as possible to prevent it from adversely affecting the photoresist performance: Excessive solvents can make the photoresist stick to the photomask, which can contaminate the photomask and also deform the exposed patterns; it can produce higher solubility of the unexposed areas (also known as dark erosion); and

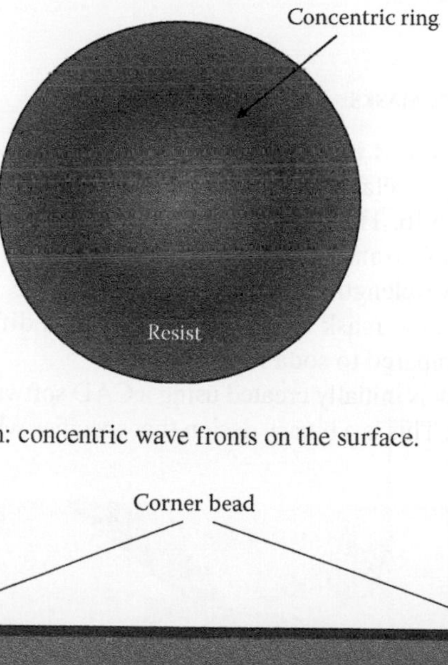

FIGURE 6.14 Common problem: concentric wave fronts on the surface.

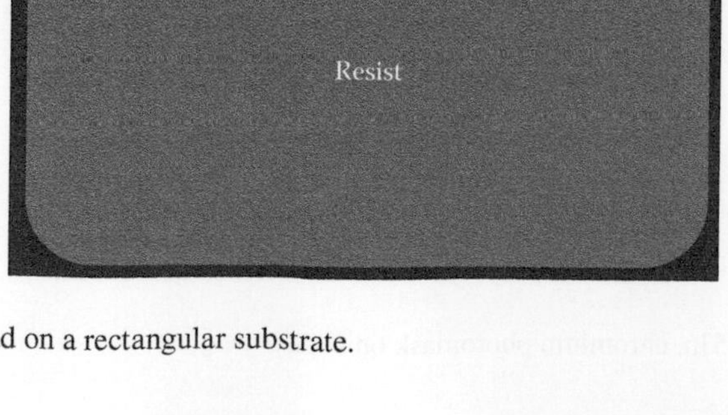

FIGURE 6.15 Edge bead on a rectangular substrate.

it can increase the diffusion rate of the photoacid that is generated during exposure, resulting in the enlargement of the exposed areas. It is, therefore, recommended to remove the remaining solvents by a process step known as solvent bake or soft bake. It involves simply raising the temperature of the substrate to allow the solvents to evaporate. Typically, the substrate is heated to about 100°C on a hot plate. Assuming a thermally conductive substrate like silicon, the bottom of the film will be close to the hot plate temperature, but the top of the film will be at a lower temperature as determined by the ambient air conditions, convective flow patterns, and exhaust flows. The diffusion coefficient of the solvents through the film is a function of the solvent content of the film. The temperature gradient will produce a gradient in solvent vapor pressure, which in turn will produce a gradient in the diffusion coefficient. The overall effect of this gradient is an acceleration of the evaporation rate. The required bake time to reach a certain minimum solvent concentration will be a function of the film thickness, but it is typically about 1–3 min.

Oven baking, instead of hot plate, is a more manufacturable process because it allows a large number of substrates to be placed on a rack and baked simultaneously. In addition to increasing throughput, it also makes it possible to have identical bake conditions and bake times to achieve run-to-run process consistency. However, the temperature gradient across the photoresist film in an oven will be nearly zero because the substrate temperature will be the same as the ambient temperature of the air above the photoresist film. The lack of a thermal gradient makes the evaporation rate much slower compared to a hot plate bake. As a result, bake times have to be significantly increased. Bake times are typically about 10 times longer in an oven compared to a hot plate.

6.3 PHOTOMASKS

6.3.1 Laser-Written Photomasks

In photolithography, a photomask is imaged onto the photoresist and then exposed with UV light. The photomask contains opaque metal traces (made of chromium or iron oxide) on a UV-transparent substrate, as shown in Figure 6.16. The substrate is typically soda lime glass, which is an inexpensive glass but it is still reasonably transparent at the commonly used UV wavelength of $\lambda = 365\,nm$. But it is not transparent for wavelengths below 300 nm. Therefore, when deep-UV (DUV) wavelengths are used for exposure, the mask has to be made from a different substrate, such as fused silica, at an increased cost compared to soda lime glass.

The design of the photomask is initially created using a CAD software similar to any printed circuit board (PCB) design procedure. This is a binary design that specifies which areas should be opaque and

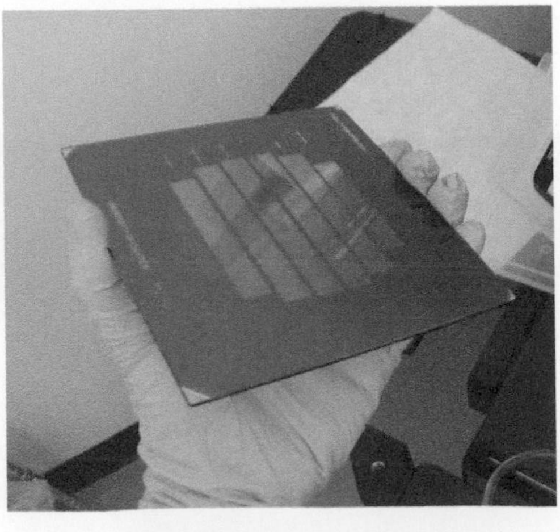

FIGURE 6.16 A 5 in. × 5 in. chromium photomask on soda lime glass.

which areas should be transparent. The metal, such as chromium, is first deposited on the mask substrate using evaporation or sputtering. A suitable photoresist is then spin coated on top of the metal layer. The design file from the CAD program is then used to drive a laser writing exposure system. This is a UV laser beam focused to a small spot, with a motorized *XY* stage to translate the substrate in the lateral plane. The stage moves the substrate line by line, and the beam is turned on and off by an acousto-optic modulator to expose the photoresist according to the design file [21]. In some systems, a separate acousto-optic deflector is also used to position the beam more accurately in addition to the mechanical translators. The photoresist is then developed to remove the exposed areas. The metal is then etched in a chemical bath. The unexposed areas of the photoresist will act as a chemical barrier to prevent etching under those areas. Then, the photoresist is stripped off to reveal the desired metal pattern on the glass substrate. Finally, a protective scratch-resistant coating is applied on top of the chromium layer.

The smallest feature that can be printed with a laser writer depends on the spot size, which in turn depends on the wavelength and numerical aperture (NA). In most laser writers, a HeCd laser at 442 nm or 325 nm wavelength is used as the illumination source. Compared to an incoherent mercury vapor lamp, a coherent laser source allows the beam to be focused to a much smaller spot. However, the definition of a "feature" is not the same as the spot size. A design feature such as a line or a square has to be multiple spots wide to achieve a reasonably straight edge and sharp corners. The straightness of an edge also depends on the scan direction of the writing laser beam. The smoothest lines are achieved when the lines are parallel to the writing direction, as illustrated in Figure 6.17. The minimum printable feature size is known as the critical dimension (CD) of the system. A small spot size will improve the CD, but it will increase the time it takes to write the entire design. Therefore, the cost of a photomask is directly related to the CD. A larger CD (of the order of 3 μm or more) can be written with a larger spot size by defocusing the beam to accelerate the writing process. These masks are typically inexpensive. As of this writing, a 5 in. × 5 in. photomask with features smaller than 3 μm could be obtained from commercial sources for less than $500. A CD smaller than 1 μm will require more writing time and greater placement tolerance, so the cost will be greater.

Chromium has excellent adhesion to glass and can be easily patterned with high fidelity. It is also very opaque across the UV and visible spectrum. An alternative to chromium is iron oxide. This is opaque to UV but it is partially transparent to visible light. The partial transparency is beneficial when aligning the photomask to the substrate. It allows one to see the substrate features that would otherwise be hidden behind the masked areas. However, iron oxide is slightly more difficult to process than chromium; therefore, the CD is slightly worse than with chromium.

Instead of fabricating a photomask and then using that photomask to replicate the pattern on the final substrates, the laser writer can also be used to directly expose the desired pattern on the final photoresist film. This process circumvents the need to make a photomask and is known as mask-less lithography, or direct-write lithography [22]. Direct-write lithography is an acceptable process when only a few substrates have to be patterned. Obviously, this requires one to have a mask writing tool in their laboratory.

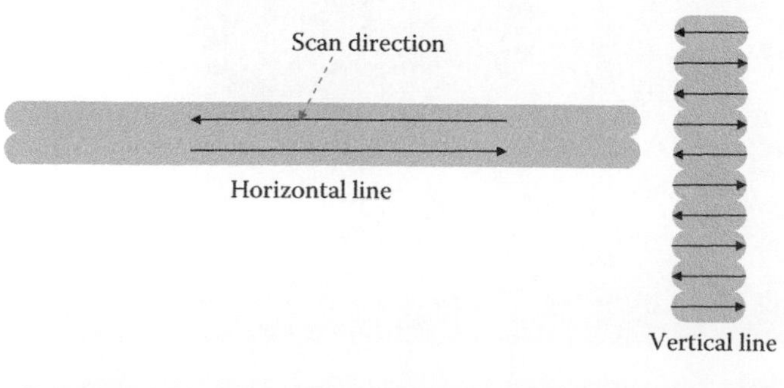

FIGURE 6.17 The effect of spot size and scan direction on horizontal and vertical lines.

6.3.2 Film Photomasks

Film photomasks are widely used in the production of printed circuit boards (PCB). The mask in this case is created on a flexible film coated with a monochromatic photographic emulsion, which is exposed with a scanning laser beam and then developed. The writing tools are very similar in operation to a traditional laser printer, and are known as photoplotters. Compared to chrome-on-glass photomasks, film photomasks have a lower contrast between the clear and opaque areas, and the CD is also significantly worse. The resolution is typically specified in dots-per-inch (dpi), with some of the highest being 50,000 dpi. The CDs are generally larger than 10 μm.

Although film photomasks are exclusively the domain of PCBs, they can be a very inexpensive alternative to chrome-on-glass masks for some limited device applications. For the same size, a film mask can be less than a tenth of the cost of a glass mask. If the device geometries are larger than 10 μm, and the line edge roughness is not a critical requirement, film masks can be used with reasonable results. Film masks are most suitable for preliminary design explorations or academic class demonstrations where performance and optimization are not the main focus. Film masks can be plotted on large sheets and then cut down to the required size. They are then mounted on glass mask blanks (the same substrates used for making chrome-on-glass masks) using a small quantity of isopropyl alcohol (IPA) as the temporary adhesive, as shown in Figure 6.18. This allows the film to be removed and repositioned easily. For geometries in the range of several tens of micrometers, this results in a reasonably good pattern definition at a small fraction of the cost of a conventional chrome-on-glass mask. Figure 6.19 shows an example of a photodetector structure fabricated using film masks, where it can be seen that the lines and structures are reasonably well defined.

6.3.3 Electron Beam–Written Photomasks

As mentioned earlier, the spot size of a UV laser writer depends on the wavelength and the NA of the system. 500 nm is generally the smallest CD that can be written with reasonable accuracy using a HeCd writer at 325 nm. For features smaller than 500 nm, electron beam writing will be required. High-energy electron beams can be focused down to a much smaller spot, of the order of 1–10 nm.

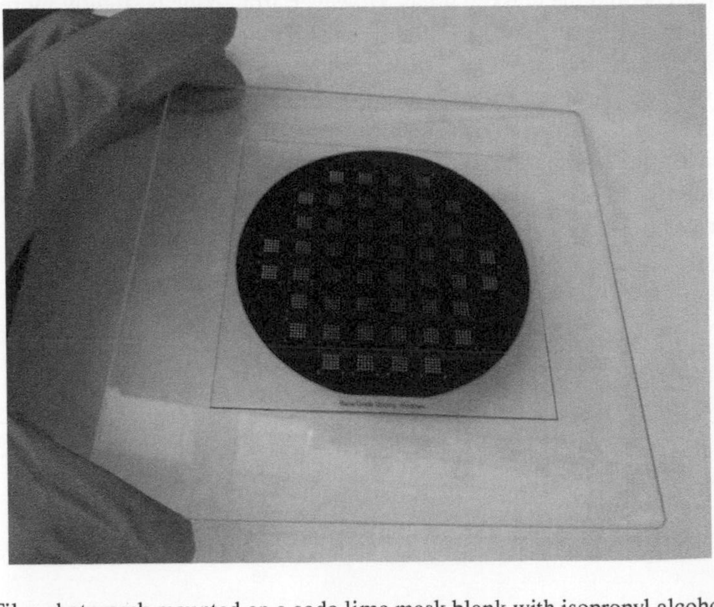

FIGURE 6.18 Film photomask mounted on a soda lime mask blank with isopropyl alcohol.

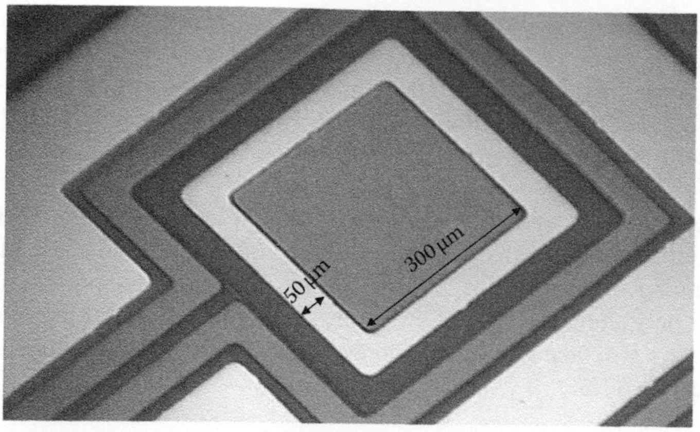

FIGURE 6.19 Example of a completed photodetector device using photoplotted film masks. For dimensional reference, the central square is 300 µm × 300 µm.

Certain photoresists are sensitive to electron beam exposure and produce a similar solubility change as with UV exposure. The e-beam writer is conceptually very similar to a laser writer, except the UV beam is replaced with an electron beam and everything takes place inside a high vacuum chamber. The write times are also significantly longer due to the small spot size. Therefore, the system is far more complex and more expensive than a laser writer.

The focused spot size of an electron beam is determined by de Broglie's wavelength of electrons, which is inversely related to the momentum of the electron as

$$\lambda = \frac{h}{\sqrt{2mE}} \tag{6.17}$$

where
E is the energy of the electron
λ is its wavelength

At 100 keV, the wavelength of the electron is only 4 pm (4×10^{-12} m). However, this does not mean it is possible to write features as small as 4 pm. The focusing system used for electron beams (which are magnetic) is much more primitive compared to optical focusing. While diffraction is the primary limitation in optical projection lithography, chromatic and spherical aberrations play a far greater role in electron beam lithography (EBL). Instead of picometers, practical values of spot sizes are in the range of 1 nm, which is many orders of magnitude greater than the diffraction limit. Once the electron beam arrives at the photoresist, scattering in the photoresist degrades the resolution even further. At low electron beam energies, most of this scattering takes place in the photoresist film. If the energy is high enough for most of the electrons to penetrate past the photoresist layer into the substrate, the scattering effect can be minimized. This is one of the reasons for why most e-beam lithography systems run at voltages around 100 keV [23]. However, if the beam energy is too high, the backscattered electrons can become dominant and produce what is known as proximity effects. The conductivity of the substrate also dictates the maximum energy that can be used. Finally, most of the resist exposure takes place through low-energy secondary electrons that are released from the primary incident electrons. These electrons successively reduce the molecular weight of the resist through a process known as scissioning. These secondary electrons can travel up to 10 nm in random directions due to multiple collisions. Combining all of these factors, minimum achievable features in EBL are of the order of 10 nm, which is a sobering number compared to the electron wavelength of 4 pm [24].

Electron beam writing requires the substrate to be electrically conductive so that the incident electrons can have a discharge path to ground. Insulating or semi-insulating substrates will cause the photoresist to accumulate a negative charge and deflect the incident electrons. This will result in pattern distortion and loss of resolution. For writing photomasks, although the glass substrate is electrically insulating, the presence of the chromium metal layer provides an adequate discharge path.

Due to the long write times and high equipment costs, the cost of e-beam-written photomasks is significantly greater than laser-written photomasks. The write time could take from several hours to several days depending on the spot size, pattern density, mask size, and photoresist sensitivity. The write times are not only a function of the CD but also a function of the density of the written traces. A highly dense pattern will take significantly longer than a sparse pattern. More details of EBL will be discussed in Section 6.13.1.

6.4 UV LIGHT SOURCES

Mercury vapor lamps are the most widely used UV light sources in photolithography. This is due to their low cost, high efficiency, and long life. High-pressure mercury vapor arc lamps have characteristic UV emission lines at a number of different wavelengths from 254 (DUV) to 579 nm (yellow). Of most interest in photolithography are the lines in the UV spectrum. The 436, 405, and 365 nm lines are the most dominant lines in the near-UV range and are referred to as g-line, h-line, and i-line, respectively, as shown in Figure 6.20. Photoresist parameters are often measured and specified at these emission wavelengths, although the photoresist can perform equally well at any intermediate wavelengths. The i-line (365 nm) is one of the most commonly used lines for photolithography because it is the shortest wavelength that still falls within the spectral sensitivity range of most resists and also has a reasonably high emission power. Besides chemical sensitivity, shorter wavelengths also have less diffraction effects and will produce better images with greater spatial resolution.

The DUV spectral range including the 254 nm line falls in a separate category of lithography because it requires different photochemistries. It also requires fused silica optics because ordinary borosilicate or soda lime glass is not transparent at this wavelength. For shorter wavelengths, excimer lasers are used in lithography systems. The most commonly used excimer lasers are the KrF laser (248 nm), ArF laser (193 nm), and the F_2 laser (157 nm).

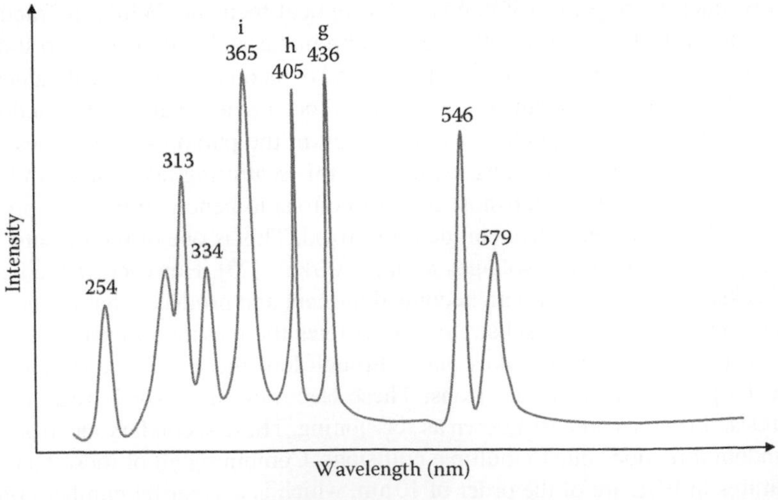

FIGURE 6.20 Emission spectrum of a high-pressure mercury vapor arc lamp.

6.5 CONTACT MASK LITHOGRAPHY

Contact mask lithography is conceptually the most straightforward system. Although it is no longer used in large-scale production environments, it is still the least complex and the least expensive method. Therefore, it is the most widely used lithography system in research laboratories.

In contact lithography, the photomask is placed in direct physical contact with the photoresist film and exposed to UV light. The primary factors that affect the image resolution are the effectiveness of the contact and the thickness of the photoresist film. If a small gap exists between the photomask and the photoresist, the image will become diffracted by the time it reaches the top surface of the photoresist. For this reason, the surface quality of the photoresist plays a critical role in the image resolution. Contamination or topographic features will significantly degrade the resolution. Even with a perfect photoresist film, any warp and bow of the substrate, which are always present to some degree in any substrate, will prevent a uniform contact from being achieved. Contact lithography systems use a variety of contact modes to overcome some of these problems. When the substrate and mask are just brought into contact without further compression, this is known as the soft contact mode (see Figure 6.21). This is ideal for fragile substrates that cannot withstand excessive contact pressures. A hard contact mode utilizes an active compression of the substrate onto the photomask to flatten out any bows or surface contaminations, as shown in Figure 6.22. Some systems also utilize a vacuum contact mode to further increase the pressure between the substrate and photomasks [25,26].

Even with a perfect contact, diffraction cannot be completely eliminated. The thickness of the photoresist film itself will present an optical path for diffraction to take place. The image will enlarge and diffract as it propagates through the photoresist film, reflect off the substrate surface, and expose the photoresist again during the return path. The calculation of the aerial image due to all of these effects is not a straightforward process. A model that includes the electromagnetic fields in two or three dimensions with all of the film interfaces will require significant computational effort using numerical techniques such as the finite-difference time-domain (FDTD) method and rigorous coupled-wave analysis (RCWA) [27–29]. However, we can get an approximate formula to predict the image resolution by

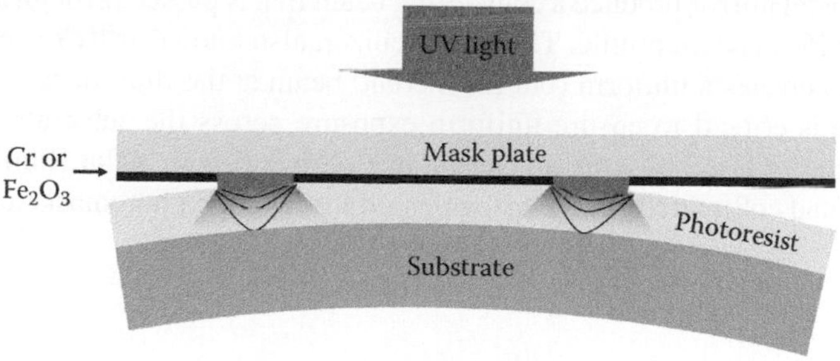

FIGURE 6.21 Soft contact mode.

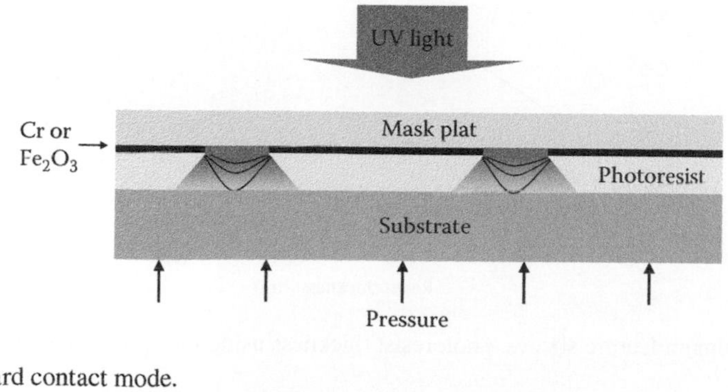

FIGURE 6.22 Hard contact mode.

using Fresnel's integrals for diffraction through a circular aperture. The smallest peak is created at the image plane when the aperture diameter is approximately one Fresnel zone wide. That is,

$$\rho^2 = \lambda z \tag{6.18}$$

where
ρ is the radius of the first Fresnel zone
z is the distance
λ is the wavelength

From this, we can get the diameter of the aperture as

$$D_{min} = 2\sqrt{\lambda z}. \tag{6.19}$$

A plot of D_{min} vs. photoresist thickness z for an illumination wavelength of 365 nm is shown in Figure 6.23. At a wavelength of 365 nm, with a zero air gap and a photoresist thickness of 500 nm, the smallest feature that we can print with reasonable sharpness using contact lithography is about 854 nm. This is the best-case scenario, and any contamination on the substrate or photoresist will introduce additional gaps between the mask and photoresist and will degrade the resolution. A dust particle 1 μm in size will create a 1 μm air gap. Using Equation (6.19), D_{min} will enlarge to 1.48 μm. This is one of the reasons why an ultraclean particle-free environment is critical for performing contact lithography.

It is possible to reduce the photoresist thickness to improve the resolution, but this is not without trade-offs. A photoresist that is too thin may not withstand the subsequent etching or pattern transfer steps. In addition, the relationship between resolution and photoresist thickness is not linear—a 50% reduction in thickness will only produce a 30% improvement in resolution. The same is true with the source wavelength as well. Therefore, 500 nm is often advertised as the resolution limit of contact photolithography using i-line mercury vapor lamps.

Figure 6.24 shows the basic components of a typical contact mask exposure system. The UV lamp and the ellipsoidal mirror produce a converging beam that is passed through an optical homogenizer to create a uniform beam profile. The homogenizer, also known as fly's eye, consists of a 2D array of lenslets that creates a uniform (but incoherent) beam at the illumination plane [25]. A flat illumination profile is critical to ensure uniform exposure across the substrate. Gaussian beams typically encountered in laser systems are unsuitable. A beam with a flat illumination profile is created, expanded, and collimated before being passed through the photomask/substrate assembly.

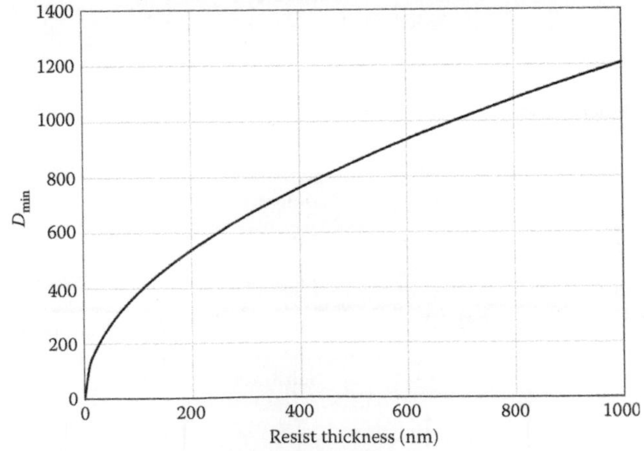

FIGURE 6.23 Minimum feature size vs. photoresist thickness using the approximate Fresnel zone formula from Equation 6.19.

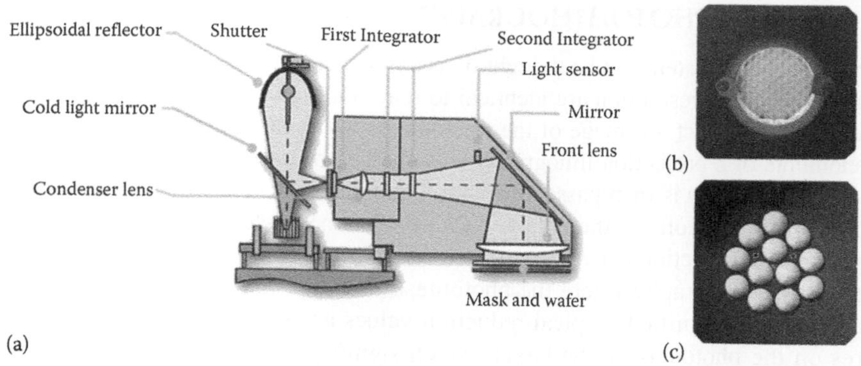

FIGURE 6.24 Components of a contact mask aligner. (a) The schematic of the illumination system, (b) first optical integrator, and (c) second optical integrator with the 2D lenslets. (Reproduced with permission from Suss MicroOptics SA., Switzerland.)

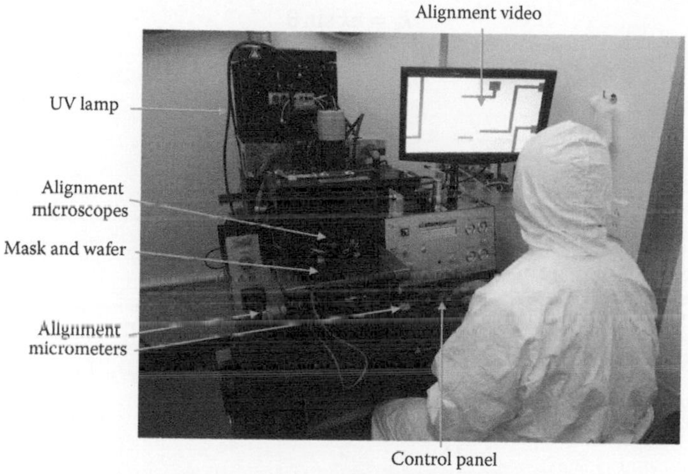

FIGURE 6.25 Contact mask aligner in operation in the author's laboratory.

Alignment of the substrate with the photomask is another critical aspect of photolithography. This is usually accomplished by holding the mask stationary and translating the substrate laterally in the X, Y, and θ directions, while also viewing the prerequisite alignment features on the substrate through the transparent portions of the photomask using a microscope. After satisfactory alignment between the photomask and substrate is obtained, the substrate is compressed against the photomask to perform the exposure. A photograph of a contact mask aligner manufactured by Suss MicroTec is shown in Figure 6.25.

The biggest advantage of contact photolithography is the simplicity of the equipment. The refractory optical components are relatively simple and are primarily used for creating a uniform illumination profile, not for imaging lithographic features. As a result, the majority of research laboratories rely primarily on contact photolithography. Its biggest disadvantage, however, is the requirement for a hard physical contact between the substrate and the photomask. This aspect makes it difficult to scale up for volume production. After each exposure, the photomask has to be removed and cleaned for any residues of photoresist. The wafer warp also makes it difficult to scale up to large wafer sizes. Even with hard contact it becomes increasingly difficult to maintain planarity over large substrate sizes. As a result, the majority of integrated circuit manufacturing has transitioned to noncontact projection lithography systems.

6.6 PROJECTION PHOTOLITHOGRAPHY

Projection lithography system can be thought of as an inverse of the optical microscope. In fact, the governing equations for resolution are identical to that of an optical microscope. It uses refractory optical elements to project an image of the photomask onto a photoresist film. Figure 6.26 shows the basic elements of a projection lithography system. The beam is produced by a UV lamp and a parabolic reflector, which is then passed through a homogenizer to create a uniform intensity profile. The condenser lens collects the light and illuminates the photomask, which is then imaged on the substrate by the projection lens.

Unlike contact lithography where the photomask to image ratio is 1:1, in projection systems, the image can be demagnified. Typical reduction values are in the range of 5–10X. This allows the features on the photomask to be larger, which significantly relaxes the design tolerances as well as the photomask cost.

The resolution limit of the system is governed by the numerical aperture (NA) of the projection lens and the illumination wavelength. Referring to Figure 6.27, if the maximum angle of the converging beam is θ, the component of the *k*-vector along the image plane will be

$$k_{\|} = nk \sin\theta \tag{6.20}$$

where

$$k = \frac{2\pi}{\lambda}$$

and n is the refractive index of the medium.

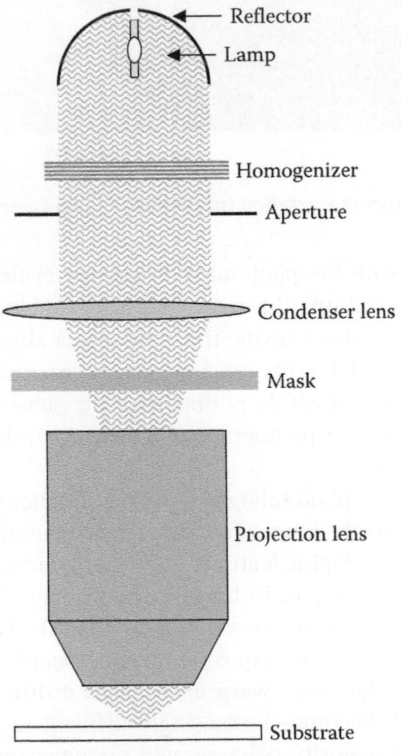

FIGURE 6.26 Basic elements of a projection lithography system.

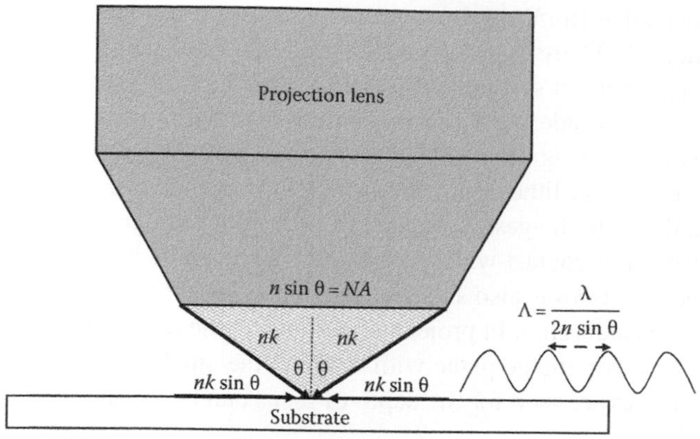

FIGURE 6.27 Relationship between the numerical aperture, wavelength, and spatial resolution.

The two counterpropagating k-vectors from the beam will create a standing wave with a spatial wave vector of

$$K = 2nk\sin\theta. \tag{6.21}$$

The pitch (the distance between two adjacent peaks) that corresponds to this spatial wave vector will be

$$\Lambda = \frac{\lambda}{2n\sin\theta} \tag{6.22}$$

which is typically defined as the maximum achievable resolution. Since the NA is defined as

$$\text{NA} = n\sin\theta \tag{6.23}$$

the smallest pitch that can be resolved is usually expressed as

$$\Lambda = \frac{\lambda}{2NA}. \tag{6.24}$$

If we assume that the pitch Λ consists of equal widths of bright and dark fringes, the width of a single printed line would be $\Lambda/2$. This is referred to as the half-pitch (HP). The HP is usually defined as the CD of the overall lithography system.

Therefore, the half-pitch, HP, becomes

$$HP = \frac{\lambda}{4NA} \tag{6.25}$$

which is more generically written as

$$HP = k_1\frac{\lambda}{NA}. \tag{6.26}$$

Comparing equations (6.25) and (6.26), the smallest possible value for k_1 is 0.25, but this will only occur in a perfect optical system where $NA = 1.0$. In general, due to imperfections in the lenses and photoresist performance, k_1 will be higher than 0.25. The image size also plays a role in the value of k_1. Larger images will suffer more from spherical aberrations and will result in a larger value for k_1.

The largest possible value for NA is 1.0 (assuming air as the incident medium). In practice, this value will be lower than 1.0. There is also a practical trade-off between spherical aberrations and NA. Nevertheless, assuming a perfect system with $k_1 = 0.25$ and $NA = 1.0$ using $\lambda = 365$ nm, we can find that the HP will be 91 nm. Considering a more realistic system with $k_1 = 0.5$ and $NA = 0.5$, the HP will be 365 nm. Although this is not significantly better than contact lithography, most manufacturing systems moved away from contact lithography and adopted projection lithography due to its increased throughput and reliability. The biggest advantage of projection lithography is its noncontact nature. The mask is never brought in contact with the substrate, so it can be kept clean, allowing for greater throughput. The reduction optics is also an advantage, allowing for lower photomask costs.

Another important consideration in projection systems is the depth of focus. This is the vertical distance above and below the image plane within which the smallest resolved feature will remain reasonably resolved. The expression for the depth of focus can be given by

$$DOF = k_2 \frac{\lambda / n}{\sin^2 \theta}. \tag{6.27}$$

Using $\lambda = 365$ nm with $NA = 0.5$ and $k_2 = 0.5$, we can get a $DOF = 730$ nm. The substrate planarity, warp, and topographic features must remain within this value for a proper image to be formed. This is not an easy criterion to satisfy. Even the best substrates have warp numbers that significantly exceed this DOF value. However, it becomes more likely to satisfy this criterion if the exposure can be limited to a small field size. As a result, projection systems use an image size that is usually no larger than about 1 in. × 1 in. This image is repeated across the whole wafer in a step-and-repeat configuration, also known as steppers, to produce multiple copies of the same design. This is illustrated in Figure 6.28. Assuming a 5× reduction optics, a 5 in. × 5 in. photomask will be projected onto a 1 in. × 1 in. image size.

The progress in lithography has not only been toward smaller and smaller device geometries but also toward larger chip sizes. Larger images require larger projection lenses, and these naturally lead to greater optical aberrations. One way to mitigate this problem is by linearly scanning each image through a slit instead of projecting the whole image. This slit can be positioned through the best parts of the optical system with the lowest aberrations. This system is known as the step-and-scan system, and it is currently the most widely used method in manufacturing systems.

Mercury vapor lamps operating at 365 nm were the dominant light sources for lithography until the late 1980s, but continuing demand for smaller device geometries has made it necessary to look for shorter wavelength sources. Deep UV (DUV) excimer lasers were the only efficient light sources at these wavelengths, so initially 248 nm KrF lasers were used and then 193 nm ArF lasers [30,31]. While the wavelength has been getting shorter, the quality of the optics has also been improving,

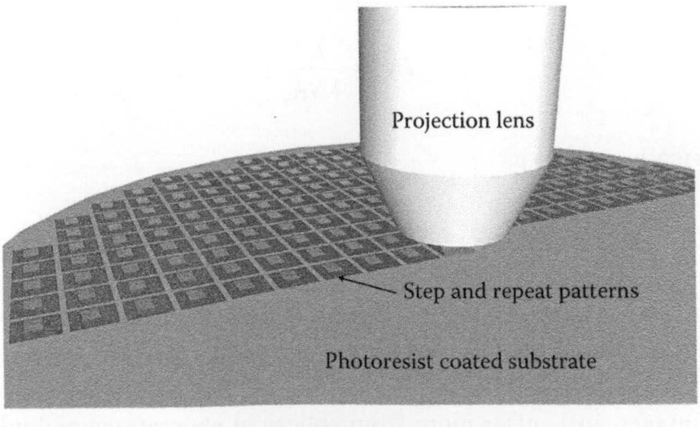

FIGURE 6.28 Step and repeat configuration used on projection lithography.

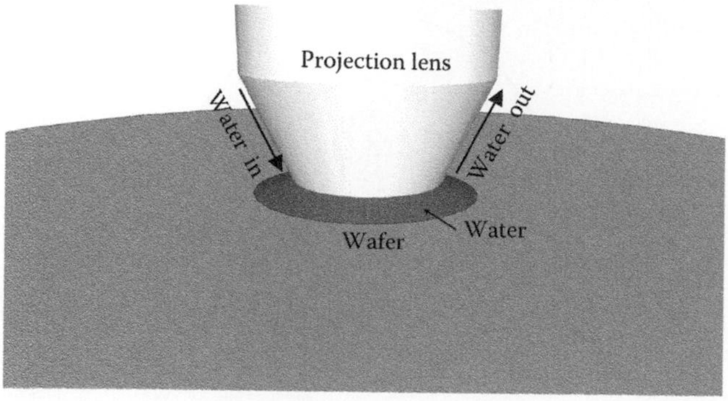

FIGURE 6.29 Water-immersion lithography.

resulting in the k_1 parameter approaching very close to the theoretical limit of 0.25. The NA has also been increasing to 0.9. Using Equation 6.26, we can verify that the HP will then become 53 nm.

Subsequently, significant effort was spent on developing a lithography system using the 157 nm F_2 laser, but this proved to be extremely difficult due to the imaging optics [32]. Fused silica is not transparent at this wavelength, and the efforts to develop high-quality CaF_2 optics turned out to be too costly and too difficult. As a result, lithography systems have remained at 193 nm for over a decade. Nevertheless, the HP has been steadily improving due to a number of other innovative techniques [33,34].

One of these techniques is immersion lithography [35–38]. The *NA* is limited to 1.0 only if air is used as the incident medium. If air is replaced with a fluid with a high refractive index, the *NA* can become larger than 1.0. High refractive index oils are routinely used in oil-immersion microscopy to achieve *NA* of 1.6, also known as hyper-*NA*. In photolithography, oils are generally incompatible with photoresists and are also not transparent to DUV wavelengths, so liquid water is used instead. At 193 nm, the refractive index of water is 1.44, which allows the *NA* to reach as high as 1.44. A film of water is continuously flowing between the projection lens and the photoresist as the wafer is repeatedly stepped and scanned, as shown in Figure 6.29. This is known as the 193 nm immersion lithography, abbreviated as 193i. Despite a number of problems, such as leaching of chemicals by the water, bubble formations, and contamination of the lenses, it has been successfully developed into a commercial system. As a result, it became possible to reach HP values of 35 nm, with a *DOF* of 38 nm.

Further refinement in lithographic performance continued to be developed, including phase-shifted masks, optical proximity corrections (OPCs), off-axis illumination, double patterning, and self-aligned double patterning. Currently, features as small as 20 nm are being produced by major electronic chip manufacturers using the 193 nm laser source. Nikon, ASML, and Canon are the three major manufacturers of semiconductor lithography equipment, with the first two still producing DUV projection lithography systems.

6.7 BASIC PROPERTIES OF PHOTORESISTS

6.7.1 Components of Photoresists

Photoresists contain three major components: (1) solvent, (2) resin, and (3) photoactive compound (PAC), as shown in Figure 6.30. The main purpose of the solvent is to allow liquid dispensing and modification of the viscosity to allow spin coating. After spin coating, solvents are usually expelled during the soft bake step. The resin forms the bulk of the remaining component and is the main structural element of the photoresist. This is the "resist" part of the photoresist. Novolac is the most commonly used resin, which is part of a family of phenolic resins also used in PCBs and adhesives. Other resins such as poly methyl methacrylate (PMMA) are also used in some photoresists.

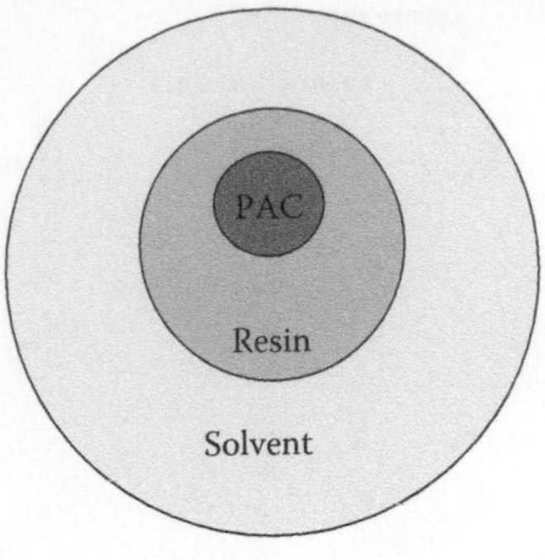

FIGURE 6.30 Components of a photoresist.

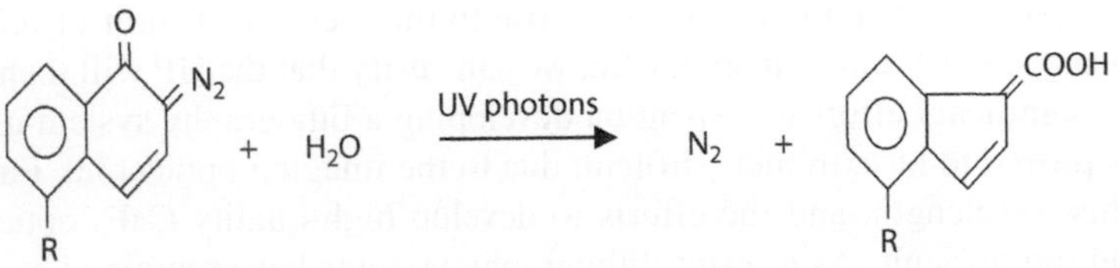

FIGURE 6.31 Reaction of the photoactive compound during exposure to UV.

The PAC is the UV-sensitive component of the photoresist. This is the "photo" part of the photoresist. Diazonaphthoquinone (DNQ) is the most commonly used PAC. Novolac has a missing formaldehyde group that prevents complete polymerization. It is therefore slightly soluble in weak alkaline base solutions (developer). When DNQ is introduced in the resin, it binds to these missing sites and reduces the solubility of the resin by two or more orders of magnitude. Therefore, unexposed photoresist exhibits an extremely low solubility in the developer solution. When exposed to UV light, the DNQ releases itself from the resin and combines with water to produce carboxylic acid and nitrogen, as shown in Figure 6.31. The carboxylic acid binds with the Novolac resin and increases its solubility by about two orders of magnitude. As a result, the solubility of the resin changes by about four orders of magnitude after UV exposure. Photoresists that exhibit an increase in solubility after exposure are known as positive-tone photoresists [4].

DNQ is sensitive to a broad UV spectral range, from about 320 nm to 400 nm. At longer wavelengths, the photon energy becomes too small to catalyze the photoacid reaction, and the transmission through the photoresist will increase; at shorter wavelengths, the resin starts to absorb most of the radiation. Although the optical absorption is large at short wavelengths, they do not necessarily result in a photoacid reaction.

After a prolonged exposure to UV, all of the DNQ will be consumed and converted to carboxylic acid. The absorption spectrum before and after UV exposure can be used to separate the DNQ's absorption from the resin's absorption. Figure 6.32 shows an absorption spectrum of a representative photoresist before and after UV exposure. The unbleached (unexposed) photoresist exhibits a higher optical absorption than the fully bleached (exposed) photoresist. The residual absorption in a fully exposed photoresist is due to the resin. In this plot, the absorbance is defined as the optical density $\left(\log_{10}(1/T)\right)$, where T is the optical transmission through the photoresist film.

Photoresists are developed and optimized for specific spectral bands. Every time the industry transitioned to a shorter wavelength, such as from near UV to DUV and then from DUV to extreme UV, new photoresists had to be developed, which became major bottlenecks in the technology. In the DUV wavelengths, the background optical absorption in the resists was too high, resulting

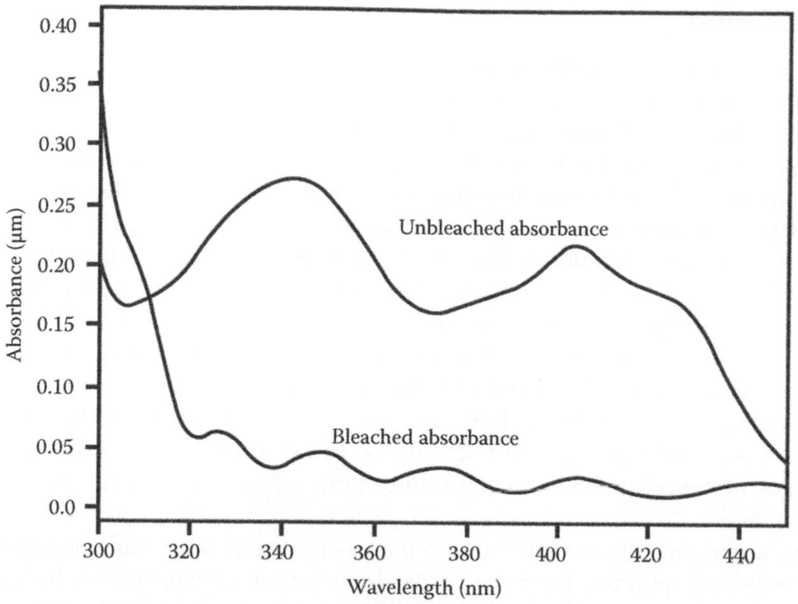

FIGURE 6.32 Absorption spectrum of the photoresist before and after UV exposure for the Rohm and Haas SPR-955 photoresist. (Courtesy of Dow Electronic Materials.)

in poor optical sensitivity. To overcome this problem, a process known as chemical amplification was developed, where a single photoacid molecule is used as a catalyst to induce a change in the solubility of a large number of resin molecules. This is a thermally activated process, so a bake step is required after the initial exposure, known as the postexposure bake (PEB) [39,40].

6.7.2 Effects of Moisture on Photoresist Performance

Interestingly, water is a necessary ingredient for the exposure reaction [41] and is shown in Figure 6.31. This water primarily comes from the ambient humidity of the air. Unfortunately, the soft bake step will remove not only the solvents but also the moisture in the film. Therefore, before exposing the photoresist, it has to be allowed to sit in ambient air for a sufficient length of time for moisture to permeate and rehydrate the film. The required rehydration time increases exponentially with film thickness. Film thicknesses of the order of 1 µm will require less than 1 min to rehydrate, but films thicker than 10 µm may require 30 min or even longer. In dry ambient conditions, the duration will have to be increased even more [42].

In inadequately rehydrated films, a gradient in moisture concentration will exist along the film thickness. After exposure, the bottom portions of the photoresist with a lower moisture content will not respond to light exposure and will fail to develop properly. This effect is sometimes misinterpreted as arising due to insufficient exposure dose. But increasing the UV dose will not improve the performance because no amount of exposure will increase the solubility if the film has not been adequately rehydrated.

For this reason, lithography areas must be maintained at a constant humidity level, typically around 45%. Without active humidity control, laboratory air becomes extremely dry in the winter months and extremely humid in the summer, resulting in widely varying lithography performances. While inadequate moisture is usually the common problem, excessive moisture can also affect the photoresist performance. Condensation of moisture on the substrate surface prior to spin coating weakens the adhesion of the photoresist to the substrate and causes the photoresist features to become detached during the develop step.

6.7.3 Development

Exposed photoresist becomes highly soluble in the developer. The difference in solubility between an exposed and unexposed photoresist can be several orders of magnitude. This property enables sharp vertical walls to be created even from diffused exposure profiles. The plot of solubility vs. exposure dose has a steplike character, as shown in Figure 6.40. This is very similar to the optical density vs. exposure dose curve used in photography, known as the Hurter–Driffield curve [43].

The developer is typically a weak solution of potassium hydroxide (KOH), sodium hydroxide (NaOH), or tetramethylammonium hydroxide (TMAH) along with other ingredients that act as surfactants to improve the wetting properties. The metal hydroxide developers can be buffered to maintain a constant pH during usage. They also have a wide process margin with a lower sensitivity to concentrations and temperature. However, metal ions can be a source of contamination in complementary metal oxide semiconductor (CMOS) electronic devices, and they can also form salt deposits in the developer bowl. TMAH developers do not contain any metal ions; hence, they are also known as metal-ion-free (MIF) developers. But their effectiveness is much more sensitive to changes in concentration and temperature. Both types of developers are widely used in photolithography.

Although a wafer can be developed by simply immersing it in a beaker, more consistent results can be obtained by using a spray developer or a puddle developer, or a combination of both, in a configuration that is very similar to a spin coater. In puddle development, a predefined quantity of developer is dispensed onto the wafer and allowed to wet the surface and form a stationary puddle, as shown in Figure 6.33a. However, the developer becomes saturated with the dissolved photoresist, so it has to be spun off and a new puddle formed to keep the development rate consistent. The spray method uses a fine mist of developer continuously dispensed from a nozzle while spinning the wafer, as shown in Figure 6.33b. This uses less developer and produces a more uniform coverage. In the spray–puddle development, both methods are used alternately—first, a puddle is formed, followed by a spray cycle, and then a new puddle is formed, and the process is repeated until the process is complete.

6.7.4 Modeling the Optical Performance of Photoresists

6.7.4.1 Dill Parameters

The optical exposure properties of photoresists are specified using three parameters, *A*, *B*, and *C*, known as the Dill parameters [44]. These can be treated as optical constants of the photoresist, just like refractive index for dielectric materials. In this model, the DNQ concentration in the photoresist film at time *t* is expressed as a normalized value M that spans between 0 and 1.0. As a result, the value of *M* will be 1.0 before the photoresist is exposed. As the film undergoes exposure, *M* will decline toward zero. However, due to the optical absorption in the film, the top portions of the film

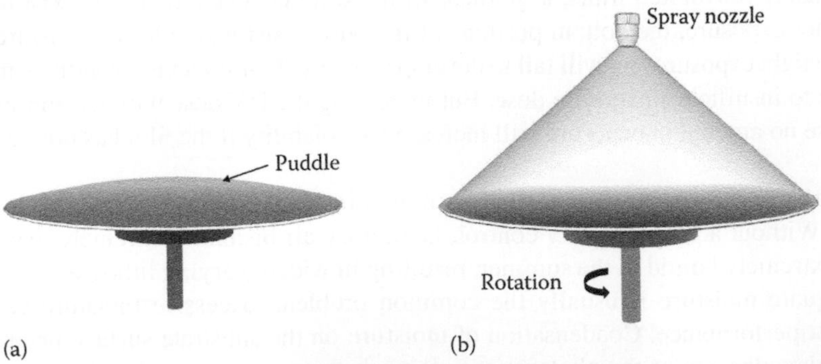

FIGURE 6.33 (a) Puddle develop system; (b) spray develop system.

will absorb more energy than the bottom portions. Therefore, in general, M will become a function of position inside the photoresist film, as well as exposure time. Therefore, we will represent the DNQ concentration as $M(r,t)$, where r is a position vector and t is time. When the photoresist is unexposed, the DNQ distribution will be $M(r,0)=1.0$. After it is fully exposed for a very long time, and all the DNQ is entirely depleted, the DNQ distribution becomes $M(r,\infty)=0.0$.

In the Dill model, the optical absorption coefficient of the photoresist is expressed as a function of the A and B Dill parameters as

$$\alpha(r,t)=AM(r,t)+B \tag{6.28}$$

and the depletion rate of DNQ is written as

$$\frac{dM(r,t)}{dt}=-I(r,t)M(r,t)C \tag{6.29}$$

where $I(r,t)$ is the optical intensity inside the photoresist film.

The refractive indices of the photoresist, substrate, and other materials in the system are also assumed to be known. Initially, at $t=0$, we will have $M(r,0)=1.0$; therefore, from Equation 6.28, we can get $\alpha(r,0)=A+B$. Using this absorption coefficient and the incident optical intensity profile $I(r_0,0)$, where r_0 is the top surface of the photoresist, the optical intensity distribution $I(r,0)$ inside the photoresist film can be calculated. This requires solving the Maxwell's electromagnetic wave equations. Using this intensity distribution $I(r, 0)$, we can find $\frac{dM(r,t)}{dt}$ by using Equation 6.29. This allows us to find the DNQ distribution $M(r,\Delta t)$ for the next time step $t=\Delta t$. Then we can return to Equation 6.28 and calculate the new value of $\alpha(r,\Delta t)$, which can then be used to find the new intensity distribution at the time step $t=\Delta t$. This process has to be repeated incrementally for each Δt until the exposure time t_{exp} is reached. The end result of this calculation will be a DNQ distribution $M(r,t_{exp})$. This value can then be used in a solubility model to predict how the photoresist will dissolve in a developer.

It can be seen from Equation 6.28 that A represents the functional absorption coefficient of the DNQ molecules and B represents the nonfunctional absorption coefficient. The latter arises primarily due to the absorption by the resin. The C parameter represents the optical sensitivity of the photoresist—that is, it specifies how fast the photoresist gets exposed. Additionally, it should be noted that the A, B, and C parameters are wavelength-specific. Generally, these parameters are specified at the wavelength for which the photoresist was manufactured, although it can be measured for any wavelength.

$I(r_0,0)t_{exp}$ is the total energy per unit area (optical dose) supplied to the photoresist by the incident UV light. The majority of this energy will be absorbed by the DNQ, but some of it will be lost due to reflection, transmission, and scattering. The spatial distribution of $M(r,t_{exp})$ will be linearly related to the spatial distribution of the absorbed energy.

The solution of Equations 6.28 and 6.29, along with the optical intensity distribution $I(r,t)$, can only be performed numerically. The most difficult part of this calculation is to solve Maxwell's equations. This has to include the effects of the substrate, any films on the substrate, the photoresist film, and the air interface above the photoresist. Hence, this is a classical multilayer optical problem, very similar to the problems discussed in Chapter 4: Thin Film Optics except in this case, we also have to include the effects of photoresist absorption and diffraction. Furthermore, unlike the multilayer dielectric structures where we only had discrete interfaces, in this case, we also need to account for arbitrary refractive index profiles and continuously varying absorption coefficients. This is a difficult computational problem in three dimensions and nontrivial even in two dimensions. However, if we consider it as a 1D problem by ignoring the effects of diffraction, it becomes easier to solve.

6.7.4.2 Diffusion

The carboxylic acid generated from the DNQ can diffuse over time, especially at elevated temperatures. On highly reflective substrates, this can be beneficial because it will wash out the effects of the optical standing waves. Standing waves are ripples in the optical intensity pattern due to the counterpropagating waves from the incident and reflected waves. To avoid these ripples in the developed photoresist profile, a PEB is often recommended in many photoresist processes. However, an excessive bake can also produce too much lateral diffusion and adversely affect the pattern resolution. The diffusion is governed by Fick's equation as follows:

$$\frac{\partial M(r,t)}{\partial t} = \nabla(D\nabla M(r,t)) \tag{6.30}$$

where D is the diffusion coefficient. If we limit ourselves to one dimension only, assuming D does not vary with position, we can rewrite Equation 6.30 as

$$\frac{\partial M(z,t)}{\partial t} = D\frac{\partial^2}{\partial z^2}M(z,t). \tag{6.31}$$

Furthermore, we can designate two different diffusion coefficients: D_0 for diffusion that takes place at ambient temperature during the exposure and D_1 for diffusion that takes place at an elevated temperature during the PEB.

Since the diffusion takes place simultaneously with the exposure process described in Equation 6.29, we can combine them together into a single equation:

$$\frac{\partial M(z,t)}{\partial t} = D\frac{\partial^2}{\partial z^2}M(z,t) - I(r,t)M(r,t)C. \tag{6.32}$$

6.7.4.3 Numerical Shooting Method for Modeling the Optical Field

The numerical shooting method is a computational technique for solving one-dimensional wave equations. It is similar in many respects to the transfer-matrix method (TMM) as described in Chapter 4: "Thin Film Optics," except while TMM is best suited for computing the transmission and reflection from discrete layers of optical films, the optical shooting method can handle continuously varying quantities. This comes at the cost of a slightly higher computational burden than TMM, but for one-dimensional problems, this is not usually a significant penalty.

In this section, we will derive the numerical shooting method for Maxwell's equation and apply it to solve the Dill parameter problems. Compared to other numerical techniques for solving Maxwell's equations, this is an extremely trivial method to understand and implement. The limitation, however, at least for now, is that it can only be applied to one-dimensional problems. As a result, it does not include diffraction effects. Admittedly, diffraction is a very important consideration in photolithography, but unfortunately it will require a significant computational effort that is beyond the scope of the discussion in this chapter. Interested readers are directed to more rigorous computational models described in references [27–29,45,46].

We start with the 1D plane wave propagation equation:

$$\frac{\partial^2 E(z)}{\partial z^2} = -k_0^2 n(z)^2 E(z) \tag{6.33}$$

where

z is the direction of propagation

k_0 is the free-space wave vector $\frac{2\pi}{\lambda}$

$n(z)$ is the refractive index distribution (which can be complex)
$E(z)$ is the electric field

All of the z-dependent functions in Equation 6.33 will be discretized on a finite-difference grid as

$$E(z) = E(m\Delta z) \tag{6.34}$$

and

$$n(z) = n(m\Delta z) \tag{6.35}$$

where
m is an integer
Δz is the grid size

The second-order derivative at the grid point n can be expressed in finite-difference discretization as

$$\left.\frac{\partial^2 E(z)}{\partial z^2}\right|_m = \frac{\left(\left.\frac{\partial E(z)}{\partial z}\right|_{m-\frac{1}{2}} - \left.\frac{\partial E(z)}{\partial z}\right|_{m+\frac{1}{2}}\right)}{\Delta z} \tag{6.36}$$

$$\left.\frac{\partial^2 E(z)}{\partial z^2}\right|_m = \frac{\left(\frac{E((m+1)\Delta z) - E(m\Delta z)}{\Delta z} - \frac{E(m\Delta z) - E((m-1)\Delta z)}{\Delta z}\right)}{\Delta z} \tag{6.37}$$

$$\left.\frac{\partial^2 E(z)}{\partial z^2}\right|_m = \frac{E((m+1)\Delta z) - 2E(m\Delta z) + E((m-1)\Delta z)}{\Delta z^2}. \tag{6.38}$$

Substituting Equation 6.38 into Equation 6.33 allows us to express the one-dimensional electromagnetic wave equation in finite-difference notation as

$$\left[\frac{E((m+1)\Delta z) - 2E(m\Delta z) + E((m-1)\Delta z)}{\Delta z^2}\right] = -k_0^2\,[n(m\Delta z)]^2\, E(m\Delta z). \tag{6.39}$$

If we know the first two values of the electromagnetic field $E((m-1)\Delta z)$ and $E(m\Delta z)$, we can express the subsequent value $E((m+1)\Delta z)$ by rearranging Equation 6.39 as

$$E((m+1)\Delta z) = \left(2 - [n(m\Delta z)]^2\, k_0^2 \Delta z^2\right) E(m\Delta z) - E((m-1)\Delta z). \tag{6.40}$$

Now consider the structure as shown in Figure 6.34, where the incident field is on the right of the structure and the transmitted field is on the left of the structure. Because the transmission side contains just one wave component and the incident side contains two wave components, the numerical procedure is actually easier if we start from the transmission side and work our way toward the incident side. We will start the computation by assuming the transmitted field amplitude to be unity, that is, $A_{zt} = 1$. This value is of course completely arbitrary since the entire field distribution that results from this initial assumption can be normalized to any value after the computation.

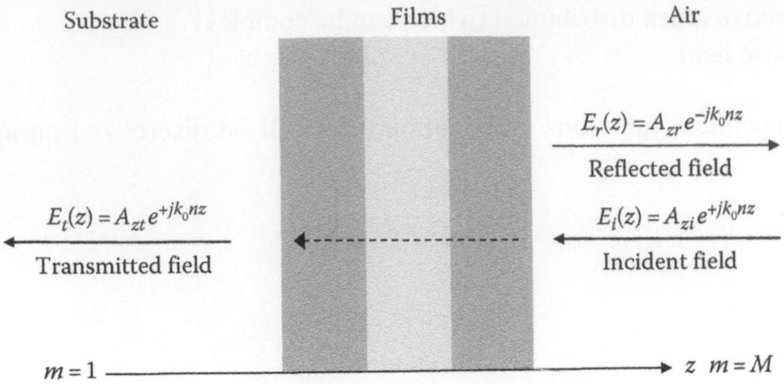

FIGURE 6.34 Optical layer structure for the numerical shooting method.

Therefore, on the transmission side, we will have

$$E_t(z) = e^{+jk_0nz} \tag{6.41}$$

and on the incident side

$$E_i(z) = A_{zi}e^{+jk_0nz} \tag{6.42}$$

$$E_r(z) = A_{zr}e^{-jk_0nz}. \tag{6.43}$$

The initial conditions will therefore be

$$E(\Delta z) = 1 \tag{6.44}$$

$$E(2\Delta z) = e^{+jk_0n(2\Delta z)\Delta z}. \tag{6.45}$$

Substituting these two initial conditions into Equation 6.40 will result in the value for the field at $E(3\Delta z)$. Then $E(2\Delta z)$ and $E(3\Delta z)$ can be used to calculate the field value at $E(4\Delta z)$, and so on up to the last point on the right-hand side $E(M\Delta z)$. This way, we can calculate the entire field distribution in the structure, starting from the substrate and ending at the outer medium (which is assumed to be air).

The next step is to calculate the amplitudes A_{zi} and A_{zr}. Since the calculated field profile contains the sum of the incident and reflected waves, a few extra steps are necessary to separate these amplitudes. If the numerical values at the last two points on the incident side are $E(M\Delta z)$ and $E((M-1)\Delta z)$, these can be expressed as

$$E(M\Delta z) = A_{zr}e^{-jk_0n(M\Delta z)M\Delta z} + A_{zi}e^{+jk_0n(M\Delta z)M\Delta z} \tag{6.46}$$

$$E((M-1)\Delta z) = A_{zr}e^{-jk_0n((M-1)\Delta z)(M-1)\Delta z} + A_{zi}e^{+jk_0n((M-1)\Delta z)(M-1)\Delta z} \tag{6.47}$$

The values of k_0 and the refractive index distribution n are known. Therefore, Equations 6.46 and 6.47 contain only two unknowns, A_{zr} and A_{zi}, which can be calculated by simple substitution. Furthermore, we will assume that the refractive index values $n(M\Delta z)$ and $n((M-1)\Delta z)$ are equal, that is, the refractive index is nonvarying at the edge of the computation window. This results in

$$A_{zr} = \frac{E((M-1)\Delta z)\, e^{+jk_0n((M-1)\Delta z)\Delta z} - E(M\Delta z)}{\left(e^{-jk_0n((M-1)\Delta z)(M-2)\Delta z} - e^{-jk_0n(M\Delta z)M\Delta z}\right)} \tag{6.48}$$

and

$$A_{zi} = e^{-jk_0 n(M\Delta z)M\Delta z}\left(E(M\Delta z) - A_{zr}e^{-jk_0 n(M\Delta z)M\Delta z}\right). \tag{6.49}$$

From this, we can get the field transmission coefficient as

$$t = \frac{A_{zt}}{A_{zi}} = \frac{1}{A_{zi}} \tag{6.50}$$

and the field reflection coefficient as

$$r = \frac{A_{zr}}{A_{zi}}. \tag{6.51}$$

The intensity reflection and transmission coefficients will be

$$T = |t|^2 \tag{6.52}$$

and

$$R = |r|^2. \tag{6.53}$$

Additionally, the local intensity at any given point is

$$I(m\Delta z) = \frac{c\varepsilon_0 n(m\Delta z)}{2}|E(m\Delta z)|^2. \tag{6.54}$$

If the incident intensity is I_{in}, the local intensity becomes

$$I(m\Delta z) = I_{in}\frac{\text{Re}\{n(m\Delta z)\}}{n(0)}|E(m\Delta z)|^2 \tag{6.55}$$

where $\text{Re}\{n(m\Delta z)\}$ is the real part of the refractive index.

The diffusion equation (6.32) can be discretized as follows:

$$\frac{M(m\Delta z, t+\Delta t) - M(m\Delta z, t)}{\Delta t} = D\frac{M((m+1)\Delta z, t) - 2M(m\Delta z, t) + M((m-1)\Delta z, t)}{\Delta z^2}$$
$$-I(z,t)M(m\Delta z, t)C \tag{6.56}$$

which can be rearranged in a more convenient form as

$$M(m\Delta z, t+\Delta t) = M(m\Delta z, t) + D\Delta t\left(\frac{M((m+1)\Delta z, t) - 2M(m\Delta z, t) + M((m-1)\Delta z, t)}{\Delta z^2}\right)$$
$$-I(z,t)M(m\Delta z, t)C\Delta t. \tag{6.57}$$

The solution method proceeds as follows. First, at time step $t = 0$, Equation 6.28 is used to calculate the absorption coefficient $\alpha(z,0)$ assuming $M(z,0) = 1$, from which we can evaluate the imaginary part of the refractive index as

$$\text{Im}\{n(z,0)\} = \frac{\alpha(z,0)\lambda}{4\pi}. \tag{6.58}$$

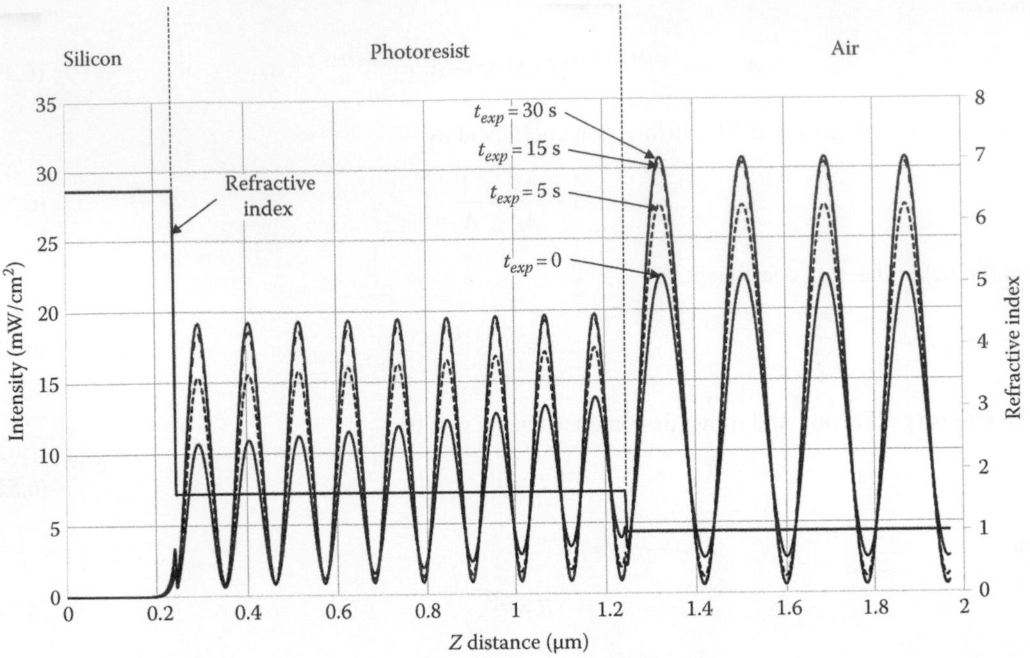

FIGURE 6.35 Intensity profile through the photoresist film on a silicon substrate as a function of exposure time.

Along with the real part of the refractive index, this is used in Equations 6.40, 6.44, and 6.45 to evaluate the field profile $E(z)$. Then the intensity profile $I(z)$ is calculated using Equation 6.55. This intensity is then used in Equation 6.57 to calculate the normalized DNQ concentration $M(z,\Delta t)$ at the next time step $t = \Delta t$. This profile is used in Equation 6.28 to calculate the new value of $\alpha(z,\Delta t)$, and the whole process is repeated until the exposure time $t = t_{exp}$ is reached.

Figure 6.35 shows the time evolution of the intensity profile through a photoresist film on a silicon substrate. The parameters used for this calculation are shown in Table 6.1.

The ripples in Figure 6.35 are due to the reflection from the various interfaces creating standing waves, primarily from the silicon/photoresist interface where the refractive index difference is greatest. We can also see that the absorption in the unexposed photoresist produces a declining intensity in the film, and it slowly becomes more transparent as the exposure continues. Figure 6.36 shows the normalized DNQ concentration in the photoresist film as a function of exposure time. At $t = 0$, the value of $M(z,0) = 1$, and then it declines toward zero as exposure dose increases. The standing wave ripples have a significant effect on the value of $M(z,t)$. They produce alternating layers of photoresist that is exposed and underexposed. This effect ultimately results in the appearance of ripples along the sidewalls of the developed photoresist profile, as will be described later. Figure 6.37 shows the effect of diffusion during an exposure time of 30 s. The diffusion coefficient used here is very small, $D_0 = 10^{-6}$ μm^2/s, which results in the peak values of $M(z,t)$ becoming slightly smaller. However, the diffusion coefficient could be higher under elevated temperatures used during the PEB.

Figure 6.38 shows the same plot as Figure 6.35, except it is for a silica glass substrate instead of silicon. The most notable feature is the significantly smaller standing wave effect. However, we can also see that despite the smaller reflectivity of the substrate, the exposure times required to deplete $M(z,t)$ are not necessarily longer. This arises due to the standing waves which produces alternating areas of low and high intensity that counteracts the benefits of the increased substrate reflection from silicon.

The computed reflection coefficient as a function of exposure time is shown in Figure 6.39 for both the silicon and silica glass substrates. Note that the plot uses two separate axes for the data

TABLE 6.1
Photoresist Parameters Used for the Calculations Shown in Figure 6.35

Wavelength	365 nm
Incident intensity	10 mW/cm^2
Photoresist thickness	1.0 μm
Real part of the refractive index of photoresist	1.65
Dill A	0.75 μm^{-1}
Dill B	0.02 μm^{-1}
Dill C	0.02 cm^2 · mJ^{-1}
Substrate index	6.55 − *j*2.75
Ambient diffusion coefficient D_0	10^{-6} μm^2/s

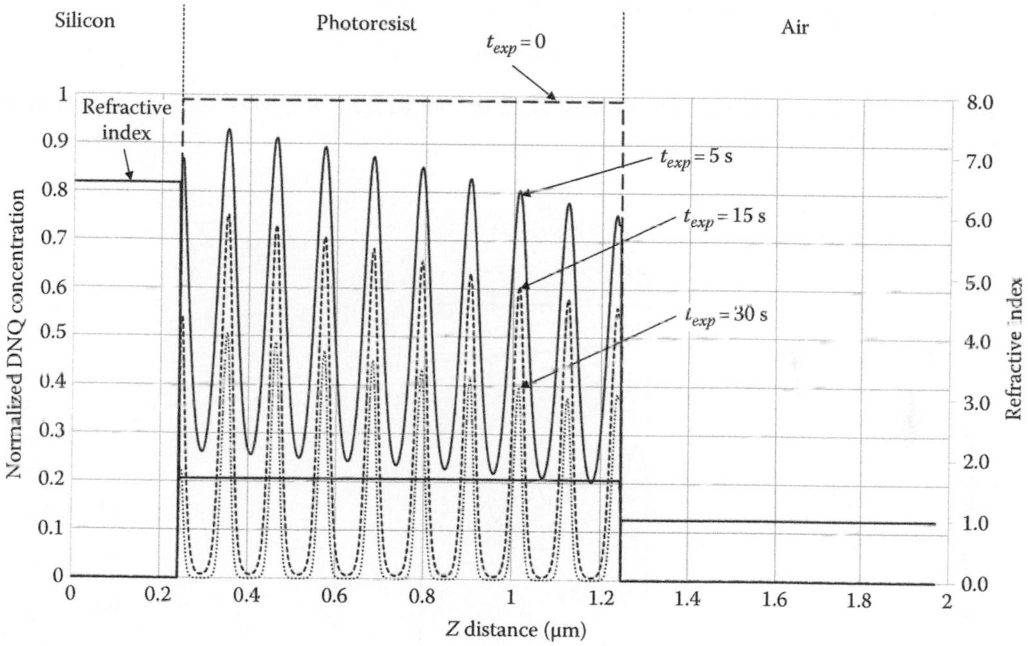

FIGURE 6.36 Normalized DNQ concentration *M* as a function of exposure time for a photoresist film on a silicon substrate.

because the reflection is significantly different between the two substrates. In the case of silicon, the reflection increases with time as the photoresist film becomes more transparent and the silicon substrate is able to reflect more of the incident light. With the glass substrate, it shows a slight decline with exposure because it is the photoresist/air interface that produces most of the reflection (instead of the photoresist/glass interface), and this reflection declines as the film becomes less absorbing. Optical measurements of reflection (or transmission) can be useful in measuring the Dill parameters. One can perform a series of reflectance (or transmission) measurements at different exposure doses and then fit the data with the model to extract all three Dill parameters. In this model, the reflection from the backside of the glass substrate (backside glass/air interface) has been ignored, but it has to be included with appropriate modifications to the model before comparing the results with measured data.

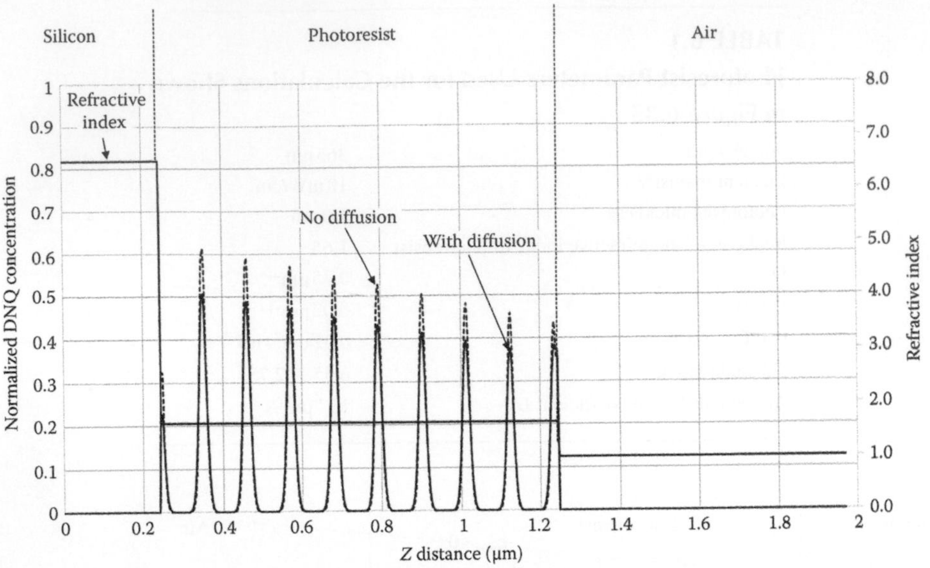

FIGURE 6.37 Effect of diffusion during exposure on the normalized DNQ concentration M for a fixed exposure time of 30 s.

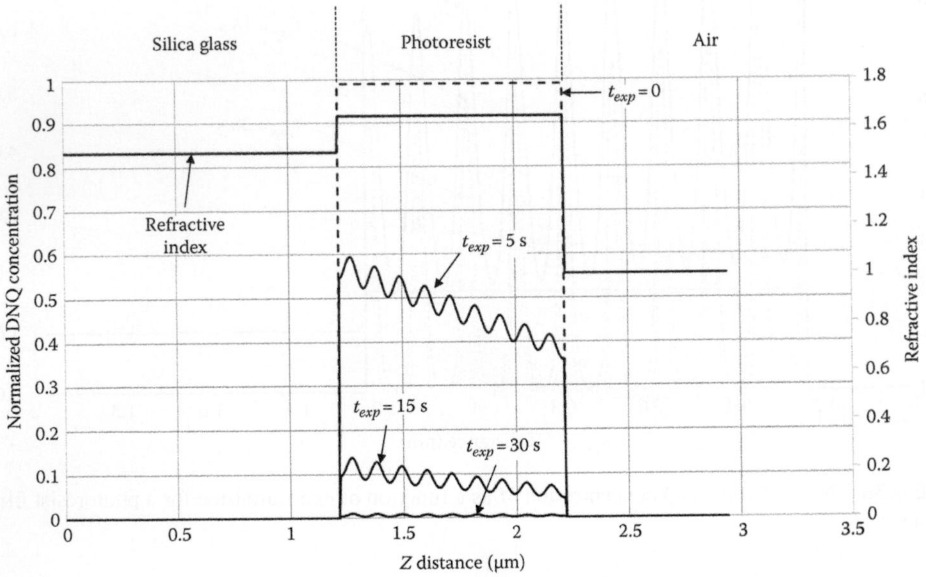

FIGURE 6.38 Normalized DNQ concentration M as a function of exposure time for a photoresist film on a silica glass substrate.

6.7.4.4 Solubility Model

The solubility of the photoresist increases by several orders of magnitude after exposure to UV light. All of the models used for this are empirical in nature. The following function is a simple way to empirically represent the solubility vs. the normalized DNQ concentration:

$$S(r) = \frac{S_1 - S_0}{1 + e^{\gamma(M(r,t_{exp}) - M_t)}} + S_0 \tag{6.59}$$

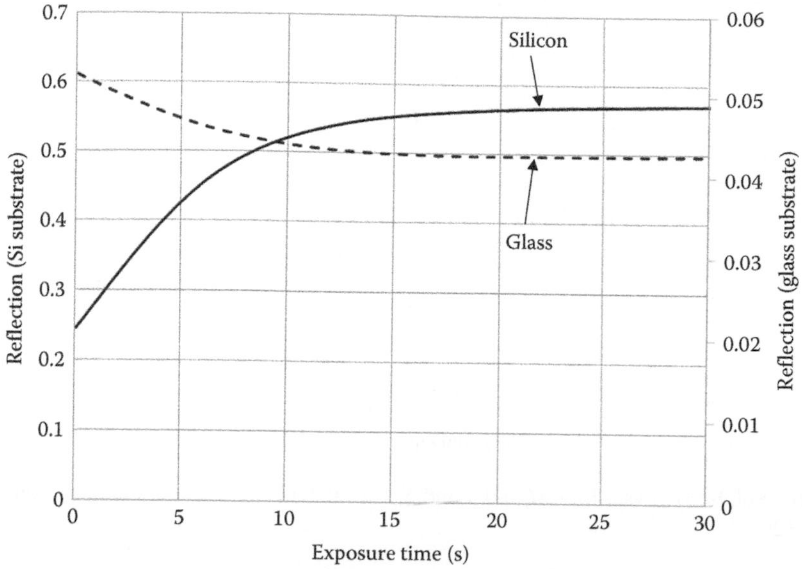

FIGURE 6.39 Reflection coefficient as a function of exposure time for a photoresist film on a silicon substrate (left axis) and silica glass substrate (right axis).

where

γ can be interpreted as the photoresist contrast
M_t as the solubility threshold
S_1 is the maximum solubility of the fully exposed photoresist
S_0 is the minimum solubility of the unexposed resist

S_0 is also known as the dark erosion rate of the photoresist. Based on this model, we can make the following three observations:

- When $M(r,0)=1.0$ (unexposed), the solubility is $(S_1+S_0)/(1+e^{\gamma(1-M_t)})+S_0$. When γ is large, this reduces to S_0.
- When $M(r,t_{exp})=0.0$ (fully exposed), the solubility is $(S_1+S_0)/(1+e^{-\gamma M_t})+S_0$. When γ is large, this reduces to S_1.
- When $M(r,t)=M_t$, the solubility is $(S_1+S_0)/2$.

Figure 6.40 shows a plot of solubility S vs. the normalized DNQ concentration M for a threshold value of $M_t=0.3$, minimum solubility $S_0=0.01$ μm/min and the maximum solubility $S_1=2$ μm/min. It can be seen that the solubility increases rapidly as DNQ concentration falls below M_t in a highly nonlinear fashion. This is fairly typical of most photoresists. This is an important result and is utilized in some of the resolution-enhancement techniques. This is also the reason why photoresists often display steep sidewall profiles even when exposed with a diffused intensity profile.

6.7.4.5 Quasi-Two-Dimensional Model

The optical model demonstrated in the previous section was strictly one-dimensional and did not include any diffraction effects in the lateral directions. Extending it to two or three dimensions with diffraction is possible, but it is significantly more complicated. However, we could compute the diffracted beam profile separately and consider it as the input to the one-dimensional model. This would allow us to develop a *qualitative* understanding of how the sidewall profiles of photoresist emerge as a function of exposure dose. Such a model is illustrated in Figure 6.41. We will assume

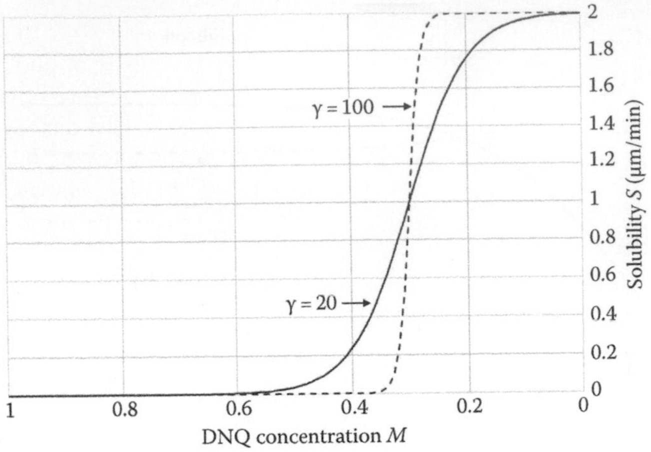

FIGURE 6.40 Solubility S vs. M for $M_t = 0.3$ and $S_0 = 0.01$ for two different values of contrast values of $\gamma = 100$ and $\gamma = 20$.

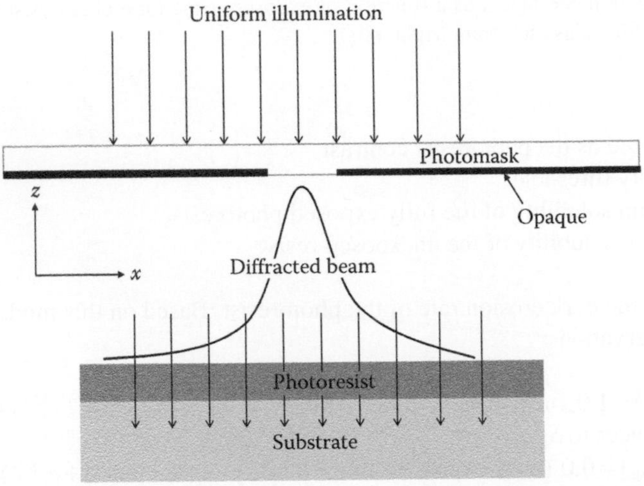

FIGURE 6.41 Quasi-two-dimensional model of the optical wave propagation inside a photoresist film.

that a diffracted beam profile as shown in the Figure 6.41 is representative of the average beam profile throughout the film. We can then take vertical slices through the photoresist film and calculate the field profile through each slice using the one-dimensional optical model. Within each slice, the optical intensity distribution $I(z,t)$ and the DNQ distribution $M(z,t)$ can be solved as a function of exposure time using the numerical shooting method and the Dill parameters, as described in the previous section. The only difference from one slice to another will be the incident optical intensity. This model effectively ignores any diffraction through the photoresist film and only includes the lateral intensity variation of the incident beam. As a result, this can be thought of as a quasi-two-dimensional model rather than a full two-dimensional model.

Consider the same photoresist as before on a silicon substrate with the additional parameters as shown in Table 6.2.

The incident beam profile through a 3 μm wide slit in the photomask, after propagating roughly halfway through the photoresist, is assumed to be as shown in Figure 6.42. The integrated value of this intensity divided by 3 μm is 10 mW/cm^2, which is the incident intensity of the optical beam at the photomask.

TABLE 6.2
Additional Parameters Used for the Calculations Shown in Figure 6.43 & 6.44

Post-exposure bake diffusion coefficient D_1	10^{-5} μm²/s
γ	20
Threshold DNQ concentration M_t	0.3
Dark erosion rate S_0	0.01 μm/min
Maximum solubility rate of exposed resist S_1	2.0 μm/min

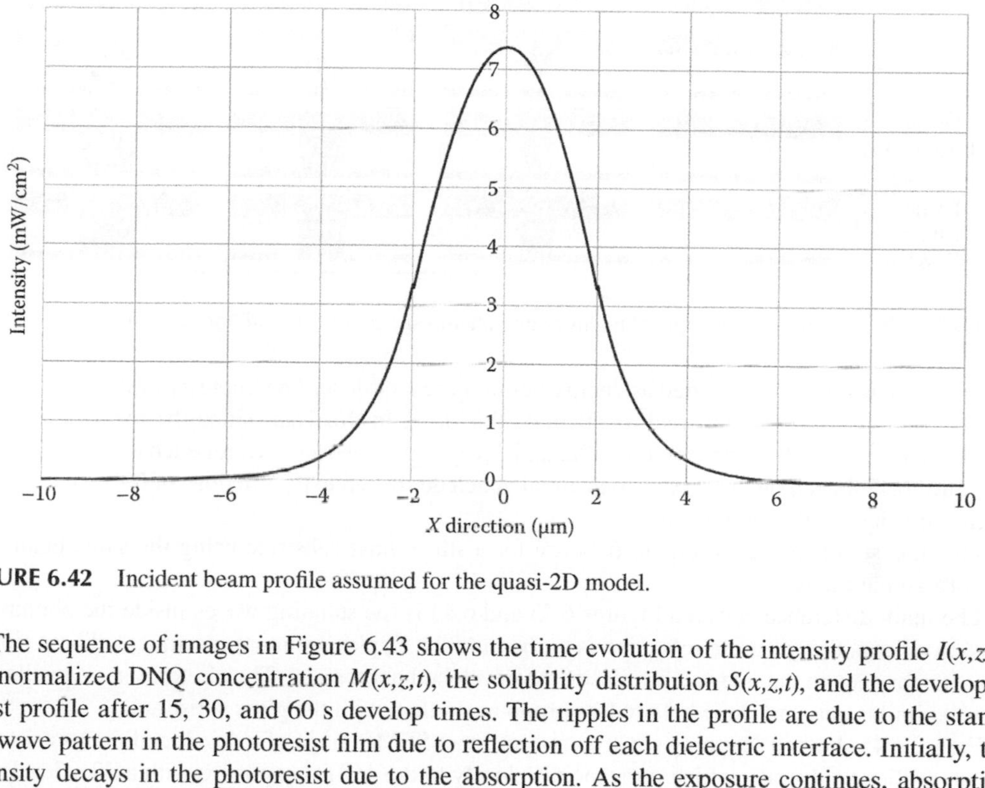

FIGURE 6.42 Incident beam profile assumed for the quasi-2D model.

The sequence of images in Figure 6.43 shows the time evolution of the intensity profile $I(x,z,t)$, the normalized DNQ concentration $M(x,z,t)$, the solubility distribution $S(x,z,t)$, and the developed resist profile after 15, 30, and 60 s develop times. The ripples in the profile are due to the standing wave pattern in the photoresist film due to reflection off each dielectric interface. Initially, the intensity decays in the photoresist due to the absorption. As the exposure continues, absorption decreases due to the declining DNQ concentration. Similarly, $M(x,z,t)$ starts at a value of 1.0 and quickly declines as the photoresist is exposed. It can be seen that the top portion of the photoresist is exposed more than the bottom portion. The process also includes a PEB for 60 s during which the diffusion becomes pronounced.

The solubility plots are the most important, as they directly correspond to the photoresist profiles one would obtain after the develop step. The develop step proceeds from the top portions of the photoresist toward the bottom. Insufficient develop will result in incomplete removal of the exposed photoresist. Excessive develop can erode the unexposed areas of the photoresist due to the dark erosion rate S_0.

The outwardly sloping sidewall angle seen in these plots is characteristic of all positive-tone photoresists, and this has important implications in subsequent processing. Also, notice that the sidewall angle decreases with increasing exposure, becoming more and more vertical. Another effect is the widening of the exposed areas, or narrowing of the unexposed areas. It should be noted that these effects are present even without diffraction through the photoresist (which has been ignored in this model). With diffraction effects included, the sidewall angle and the widening of the lines with exposure dose will become even more pronounced.

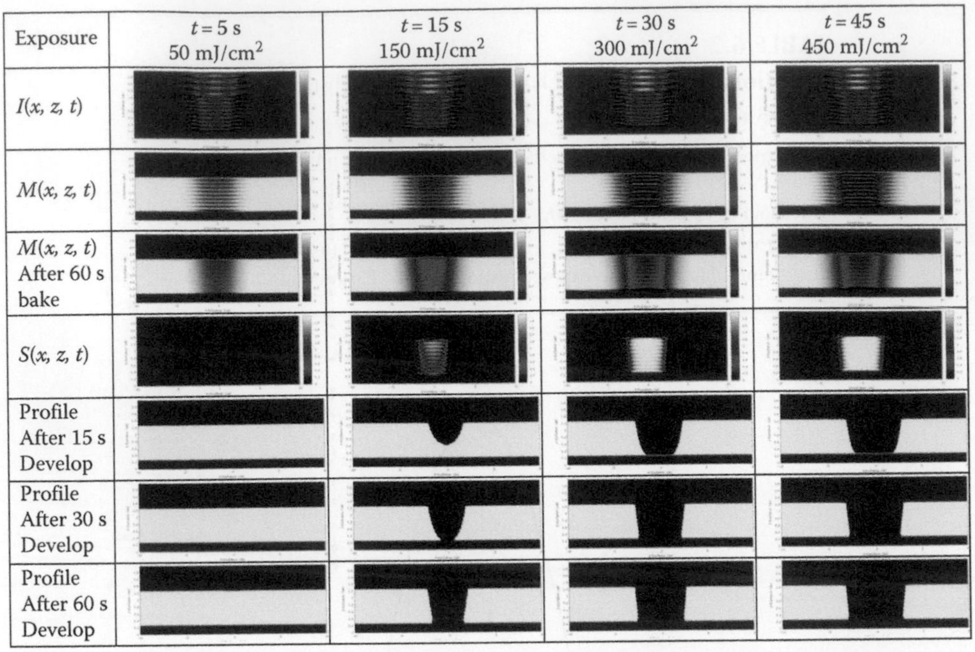

Exposure
t = 5 s
50 mJ/cm2
t = 15 s
150 mJ/cm2
t = 30 s
300 mJ/cm2
t = 45 s
450 mJ/cm2
I(x, z, t)
M(x, z, t)
M(x, z, t) After 60 s bake
S(x, z, t)
Profile After 15 s Develop
Profile After 30 s Develop
Profile After 60 s Develop

FIGURE 6.43 Evolution of the optical intensity and photoresist profile on a silicon substrate.

Exposure is typically specified as energy per unit area, or dose. The intensity of the beam reaching the photomask is usually used to calculate this dose. In this case, since the incident intensity was 10 mW/cm^2, the doses are 50, 150, 300, and 450 mJ/cm^2, respectively, for each exposure column in Figure 6.43. Based on these plots, one could conclude that an exposure dose of 300 mJ/cm^2 is the optimum value for this photoresist.

The next set of images in Figure 6.44 are for a silica glass substrate using the same beam and photoresist parameters.

The main difference between Figures 6.43 and 6.44 is the standing waves inside the photoresist film. In silicon, the substrate reflection is higher and the ripples are more prominent. These ripples can often be seen in the sidewall profiles of the developed photoresist. The photoresist sidewalls also show a similar outward slope but are less pronounced than with the glass substrate.

Figure 6.45 shows a scanning electron microscope (SEM) image of a developed photoresist profile on a silicon substrate, which corresponds closely to the 150 mJ exposure with a 30 s develop in Figure 6.43. We can also see the ripples along the sidewalls arising from the substrate reflection.

6.7.4.6 Bottom Antireflection Coatings

Silicon has a high reflection coefficient for UV wavelengths between 60% and 70% in the 365–193 nm wavelength range. As we just demonstrated in the previous section, this reflection will produce strong interference fringes inside the photoresist, which will be exhibited as ripples along the sidewalls. These could then lead to line edge roughness and problems with lift-off patterning of thin films. Furthermore, if an anti-node of the interference pattern coincides with the substrate/photoresist interface, it could lead to poor attachment of the photoresist to the substrate. Due to the short propagation distances, interference effects can be seen even with spatially incoherent light sources such as the mercury vapor lamp, but it gets more pronounced with laser light sources, such as the 193 nm ArF or 248 nm KrF lasers. Although a PEB will significantly reduce this interference effect, the diffusion that occurs during the bake step can also cause lateral diffusion, which can compromise the spatial resolution of the exposures. A better method is to eliminate the reflection in the first place by using an antireflection film between the substrate and the photoresist.

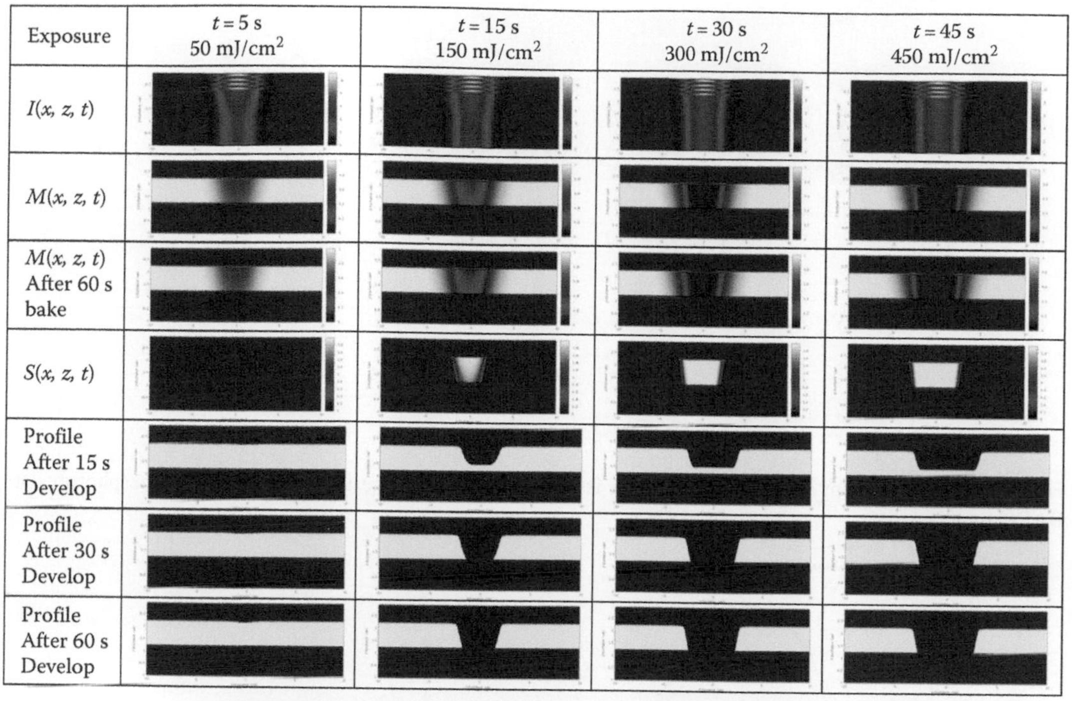
Exposure
t = 5 s
50 mJ/cm2
t = 15 s
150 mJ/cm2
t = 30 s
300 mJ/cm2
t = 45 s
450 mJ/cm2
I(x, z, t)
M(x, z, t)
M(x, z, t) After 60 s bake
S(x, z, t)
Profile After 15 s Develop
Profile After 30 s Develop
Profile After 60 s Develop

FIGURE 6.44 Evolution of the optical intensity and photoresist profile on a silica glass substrate.

FIGURE 6.45 Scanning electron microscope image of a developed photoresist profile on silicon showing the sidewall taper angle and the ripples along the sidewall due to the standing waves.

The antireflection film can be a vacuum-deposited inorganic film, or a more convenient spin-coated organic film. These are commonly known as bottom antireflection coating (BARC) films [47–49]. There is also another important distinction between general antireflection concepts and this case. The antireflection condition we are trying to achieve is not to eliminate the total reflection from the front surface of the photoresist; instead, we want the reflection into the photoresist/BARC interface to be zero. This is illustrated in Figure 6.46.

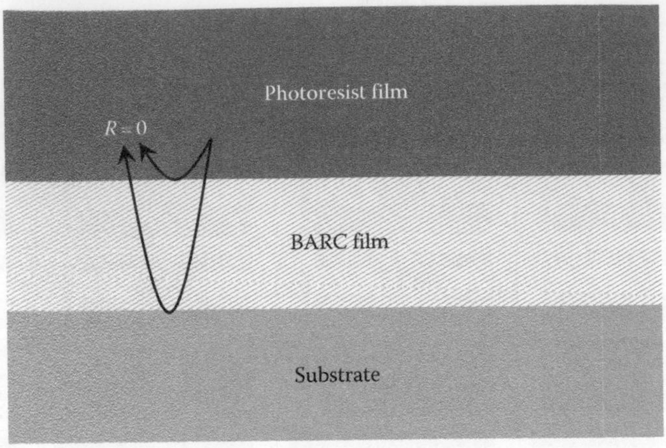

FIGURE 6.46 Bottom antireflection coating is designed to eliminate the reflection from the photoresist/BARC interface.

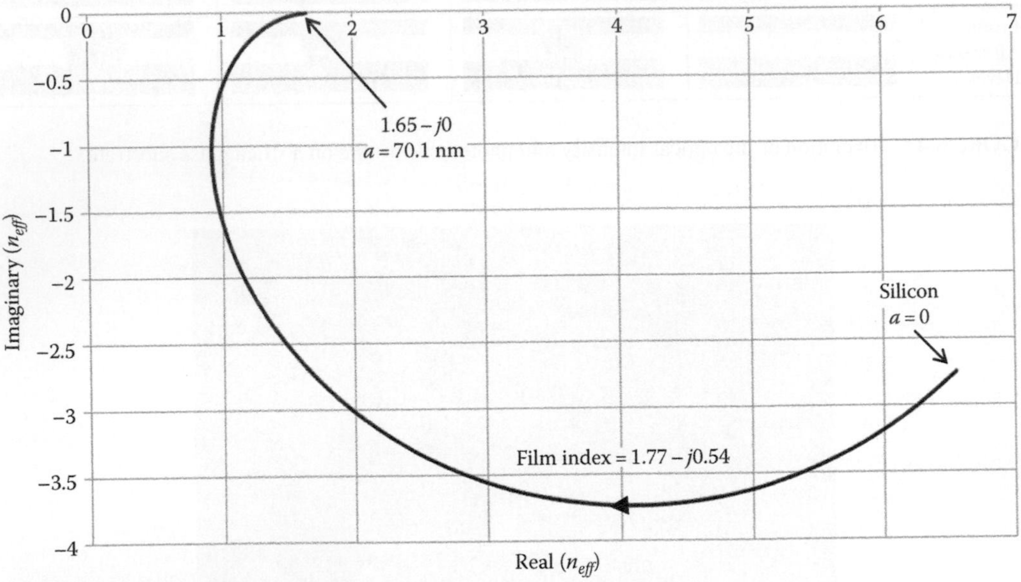

FIGURE 6.47 Contour of the complex n_{eff} for a single layer antireflection film on silicon at $\lambda = 365$ nm. Please refer to Chapter 4 for details on this method of calculation.

The refractive index of silicon is a complex number. Therefore, finding the antireflection condition is not as straightforward as it is with pure dielectrics. Nevertheless, the required film thickness and refractive index can be computed by using the multilayer effective index method discussed in Chapter 4. Figure 6.47 shows the effective index contour for a film on a silicon substrate at $\lambda = 365$ nm. The required film index to produce anti-reflection in the photoresist is $1.77 - j0.54$ and the required thickness is 72.1 nm. This produces a contour that starts at $6.55 - j2.75$ (refractive index of silicon) and ends at 1.65 (refractive index of the photoresist).

Figure 6.48a shows the calculated photoresist profile with the BARC layer after exposure and develop, and Figure 6.48b shows the same photoresist without the BARC layer. Both of these were calculated without any diffusion (D_0 and D_1 were assumed to be zero), so the only effect that is responsible for reducing the sidewall ripples can be attributed to the BARC layer.

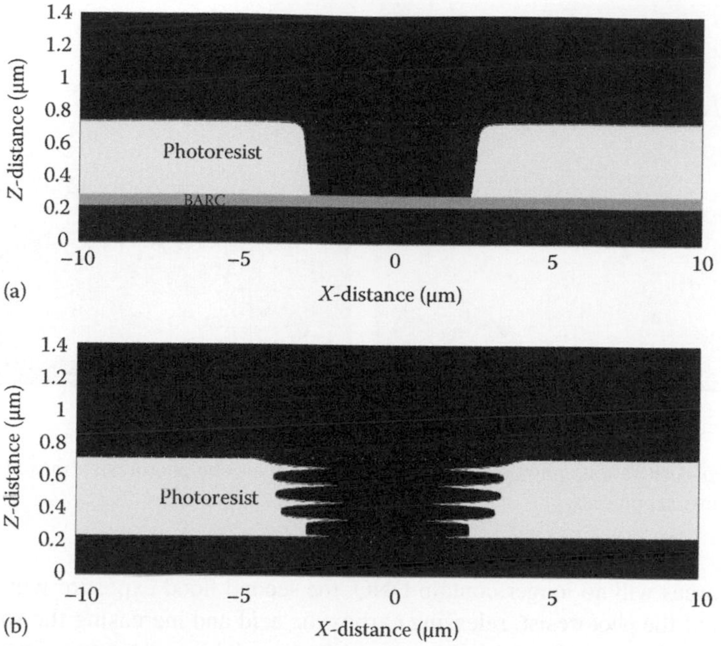

FIGURE 6.48 (a) Computed photoresist profile with the BARC layer and (b) computed photoresist profile without the BARC layer.

6.7.5 Negative-Tone Photoresists

Negative-tone photoresists are chemically very different than positive-tone photoresists. The UV exposure in negative photoresists produces a polymer cross-linking effect that increases the molecular weight of the resin [50]. This results in a reduced solubility of the resin after exposure. In Novolac + DNQ positive tone photoresists, the solubility moves from an inhibited state to an enhanced state, bringing several orders of magnitude change in solubility. In negative photoresists, the solubility moves from a moderately high value to a low value. The change in solubility is not as large as they are with positive photoresists. Therefore, positive photoresists are preferred for high fidelity and high image contrast compared to negative photoresists. Nevertheless, negative photoresists do have other advantages, such as better adhesion, better structural integrity, and lower plasma erosion rates. Most importantly, they have inwardly sloped sidewall angles that are more favorable for lift-off patterning processes.

In some cases, negative-tone photoresists, due to their cross-linking ability, are used as permanent structures rather than as a sacrificial film. An example is the SU-8 photoresist. This is an epoxy that is cured by UV and can be used just like any other photoresist. However, once it is exposed and cross-linked, it becomes a hard plastic with good structural properties. Hence, it is most often used as a permanent structure in MEMS and for the fabrication of microfluidic devices.

6.7.6 Image Reversal

In some cases, it is necessary to create a negative image on a positive photoresist without changing the polarity of the photomask or switching to a negative-tone photoresist. Several manufacturers offer photoresists that are capable of reversing the polarity of the image by performing a few additional steps. After the initial exposure (with a mask), the wafer is baked at an elevated temperature. The photoresist is designed with amine salts that decompose at the elevated bake temperature and neutralize the carboxylic acid that was produced during the exposure. This reverts the solubility of the resin back to its low value. Then, a flood exposure is performed (without a photomask). Since

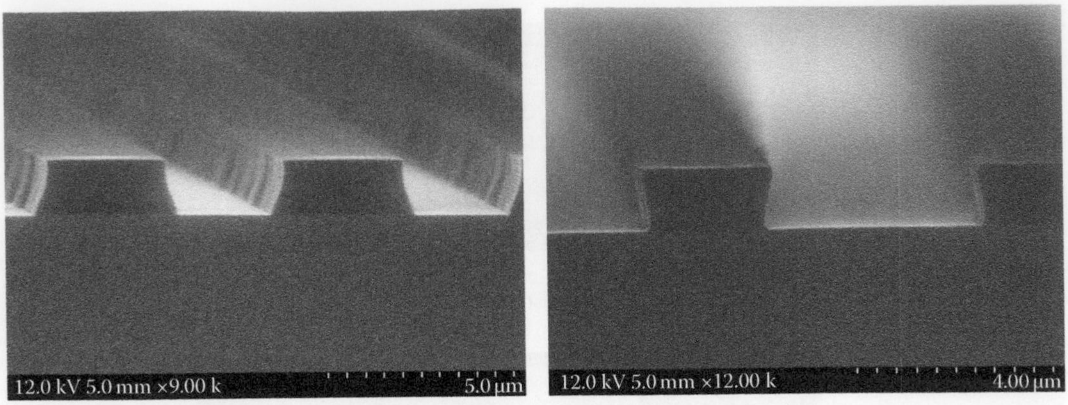

(a) (b)

FIGURE 6.49 (a) Positive-tone photoresist lines and (b) positive-tone photoresist lines after the anhydrous ammonia image reversal process.

the neutralized areas will no longer contain DNQ, the second flood exposure will only expose the remaining areas of the photoresist, releasing carboxylic acid and increasing the solubility in those areas. This renders the areas that initially exposed first insoluble and the remaining areas soluble. The end result is a pattern that will have the opposite polarity to a normal exposure process using a positive-tone photoresist.

This decarboxylation process can also be done with a conventional positive-tone photoresist without the embedded amine salts. The substrate, after the first UV exposure, is placed in a heated oven with anhydrous ammonia gas. The ammonia diffuses into the photoresist film and neutralizes the carboxylic acid. A subsequent flood exposure of UV is then performed to expose the remaining areas and achieve the negative-tone image [51,52].

Figure 6.49a shows a positive-tone photoresist profile, and Figure 6.49b shows the same photoresist with the same photomask after the anhydrous ammonia image reversal process.

6.7.7 Substrate Priming

Priming is the process of preparing the substrate for the application of photoresist [53–56]. The main consideration is the adhesion of the photoresist to the substrate. Organosilanes are commonly used as "adhesion promoters." They contain an inorganic reactive component at one end of the molecule and an organic reactive component at the other end. Hence, they act as a coupling agent that turns an inorganic surface such as silicon or a metal to appear as if it is at least partially organic. This organic surface can perform a number of functions depending on the application, including water repellency, improved wettability of certain resins, and ability to form chemical bonds with certain organic molecules. Organosilane coupling agents are widely used in industrial processes involving epoxies, resins, metals, and glasses.

In photolithography, the most common organosilane is hexamethyldisilazane (HMDS), and it is used to increase the water repellent property of the substrate surface. In fact, the term "adhesion promoter" is a misnomer because it does very little to increase the adhesion between the photoresist and the substrate. It actually *decreases* the adhesion strength [53]. What it does, however, is to prevent the aqueous developer from penetrating between the photoresist and the substrate. This is illustrated in Figure 6.50, where the left figure shows a properly treated silicon surface and the right figure shows a hydrophilic silicon surface that allows the penetration of the aqueous developer. Excessive penetration will result in patterns lifting off in the developer solution, which can sometimes be misinterpreted as being due to an excessive exposure dose.

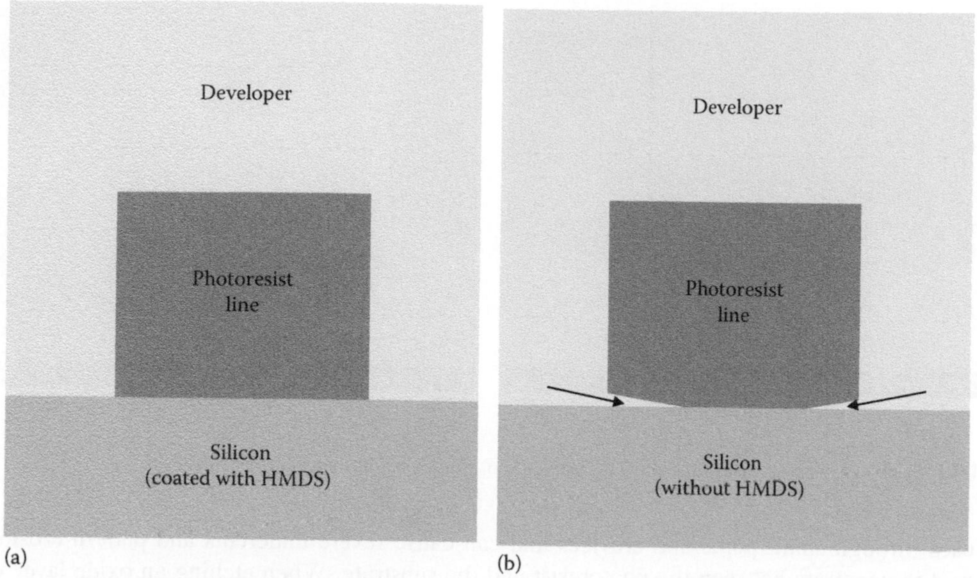

FIGURE 6.50 Effect of substrate priming with HMDS. (a) The development of a properly treated silicon surface and (b) the effect of an excessively hydrophilic surface.

Since HMDS molecules bind to SiO_2 more than to silicon, a slightly oxidized surface tends to work better than silicon that has been cleaned with BOE. An oxygen plasma cleaning step (which leaves an oxidized surface) followed immediately with the application of HMDS works well. Since ambient moisture will quickly bond to the energetic surface, the plasma-treated wafer should not be allowed to sit in air for too long. The treatment works best if the surface is fully dehydrated before the HMDS is applied. Alternatively, the surface can also be dehydrated by baking at an elevated temperature or rinsing it in IPA or performing the AMI cleaning process.

HMDS is a liquid at room temperature. It is most commonly applied by spin coating, followed by a bake step to remove excess HMDS from the surface. However, when done in ambient air, the entrapment of moisture between HMDS layer and the substrate is inevitable. This will compromise the attachment of the organosilane molecules to the substrate. This problem is mostly evidenced by inconsistent photoresist adhesion and reproducibility problems. A better technique is to apply the HMDS in the vapor phase in a vacuum oven. The vacuum in combination with the high temperature can be used to completely dehydrate the substrate. The dehydration can be enhanced by purging with dry nitrogen and then drawing a vacuum and repeating this pump–purge cycle a few times. Then, the HMDS is released into the chamber without exposing the substrate to air [56]. HMDS will reach a vapor pressure of about 20 Torr when the liquid source is held at room temperature. The length of the exposure time will affect the number of monolayers that settle on the wafer. The goal is to achieve less than one monolayer of accumulation that results in a slightly hydrophobic surface, with a water droplet contact angle of 50°–70°. Excessive accumulation of HMDS will result in the surface becoming too hydrophobic. Since HMDS actually reduces the adhesion strength between the substrate and photoresist, this will result in poor wetting of the substrate during spin coating, and in some extreme cases, no film will form during spin coating. Therefore, the wafer priming process needs to be optimized for each substrate type to achieve a consistent water droplet contact angle. For instance, silicon, silicon dioxide, and silicon nitride all require slightly different treatment times to achieve the same level of hydrophobicity. Metals can also be treated with HMDS, but the attachment properties of the organosilane to each metal type have to be individually examined.

Surface priming is important not only to ensure the survivability of the photoresist during development but also during subsequent wet processing. For example, hydrofluoric (HF) acid quickly

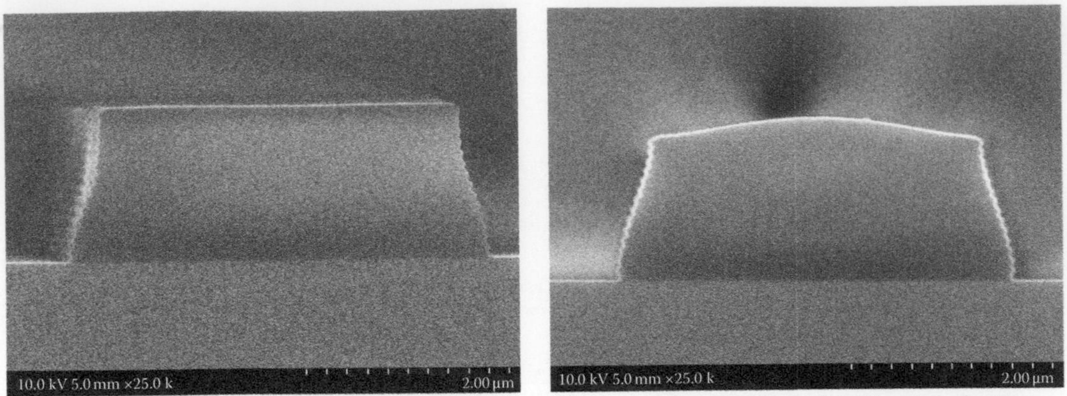

(a) Original resist pattern (b) After 120°C Hard Bake for 90s

FIGURE 6.51 A photoresist pattern (a) before and (b) after hard bake at 120°C for 90 s.

diffuses through small pores and crevices and can cause severe undercuts and pattern erosion if allowed to penetrate between the photoresist and the substrate. When etching an oxide layer with BOE with a patterned photoresist layer as an etch mask, a properly treated hydrophobic oxide surface will ensure that the HF will not penetrate and attack the oxide under the photoresist film.

It should also be noted that HMDS treatment does very little to improve the survivability of photoresists that do not use aqueous developers. For instance, SU-8, which is a negative-tone epoxy resist, uses a mixture of acetates as the developer. The application of HMDS in this case does not improve the survivability of the resist in the acetate developer. In fact, HMDS treatment has been noted to worsen the problem when used with SU-8. Instead, it is better to treat the substrate with an oxygen or argon plasma and then immediately apply the SU-8 film to improve the adhesion between SU-8 and the substrate.

6.7.8 Hard Bake

Hard baking will increase the chemical resistance and structural strength of the photoresist after the pattern has been fully developed. It is a baking step typically done at a temperature of 110°C–130°C. This is an optional step depending on what is to be done to the photoresist afterward. Hard baking will reduce the erosion rate of the photoresist during a plasma etch. It will also increase the surface adhesion and improve its performance during a wet chemical etch. However, baking close to the glass transition temperature of the resist will result in some reflowing and rounding of the edges. This is shown in Figure 6.51. This can, in some cases, be beneficial for reducing the sidewall roughness, but it will result in some loss of resolution. Hard baking removes all the remaining solvents in the photoresist and will significantly reduce outgassing in a vacuum. Instead of hard baking, the photoresist can also be cured under a DUV light source (such as a low-pressure mercury lamp emitting at 254 nm) to achieve a low plasma erosion rate and greater thermal stability [57,58]. After the hard bake, the photoresist becomes very difficult to remove, and aggressive organic removal methods may have to be employed.

6.8 SU-8 PHOTORESIST

SU-8 is a negative-tone chemically amplified photoresist, developed by IBM, that cross-links and hardens when exposed to UV light followed by a heating step [59]. It is based on a chemical known as EPON resin SU-8. It is combined with a photoacid generator and is dissolved in an organic solvent such as cyclopentanone [60–62]. The quantity of the solvent determines the viscosity of the film and hence the thickness of the spin-coated film. Upon exposure to UV light, the photoacid generator produces an acid in the exposed areas of the film. During the heating step, this acid acts as a catalyst for the cross-linking reaction of the resin. The heating step also regenerates more acid in those areas through a chemical amplification process, resulting in a significantly higher sensitivity.

SU-8 is used to make patterned structures with film thickness ranging from a few nanometers to as thick as 1 mm. One of the distinguishing characters of SU-8 compared to other photoresists is that it can be used as a structural film. Most other photoresists are used only as a sacrificial film. Fully cross-linked SU-8 is resistant to many chemicals and solvents. It is also mechanically strong and has low friction. As a result, SU-8 is widely used in MEMS and microfluidic device fabrication.

SU-8 with different viscosity levels are sold as different products. SU-8-2015 spins to a nominal thickness of 15 μm, SU-8-2050 spins to 50 μm, and SU-8-2100 spins to 100 μm. Obviously the spin speed can be adjusted to increase or decrease the thickness from these nominal values.

The optical absorption of SU-8 is very low compared to other photoresists. Despite its low absorption, the chemical amplification allows the overall sensitivity of the resist to be high. This makes it possible to expose very thick layers without suffering from the pronounced sidewall taper that would otherwise affect the profile of the photoresist.

The development is performed in an organic solvent propylene glycol monomethyl ether acetate (PGMEA). This dissolves the un-cross-linked portions of the SU-8, leaving the exposed pattern intact. Since this is not an aqueous-based developer, priming the substrate with HMDS does not necessarily lead to an increased survivability in the developer. In fact, results seem to suggest a degradation in adhesion with HMDS treatment [63]. There does not appear to be an equivalent surface treatment for increasing the survivability of SU-8. The general technique to increase the adhesion of SU-8 is to treat the substrate with a plasma to increase its surface energy and then quickly apply the SU-8 before moisture settles on the substrate. Figure 6.52 shows a 15 μm thick SU-8 that has been patterned with an array of posts and then deposited over with a thin layer of metal. The inward sidewall taper characteristic of the negative-tone photoresists is clearly evident in these structures.

During cross-linking, SU-8 undergoes a volume contraction by as much as 6%–10% [64]. This results in a significant source of film stress and could lead to tensile delamination and stress cracks, especially around sharp corners. Therefore, the PEB has to be performed slowly. For thicker SU-8 films, the temperature needs to be ramped up and down very slowly to reduce the cross-linking rate and allow more time for the stresses to relax. Figure 6.53 shows three images of 200 μm thick SU-8 structures. The left image shows stress cracks extending a significant distance into the SU-8, the

FIGURE 6.52 SU-8 structures on a silicon substrate with a metal deposition above the SU-8.

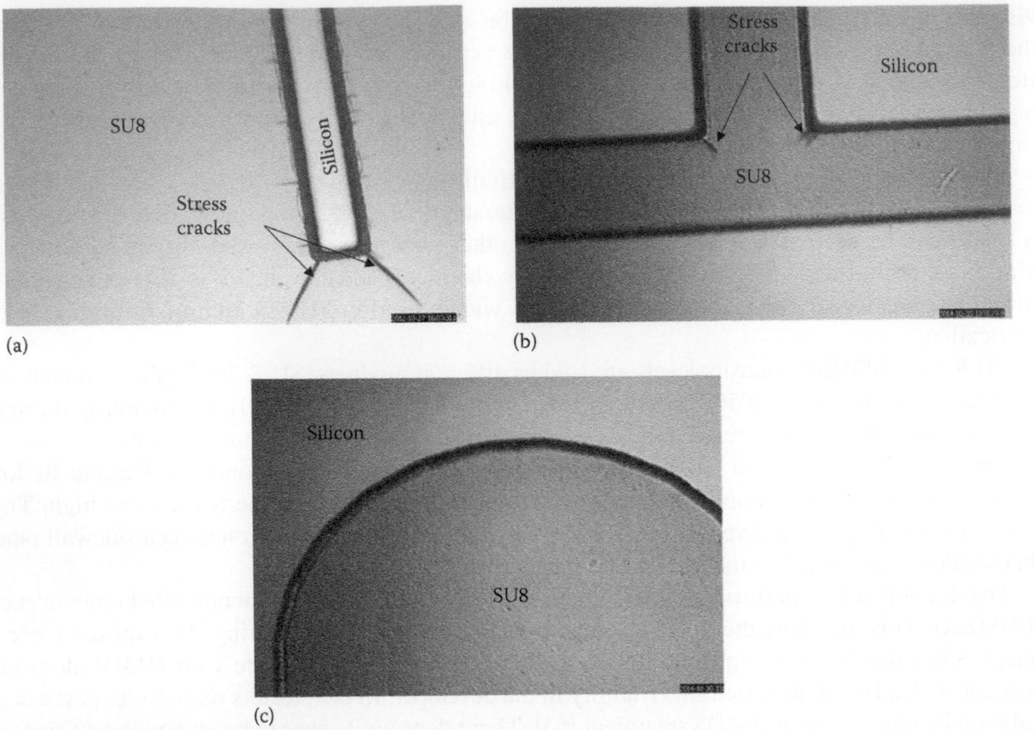

FIGURE 6.53 Appearance of stress cracks around SU-8 structures. (a) Significant stress cracks at the corners; (b) moderate cracks at the corners; (c) almost no cracks by avoiding sharp corners.

middle figures shows small cracks, and the right figure avoids most of these issues by curving the edges gradually.

The processing steps used for SU-8 can be roughly categorized as

1. Substrate preparation
2. Spin coat SU-8
3. Soft bake at 95°C to remove solvents
4. Expose with UV
5. PEB at 95°C to promote cross-linking and chemical amplification
6. Develop

A significant source of difficulty arises during the preparation of thick SU-8 layers. For thick films, of the order of 100 μm or more, the viscosity will be so great that any entrapped bubbles in the SU-8 will not naturally dissipate during spin coating process. Although it is possible to degas the SU-8 prior to spin coating, the dispensing process itself could introduce bubbles. These bubbles will usually show up during the soft bake step as they expand due to the temperature rise. Several techniques have been suggested for this, and the following is a process used in the author's laboratory based on the techniques presented in [65].

In this process, the substrate is cleaned and treated with an O_2 plasma to activate the surface. Then, without delay, the thick SU-8 (SU8-2050 or SU8-2100) is spread on the substrate with a small spatula. The viscosity of SU-8-2100 film is equivalent to that of a thick paste, so it is much easier to scoop up and spread rather than pour onto the wafer. Then the wafer is placed on a spinner and spun at the desired speed. At this stage, there will be a significant amount of trapped bubbles in the film. Then, the wafer is placed on a perfectly leveled hot plate (without heat), and the solvent

(a) (b)

FIGURE 6.54 (a) An ultrasmooth SU-8 film (200 μm thick) created using the solvent spray process. (b) A completed structure after exposure, postexposure bake, and develop.

cyclopentanone is sprayed onto the wafer from a fine mist spray bottle. The goal is to wet the surface adequately and reduce the viscosity such that the trapped bubbles will naturally rise and dissipate. The spray must be a fine mist so that the incident liquid stream does not create deformations in the film. Then, taking care not to move the wafer, the hot plate is slowly ramped up to 95°C and then ramped down. The ramp-up rate should be slow enough (typically 2–3°C/min) to evaporate the solvent without allowing it to boiling off, which could cause craters in the film. This process consistently produces ultrasmooth SU-8 films free of bubbles, as well as edge beads. In fact, the edge bead turns into a downwardly sloped profile, representative of a liquid meniscus. Representative results are shown in Figure 6.54. The left image shows an ultrasmooth SU-8-2100 spin-coated 200 μm thick on a silicon wafer. Notice the edge of the film, which has a downwardly sloped profile rather than the conventional edge bead that projects upward. The right image shows a completed structure using this SU-8 film.

6.9 PATTERNING BY LITHOGRAPHY

As stated earlier, in the majority of applications, the photoresist is used as a sacrificial film. The photoresist is exposed and developed, then the pattern is transferred to another film, and then the photoresist is discarded by rinsing off with a solvent. In some specific cases, such as SU-8, the photoresist is retained as a structural film, but these are exceptions.

There are two main approaches used in the pattern transfer process. One approach is to use the photoresist as a chemical barrier to etch an underlying film. This is how the copper traces are patterned on printed circuit boards (PCBs). The second approach is to use the photoresist as a soluble material to lift off an overlying film. These two approaches are commonly known as etch-down or lift-off methods, respectively.

6.9.1 Etch-Down Patterning

In the etch-down method, the film is deposited on the substrate first using PVD or CVD methods. The deposition method is not relevant for the subsequent photolithographic process. Then, the photoresist film is applied on top of this film, exposed, and developed. The patterned photoresist is then used as a chemical barrier to etch the underlying film through the openings in the photoresist. This is illustrated in Figure 6.55. This is typically done with a wet chemical etch. Metals and dielectrics can be etched with many acidic solutions, and their reaction chemistries can be found in

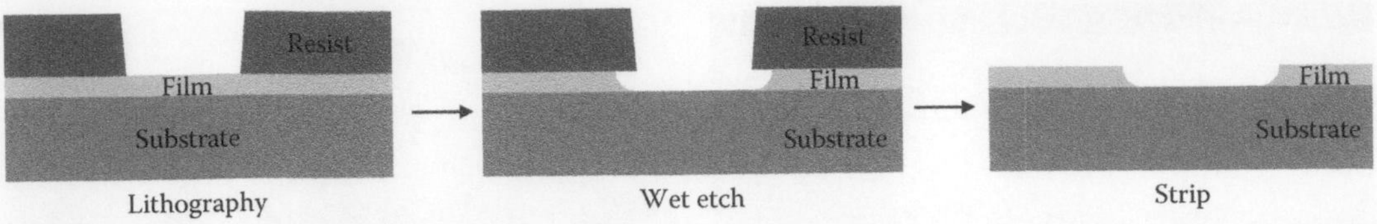

FIGURE 6.55 Etch down process with a photoresist mask.

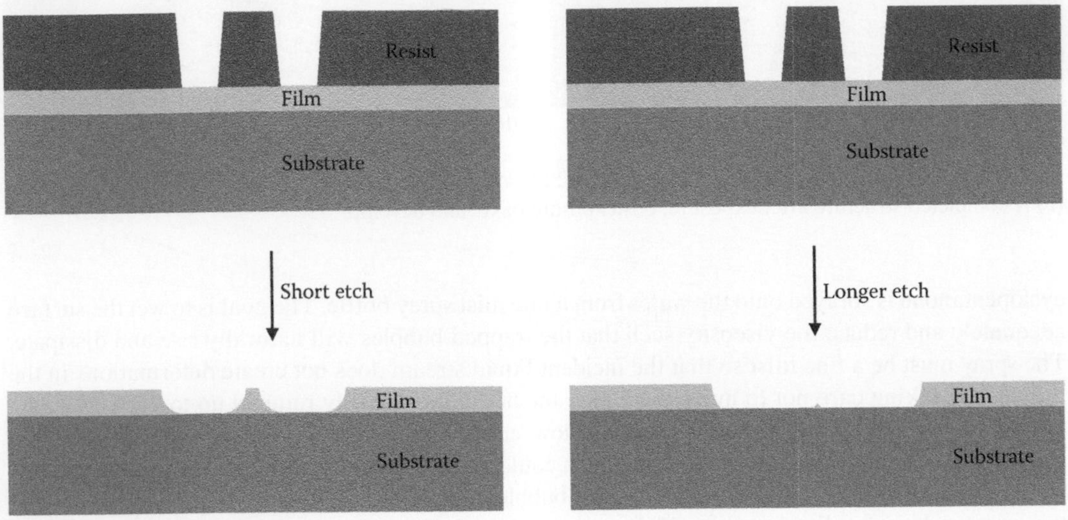

FIGURE 6.56 Lateral undercut and loss of resolution during wet etching.

the literature [66]. However, a few key aspects make this process different from a general chemical etching process. First, the etchant should not dissolve, penetrate, or react with the photoresist film. Second, the etchant should not react with the substrate. Finally, the required temperature should ideally be well below the softening temperature of the photoresist, which is normally around 120°C. Since silicon is the most commonly used substrate in micro- and nanofabrication, and Novolac is the most commonly used photoresist resin, specific chemistries have been developed by microfabrication specialists to etch various metals and dielectrics that do not affect silicon or the photoresist. For example, a ratio of $20H_2O$:$1H_2O_2$:1HF will etch titanium and 5% I_2:10% KI:85% H_2O will etch gold [66]. If a substrate other than silicon is used, or if an unconventional photoresist is used, one has to evaluate if the same etchant will work with that particular combination of substrate and photoresist. A good reference for etch chemistries commonly used in device fabrication cleanrooms can be found in [67]. These etchants could be mixed in the laboratory or purchased as premixed ready-to-use etchants [68].

Another characteristic property of wet chemical etching is the appearance of rounded edges on the sidewall profiles, as shown in Figure 6.55. This arises due to the isotropic etch rates in a wet chemical—that is, the etch proceeds at equal rates along all directions. Since the top part of the film will be exposed to the etchant longer than the bottom part, the top portions will experience a greater lateral etch than the bottom parts. As the etch process continues, the width of the film under the photoresist line will continue to shrink and may eventually disappear if it is allowed to etch for too long. This situation is illustrated in Figure 6.56. If the film thickness is t, the width of the photoresist line has to be greater than $2t$ to produce a distinct line after the etch runs to completion. In practice, the etch times are difficult to control precisely; therefore, the width of the lines has to be much larger, about $5t$–$10t$ to produce usable results.

6.9.2 Lift-Off Patterning

In the lift-off method, first, the photoresist film is spin coated, exposed, and developed. Then the film to be patterned is deposited on top of the photoresist, typically using PVD, although a CVD process such as PECVD or ALD could also be used as long as the deposition temperature is lower than the softening temperature of the photoresist, and the solvent has access to the photoresist through the sidewall or other openings. Then, the photoresist is dissolved in a photoresist stripper or simply with a solvent such as acetone. As the photoresist dissolves, the film above it will be released and float away, and the remaining areas of the film will stay attached to the substrate. This is illustrated in Figure 6.57.

The advantage of the lift-off process is that the etch chemistry of the film is no longer relevant. Any metal or dielectric film can be patterned by lift-off, even those that cannot be easily etched by chemicals. However, the solvent has to be able to penetrate the film to reach the underlying photoresist. With most thin films, the inherent voids, cracks, and defects in the deposited film will allow the solvent to penetrate and reach the photoresist layer. But this can become a problem with thicker or very conformal films. Sonication of the solvent bath can help by creating microcracks in the top film to allow the solvents to diffuse through. But sonication must be used with caution because it can also detach the film from the substrate.

The sidewall profile of the photoresist has the most significant effect on the quality of the lift-off. A clean lift-off occurs when there is a physical separation between the film above the photoresist and the film on the substrate. This leaves a gap for the solvents to reach the photoresist layer very easily and separate the two layers, as indicated in Figure 6.57. If the two layers are bridged, the lift-off will cause the film to tear. This tearing will produce irregular edges on the patterned film. Furthermore, if the film's adhesion to the substrate is poor, the tearing action can also lift the entire film off the substrate.

An inwardly tapered sidewall angle combined with a line-of-sight deposition will ensure that the sidewalls will remain uncoated. As discussed previously, positive-tone photoresists produce an outwardly tapered sidewall angle. Depending on the extent of this taper and the thickness of the film being deposited, positive resists may be unsuitable for lift-off. Negative-tone photoresist, or a positive photoresist with image reversal, works better. In general, image reversal produces better results because cross-linked negative photoresists are more difficult to dissolve in a solvent compared to unexposed positive photoresists. Example SEM images of positive- and negative-tone photoresists with a deposited metal are shown in Figure 6.58. The images in Figures 6.58a and 6.58b correspond to a positive-tone photoresist, and Figures 6.58c and 6.58d correspond to a positive-tone photoresist that has undergone an anhydrous ammonia image reversal.

Even with a proper sidewall angle, the deposition has to be nearly line-of-sight to prevent depositions from accumulating on the sidewalls. Therefore, thermal evaporation under high vacuum is the ideal choice for lift-off patterning. Sputtering can also work if the pressures are kept low enough to reduce scattering of the deposition species in the gas phase. CVD is less likely to work with lift-off because of its conformal deposition properties and also due to the high temperatures. Photoresist can reflow and become excessively hardened if exposed to temperatures higher than 120°C. ALD films can also be patterned by lift-off as long as some cracks can be induced in the film to facilitate solvent penetration.

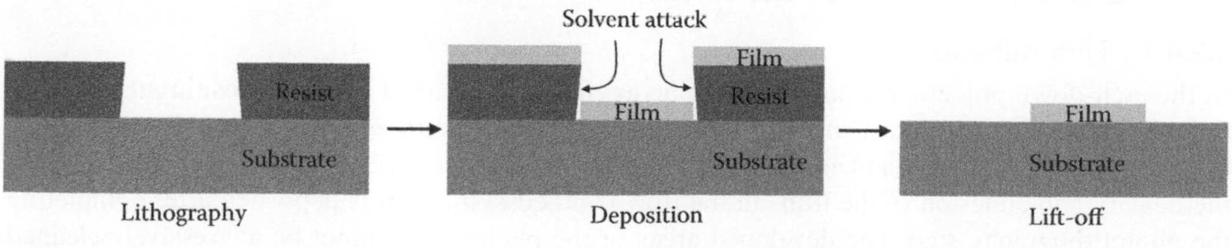

FIGURE 6.57 Lift-off patterning.

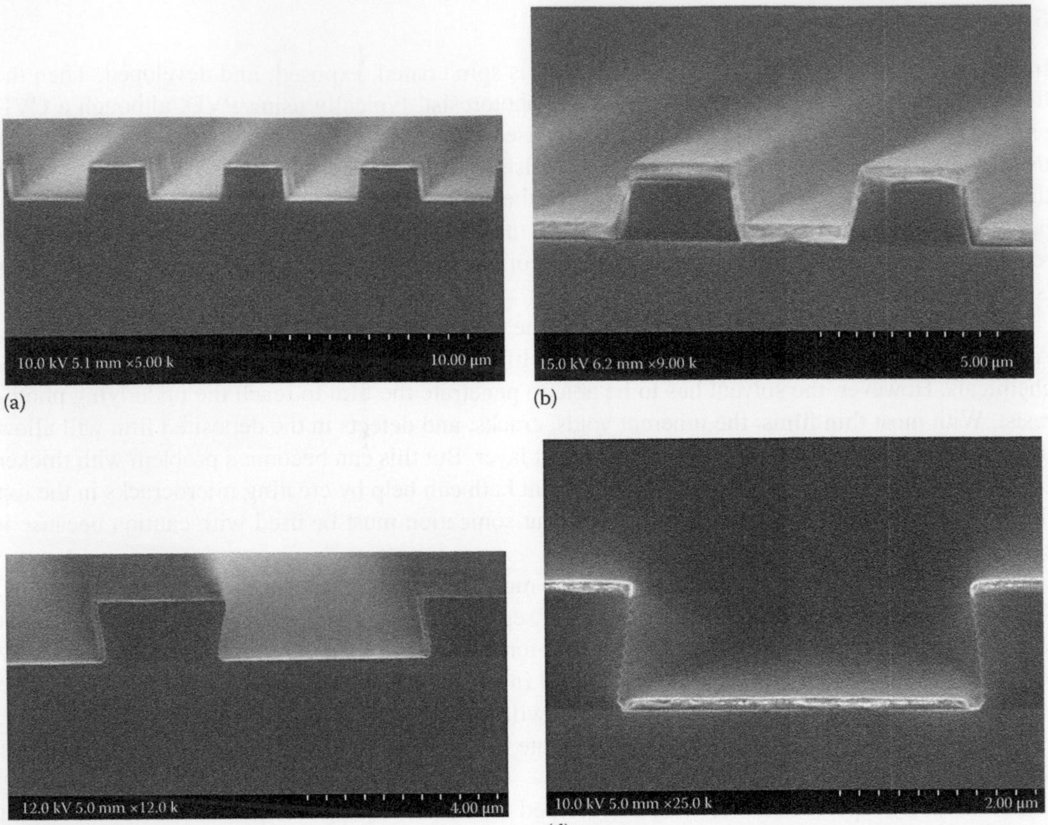

FIGURE 6.58 Metal deposition over photoresist profiles for lift-off. (a) and (b) correspond to positive tone photoresist, and (c) and (d) correspond to the same photoresist with an anhydrous ammonia image reversal.

6.9.3 Bilayer Lift-Off

In this process, two layers are spin coated. The first layer known as the lift-off resist (LOR) contains a resin without any photoactive compound (PAC). The second layer is the normal photoresist. After exposure, the photoresist layer is developed. The LOR is designed to be soluble in the same developer without requiring an optical exposure. Therefore, once the opening in the top photoresist layer fully develops, the bottom layer will dissolve isotropically creating an undercut below the photoresist layer [69]. This configuration, illustrated in Figure 6.59, is beneficial for lift-off because even if the deposition is not perfectly line-of-slight, it will still leave some areas under the overhangs uncoated. It also enables thicker films to be patterned. However, due to the isotropic undercut, the minimum achievable linewidths are somewhat compromised.

6.9.4 Etch-Down versus Lift-Off Patterning

6.9.4.1 Film Adhesion

In the etch-down process, the substrate can be aggressively cleaned before depositing the film on the substrate. The deposition can also be done under any temperature or pressure, PVD or CVD, to create the best film quality and adhesion. Ion-assisted deposition (IAD) can also be used to improve the density and adhesion of the film. In the lift-off process, the film is deposited after completing the photolithography step. The developed areas of the photoresist cannot be aggressively cleaned without destroying or compromising the photoresist structures. High temperatures or energetic ions

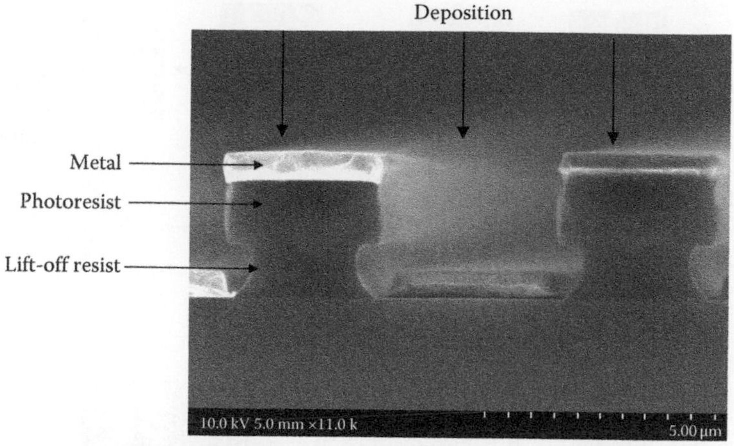

FIGURE 6.59 Bilayer lift-off process using a lift-off resist and an imaging resist with a metal that was deposited using thermal evaporation.

cannot be used during the film deposition. The pressure also has to be low to allow line-of-sight deposition. In most cases, an oxygen plasma clean (known as the ashing process) is the only cleaning step performed before the deposition. This is necessary to remove small traces of organic contamination that may be remaining in the developed areas. However, this has to be done cautiously to prevent the other areas of the photoresist from being eroded by the plasma.

6.9.4.2 Etch Chemistry

As already mentioned, the chemistry of the film is irrelevant in lift-off, except that it needs to be able to withstand the solvents used to dissolve the photoresist. This is often the deciding factor in the choice between lift-off and etch-down. For difficult-to-etch films, lift-off is an ideal choice. Another factor is the susceptibility of the substrate to the etch chemicals. Most of the etch chemistries have been developed for silicon substrate and Novolac resins. When GaAs, InP, or other substrates are used, one has to evaluate the compatibility of the etchants with these substrates. If there is any doubt of their compatibility, a lift-off process is often a safe choice.

6.9.4.3 Linewidth Control

The width of a line is difficult to maintain in a wet chemical etch. This is due to the isotropic nature of wet chemical processes. Plasma etching is an alternative to wet chemical etching because it can produce anisotropic etch profiles (Figure 6.60). For small features, this is the most common approach taken. The ions in a plasma are directed toward the cathode at nearly normal incidence. If the substrate to be etched is placed on the cathode, the etching will become directional with very little lateral etching. The details of etching processes will be discussed in Chapter 7. Lift-off is more forgiving toward line width control than etch-down. Extremely narrow lines can be lifted-off, assuming the photolithography process is able to produce lines with suitable sidewall profiles.

6.9.4.4 Film Thickness

Whether etch-down or lift-off, very thin films are much easier to pattern than thicker films. Assuming an isotropic wet etch, greater film thickness will require wider linewidths. In lift-off, the photoresist thickness places a limit on the maximum film thickness. For effective lift-off, the film thickness has to be less than about a third of the photoresist thickness. This is due to the inevitable sidewall coating that will occur due to scattering near the substrate, despite a high vacuum deposition process. For thicker films, the sidewall coatings will also increase proportionally.

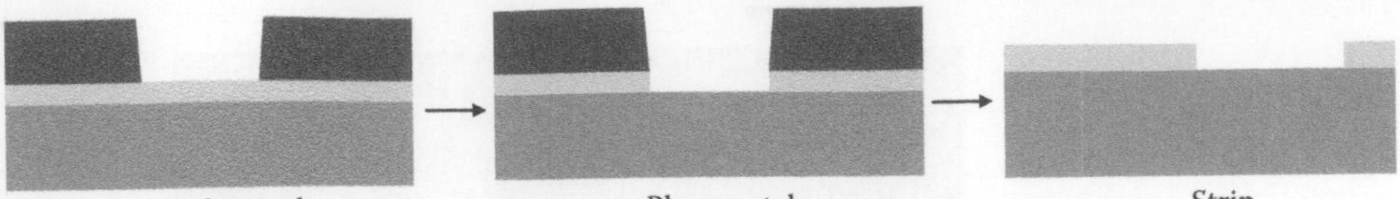

FIGURE 6.60 Plasma etching with a photoresist mask.

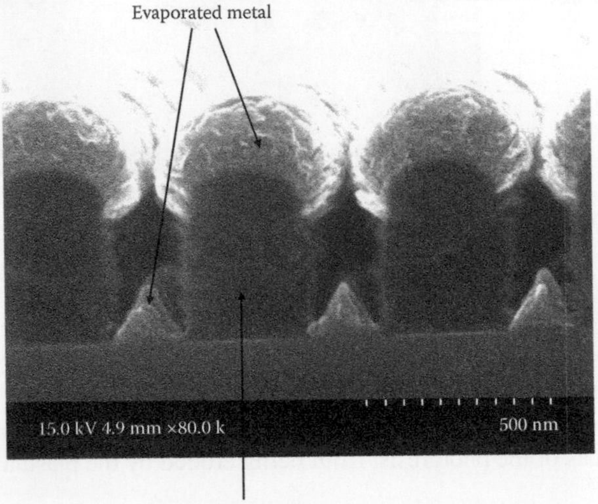

FIGURE 6.61 The effect of excessive outgassing during deposition over a pattern photoresist.

6.9.4.5 Outgassing

During vacuum deposition for lift-off, the photoresist can produce some outgassing when it is exposed to heat or plasma during the deposition. This can occur despite a beginning low base pressure in the process. The outgassing will increase the scattering rate of the depositing species and produce off-normal angular depositions. If the photoresist is not adequately hard baked, this can become a serious limitation. Increased sidewall depositions will decrease the photoresist opening sizes and eventually pinch off the openings. An example is illustrated in Figure 6.61. In this case, the lines are fairly narrow (about 250 nm), so even a small amount of sidewall build up will cause the deposition at the bottom of the trench to cease.

6.9.5 Patterning by Planarization

In addition to etch-down and lift-off, there is a third approach for patterning films. This approach is based on planarization and is also known as the Damascene process (after an ancient form of metal artwork from Damascus). In this approach, a hard dielectric such as silicon nitride is deposited and then etched down with a plasma using a photoresist layer. Then, the photoresist layer is stripped off leaving the patterned dielectric layer on the substrate. A soft metal film is then deposited on top of this dielectric to overfill the etched grooves. The exact metal thickness is not critical as long as it is greater than the groove depth. Then, the substrate is mechanically polished, with some chemicals to accelerate the process (known as chemical–mechanical planarization [CMP]). If the hardness of the dielectric is much greater than the metal, then the planarization will terminate when it reaches the dielectric layer. Only the metal inside the grooves will be left. This process is used in CMOS

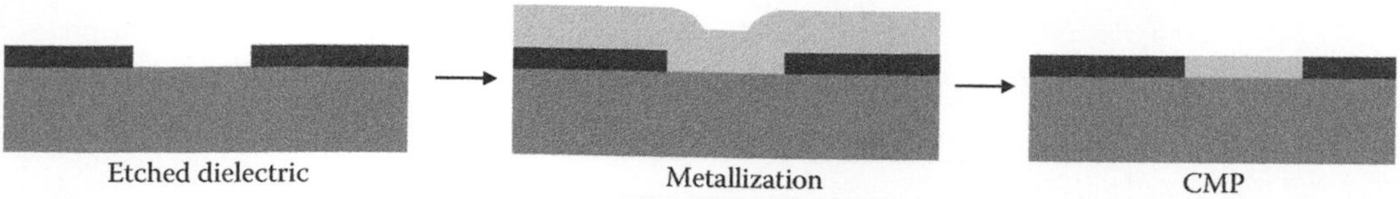

FIGURE 6.62 Patterning by chemical mechanical planarization.

electronics with copper interconnect metallization because plasma etching of copper is not as well developed as it is for aluminum [70]. The process is illustrated in Figure 6.62.

6.10 LASER INTERFERENCE LITHOGRAPHY

Laser interference lithography is sometimes described as the poor man's wafer-scale nanolithography system. It is a maskless optical projection system that can be used to expose periodic lines, grids, or lattices on a photoresist film. It can produce a half-pitch (HP) that is close to the theoretical limit of $\lambda/4n$ without requiring high-NA optics, highly flat substrates, and without any depth of focus or field size limitations. In principle, an entire wafer can be exposed with a single shot making it possible to easily scale up to large wafer sizes or high throughputs. Its main disadvantage is that it can only create repeating lines of the same structure. Therefore, it is not a replacement for a masked lithography system. But, for certain applications where periodic nanoscale structures are required, it is superior to contact or projection masked lithography. Examples of such applications include diffraction gratings, Bragg and distributed feedback reflectors, polarizers, and photonic crystals to name a few. Furthermore, since the periodic nature of the exposure arises from the coherence of the laser source, it will not have the line-to-line variations that can occur in other lithography methods due to beam spot variations or field stitching errors [71,72].

The exposure system consists of a UV laser beam that is split into two parts and recombined at the substrate at an incident angle of $\pm\theta$. The interference between the two incident beams creates periodic lines of bright and dark fringes that form the exposure pattern for the photoresist, as shown in Figure 6.63. The interference is between the horizontal components of the optical *k*-vectors, that is,

$$k_x = nk\sin\theta. \tag{6.60}$$

The two counterpropagating waves with $\pm k_x$ will produce a standing wave with a *k*-vector of

$$K = 2k_x = 2nk\sin\theta. \tag{6.61}$$

This will produce a spatial period of

$$\Lambda = \frac{2\pi}{K} = \frac{\lambda}{2n\sin\theta}. \tag{6.62}$$

In terms of HP, this can be expressed as

$$HP = \frac{\lambda}{4n\sin\theta} \tag{6.63}$$

which is identical to Equation 6.25 for a projection-masked lithography system. Unlike masked lithography systems, interference lithography relies on the spatial and temporal coherence of the beam, so only laser light sources can be used for this application. Whereas in a projection-masked system it is difficult to reach the theoretical limit of $k_1 = 0.25$ due to aberrations and nonidealities in the optical system, in interference lithography, it is always 0.25 due to the absence of any

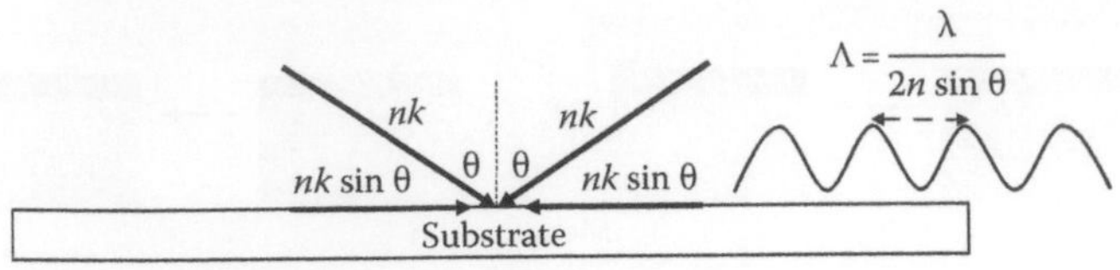

FIGURE 6.63 The interference of *k*-vectors in a two-beam interference lithography system.

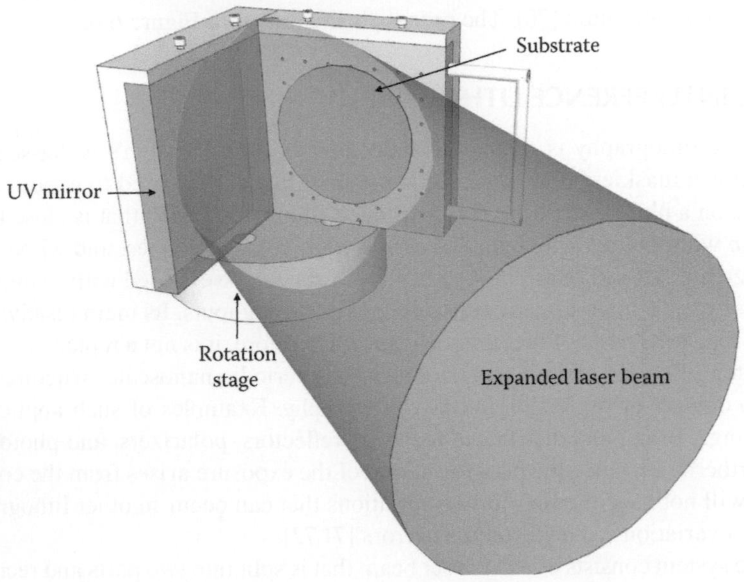

FIGURE 6.64 Lloyd's mirror configuration of an interference lithography setup.

projection lenses. From Equation 6.63, we can see that the smallest features are realized when $\theta = 90°$. However, using a plane wave illumination, referring to Figure 6.63, it is not possible to achieve this condition, so a more realistic limit is about $\theta = 75°$.

As a further enhancement, it is also possible to perform water-immersion exposure with interference lithography, which would reduce the HP by a factor of n [73,74]. In principle, using a 193 nm source and water immersion, it is possible to achieve the same ultimate HP of 33 nm as projection system, at a tiny fraction of the cost. While the lack of design flexibility may appear to be a serious limitation, it should be noted that even in masked projection systems, the best resolutions are almost always performed with a periodic array of parallel lines.

A few laser systems used in interference lithography are HeCd (325 nm), YAG (266 nm—quadrupled), KrF (248 nm), Ar (244 nm—doubled), and ArF (193 nm). Although it is possible to split the laser beam and recombine them from two separate beam paths, a more convenient and robust configuration is the Lloyd's mirror setup, as shown in Figure 6.64. It contains two rigid planes at right angles to each other, with a UV mirror on one plane and the substrate on the other plane. The laser beam is expanded and collimated to illuminate both planes simultaneously. Using simple geometry, we can show that the angle of incidence of the direct beam and the reflected beam on the substrate will be equal and opposite, and their intensities will be equal, as required for complete interference depicted in Figure 6.63. Furthermore, this angle can be controlled by rotating the whole stage around its vertical axis, which gives it the flexibility to easily change the pitch of the exposure pattern. Additionally, any platform vibrations will be common to both sides of the beam, canceling their effects.

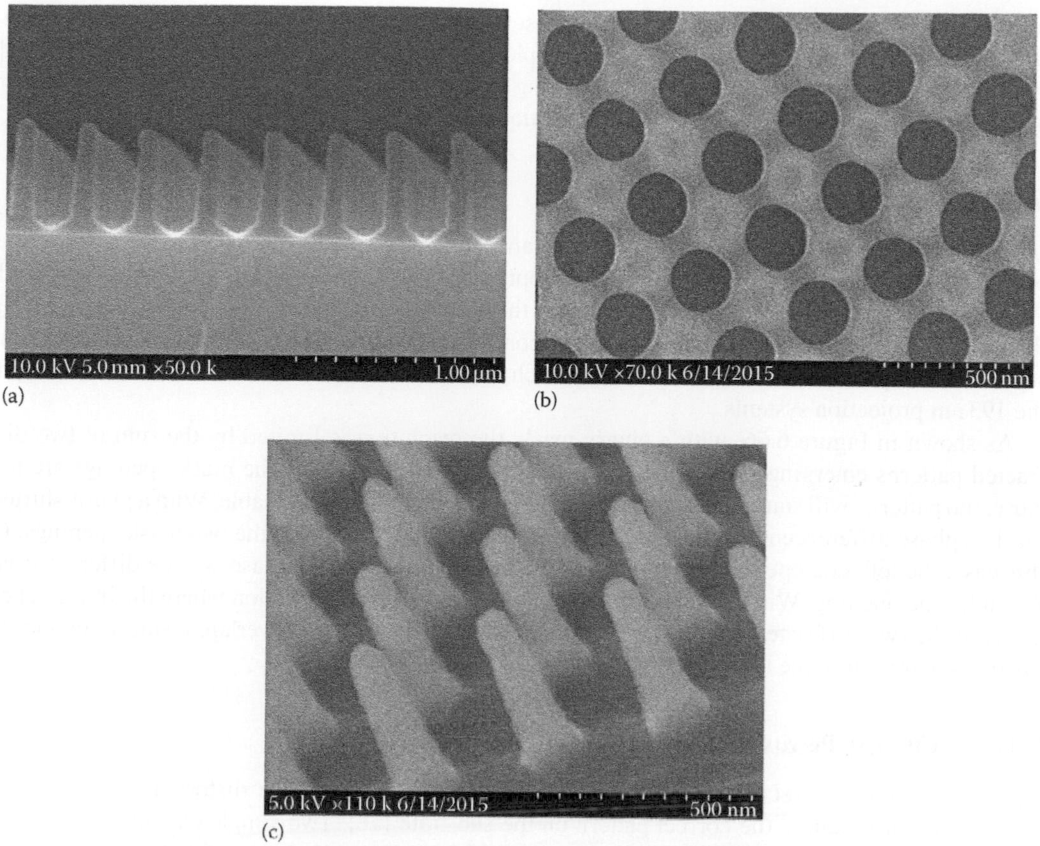

FIGURE 6.65 One-dimensional and two-dimensional photoresist patterns produced using deep-UV interference lithography system in the author's laboratory. (a) Periodic lines and spaces of 100 nm lines with 310 nm pitch, (b) 200 nm diameter holes on a 310 nm pitch two-dimensional grid, and (c) 100 nm posts on a 310 nm pitch two-dimensional grid.

In addition to the one-dimensional periodic lines, it is also possible to perform two or more consecutive exposures after rotating the substrate in-plane. This will allow two-dimensional grid patterns to be created, such as posts and holes. Figure 6.65 shows a few examples of one-dimensional and two-dimensional periodic structures on photoresist created using a 266 nm DUV laser interference lithography. The incident angle was 25°, which results in a pitch of 310 nm. By careful control of the exposure dose and resist polarity, it is possible to create isolated posts or holes. The resist thickness in this example is 500 nm with linewidths of 100 nm, which is a reasonably high aspect ratio. This is possible only because interference patterns do not have a depth of focus limitation. In principle, the depth of focus is infinite. This allows photoresist of any thickness to be exposed, making it possible to create very high aspect ratio profiles. In practice, however, the aspect ratios will be limited by the structural strength of the photoresist and the pattern's tendency to collapse or stick together during the wet development process.

6.11 RESOLUTION ENHANCEMENT TECHNIQUES

We discussed earlier that the smallest HP using 193 nm projection assuming perfect optics and water immersion is 33 nm. For over a decade, 193 nm ArF lasers have remained the mainstay of integrated circuit manufacturing. However, it does not mean the resolution has remained stagnant. In fact, the ultimate performance of the 193i system has marched on past 33 nm using a number of innovative resolution enhancement techniques. Some of these techniques include off-axis illumination, phase-shift masks, optical proximity corrections, multiple patterning and directed self-assembly [2,75,76]. Although these are used mostly in manufacturing systems, it is useful to have some knowledge of how they are implemented. A few of these techniques are briefly discussed next.

6.11.1 Phase-Shifted Masks

In a conventional mask, the pattern consists of transparent and opaque regions. These are known as binary masks, where only the amplitude of the optical wave passing through the mask is modified. Phase-shifting mask is based on modifying both the amplitude and phase of the optical field exiting the mask. This can provide an extra degree of control to manipulate the field intensity at the substrate. This requires a coherent illumination, such as a laser source, which is indeed the case with the 193 nm projection systems.

As shown in Figure 6.66, with a binary mask, the resolution is limited by the sum of two diffracted patterns emerging from adjacent openings in the mask [77]. If the mask openings are too close, the patterns will start to merge and eventually become indistinguishable. With a phase-shifted mask, a phase difference is introduced between the fields emerging from the two mask openings. In this case, the left side opening has a recess etched into it such that its phase will be different from the right side opening. With a π phase shift, it is possible to create a situation where the interference between the two diffracted patterns will subtract where the two fields overlap, resulting in a more distinct separation of the two patterns.

6.11.2 Optical Proximity Corrections

Optical proximity corrections (OPCs) involve intentionally designing a distorted pattern on the mask so that it produces the correct pattern on the substrate [78]. Two simple examples are shown in Figure 6.67, where small protrusions are introduced on the mask so that the diffracted pattern more closely approximates the desired shape. This unfortunately results in an increased resolution requirement on the photomask. These additional features on the mask are typically much smaller than the overall resolution limit of the system, and this will significantly increase the cost of the mask.

6.11.3 Self-Aligned Double Patterning

Also known as sidewall-assisted double patterning, this method is widely used in advanced lithography systems to effectively double the spatial resolution. In essence, it uses the two sidewalls of each line to create two new lines. This is done by first performing a lithography and transferring that into

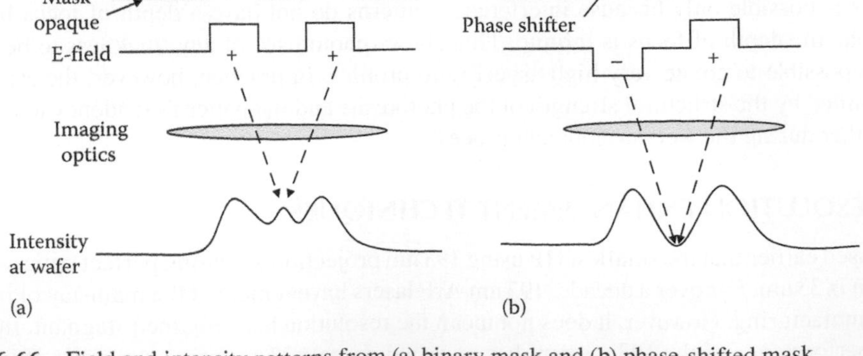

FIGURE 6.66 Field and intensity patterns from (a) binary mask and (b) phase-shifted mask.

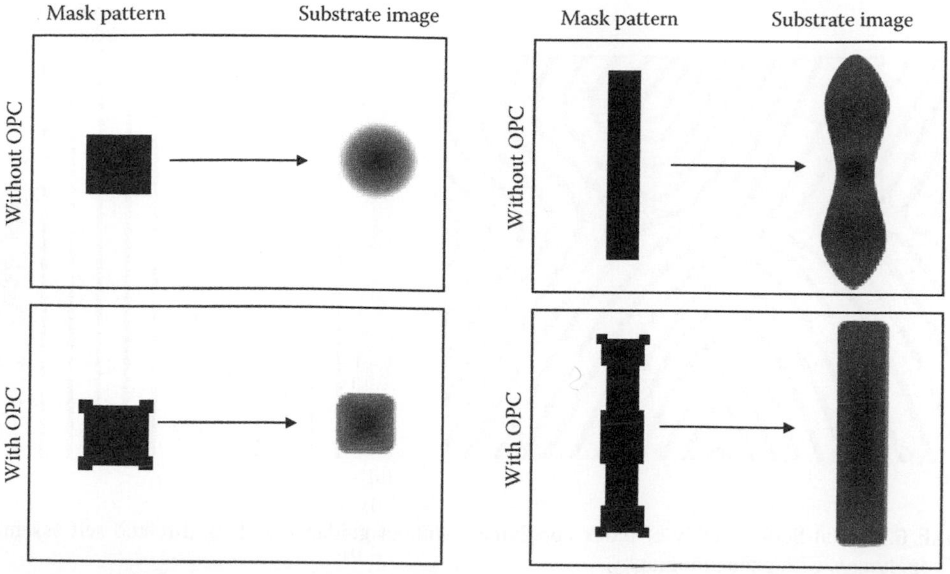

FIGURE 6.67 Examples of optical proximity corrections where the design on the mask is intentionally distorted to produce the desired representation on the substrate.

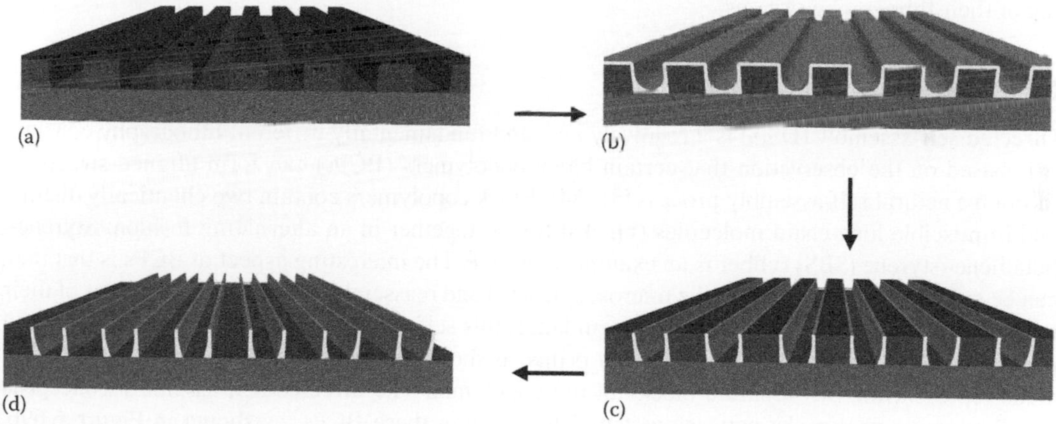

FIGURE 6.68 Self-aligned double patterning. (a) Mandrel pattern, (b) isotropic deposition, (c) removal of the mandrels, and (d) anisotropic etch.

the underlying material. This is called the mandrel. Then, an isotropic deposition is performed over that structure to coat all horizontal and vertical surfaces. This is followed by an anisotropic etch that removes all of the deposited material on the horizontal surfaces, leaving only those attached to the vertical surfaces. Finally, the first structure is selectively etched way, leaving behind a pattern that has twice the spatial frequency of the original pattern [79]. These steps are illustrated in Figure 6.68.

The advantage of this method is that the second pattern it is automatically aligned to the first pattern. It does not require an active alignment process or a second mask; hence, it is referred to as a self-aligned process. This concept can be carried forward several times. The pattern from the aforementioned process can be used as a template to double its spatial frequency once again, resulting in four times the original spatial frequency. This is referred to as the self-aligned quadruple patterning (SAQP) [80]. Yet another iteration will result in self-aligned octuplet patterning (SAOP).

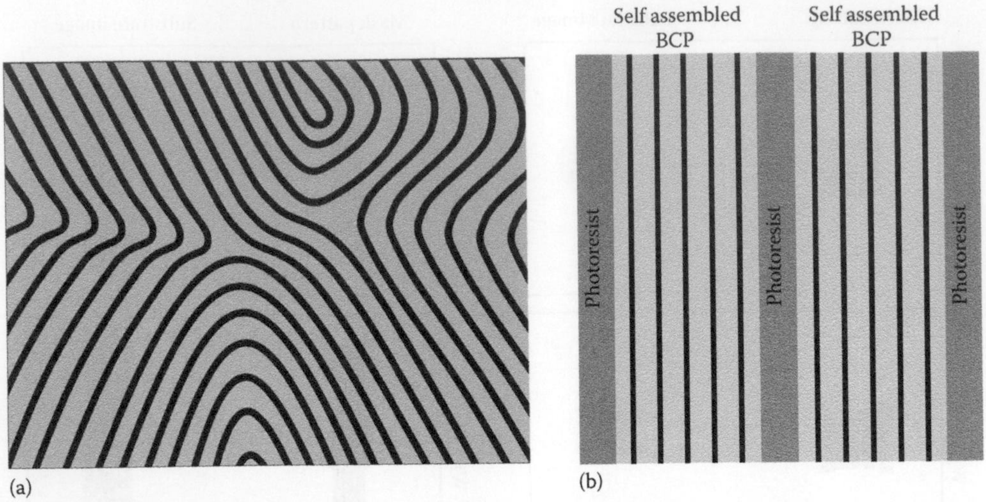

FIGURE 6.69 (a) Self-assembly of block copolymers without guidance and (b) directed self-assembly of block copolymers with photolithography.

In principle, this method enables us to achieve resolutions below 10 nm using the same 193 nm light sources. This has allowed manufacturers to push the limits further without significant retooling of their lithography systems.

6.11.4 Directed Self-Assembly

Directed self-assembly (DSA) is a relatively new and fundamentally different lithography concept. It is based on the observation that certain block copolymers (BCPs) can form aligned structures through a natural self-assembly process [81–84]. Block copolymers contain two chemically distinct and immiscible long-chain molecules (blocks) linked together in an alternating fashion. Styrene–butadiene–styrene (SBS) rubber is an example of a BCP. The interesting aspect of BCPs is that they can be made to phase separate at the microscopic level and reassemble as periodic structures of their two constituent blocks. However, without guidance, this self-assembly results in randomly oriented periodic structures. They appear like finger prints, as shown Figure 6.69a, where the dark and light areas represent the two separate blocks of the copolymer. The directed self-assembly concept is based on using photoresist patterns as templates to align these BCPs, as shown in Figure 6.69b. Since multiple periods of the copolymer blocks can fit within one period of the original lithographic pattern, this effectively results in a multiplication of the original spatial frequency. A selective etch process is then used to remove one block of the copolymer, leaving the other. This pattern is then used as the lithographic template for subsequent processing. DSA is an active area of research and is currently being developed for the next-generation lithography systems with CDs below 10 nm.

6.12 EXTREME-UV LITHOGRAPHY

Since 157 nm F_2 lasers for lithography proved to be difficult, the next practical reduction in wavelength was toward the extreme UV (EUV) range. One of these is 13.5 nm. This is more than a factor of 10 reduction in wavelength from 193 nm and represents a significant departure from nearly all aspects of DUV lithography [85–87]. To reduce optical absorption from air molecules, the entire system from the emission source to the substrate has to be in an evacuated chamber. The 13.5 nm emission is produced from a tin (Sn) plasma, which is excited by bombarding a tin surface with a high power laser (such as a CO_2 laser). This is not a very efficient process, and most of the energy is dissipated as heat. Furthermore, contamination from the sputtered tin atoms also presents a challenge to keep the optical train clean.

Due to the difficulty constructing refractory optical components, the system is designed with all-reflective optical components. All of the focusing optics, including the optical mask, are designed as reflective components with off-axis illumination. However, high reflectivity is not easy to achieve either, since most materials absorb EUV radiation. The technique currently used is a multilayer high-reflection design, similar to the ones we discussed in Chapter 4. However, this requires two materials with a large difference in refractive index. This poses a particular challenge in the EUV since most refractive indices approach 1.0 at such short wavelengths. Molybdenum and silicon are currently used as the two alternating layers. The refractive index of silicon at 13.5 nm is $0.999 - j0.0018$ and the value for molybdenum is $0.9227 - j0.0062$ [88]. This refractive index difference is extremely small compared to the designs that we studied in Chapter 4. As a result, typical reflectors use about 100 or more layers to achieve just a modest reflectivity of 50%. A large fraction of energy is still lost to absorption in the layers. Furthermore, since the layers are typically a quarter of a wavelength thick, these layers have to be about 3 nm thick each. The surface roughness also has to be extremely small to avoid scattering losses.

Nevertheless, most of these challenges have been overcome to some extent, and manufacturing systems based on EUV lithography are just starting to appear [89]. However, the acquisition cost of such systems is in excess of $100M, so one is unlikely to encounter a EUV lithography system in a research laboratory.

6.13 NONOPTICAL LITHOGRAPHY

All of the lithographic techniques discussed so far are based on illumination with light; hence, they are referred to as photolithography. It is also possible to use energy sources other than photons to illuminate a resist film and cause a chemical change similar to photolithography. Some of these are discussed in the next section.

6.13.1 ELECTRON BEAM LITHOGRAPHY

Electron beam lithography (EBL) was briefly discussed in the context of photomask manufacture in Section 6.3.3. By far the largest application of EBL is the manufacture of high-resolution photomasks. However, in most research laboratories, EBL is used as a direct write tool, where the design is written directly on the intended photoresist layer without having to make a photomask. This allows researchers to make one-off devices with features smaller than what could be achieved with contact lithography or projection lithography. This is an acceptable approach when only a few samples are needed to be made.

Both positive-tone and negative-tone photoresists are used for EBL, with positive tone being more common than negative tone. These operate on different principles as compared to optical photoresists. One of the most common positive-tone resists is polymethyl methacrylate (PMMA). This is the same material used in acrylic glass (Plexiglas). For spin coating, PMMA is dissolved in the organic solvent anisole to create different viscosities. Upon exposure to energetic electron beams, PMMA undergoes a scissioning process where its molecules are broken down into smaller fragments. These smaller molecules become soluble in a solvent such as methyl isobutyl ketone (MIBK), while the unexposed areas remain insoluble. PMMA resists have resist has poor contrast (difference in solubility between exposed to unexposed) and poor resistance to plasma etch chemistries. An alternative to PMMA is ZEP, which has a higher contrast and a similar etch resistance to optical photoresists, but it is significantly more expensive [90]. Negative-tone resists work by electron beams inducing a cross-linking reaction to decrease their solubility in the developer. At high irradiance levels, PMMA itself has been shown to have a negative-tone performance [91], and SU-8 photoresist has also been used as a negative resist with electron beams [92]. A number of other commercial negative resists have also been developed with optimized performance, such as ma-N from Micro Resist Technology, and chemically amplified resists, such as NEB-31 from Sumitomo.

As stated earlier, the wavelength of electrons can be expressed as $\lambda = h/\sqrt{2mE}$. For 100 keV, the wavelength is about 4 pm. However, practical resolution values that can be achieved is significantly larger due to spherical and chromatic aberrations in the focusing optics, as well as scatter in the photoresist. Realistic resolutions are typically on the order of 10 nm.

To fully penetrate the photoresist layer, the electron energy has to be several tens of keV. At low voltages, there will be significant forward scattering of the electrons. At higher voltages, scattering can be reduced by allowing the electrons to penetrate the underlying substrate. However, this can cause backscatter and expose the nearby areas, known as the proximity effect. As a result, the electron beam acceleration has to be optimized for each photoresist/substrate combination.

Figure 6.70 shows the simulated electron beam trajectories in a 200 nm thick PMMA film on a silicon substrate, using the software "Monte Carlo Simulation of Electron trajectory in Solids" – CASINO [93]. Each plot corresponds to a different electron energy. At 1 keV, the electron energy is insufficient to penetrate the resist film. At 10 keV, it penetrates most of the film, but the forward scattering becomes significant. At 25 and 50 keV, the forward scattering is smaller, but the proximity effects are not considered in this plot.

It is interesting to note that positive-tone electron beam resists produce an inwardly tapered sidewall profile, especially at low energies. This is opposite of the behavior usually seen in positive-tone photoresists. With UV exposure, the sidewall profile in photoresists is dominated by the optical absorption, which results in the exposed region getting smaller with increasing depth. With electron beams, the sidewall profile is dominated by forward scattering, which results in the exposed region getting larger with increasing depth. This results in a favorable sidewall profile for performing lift-off lithography.

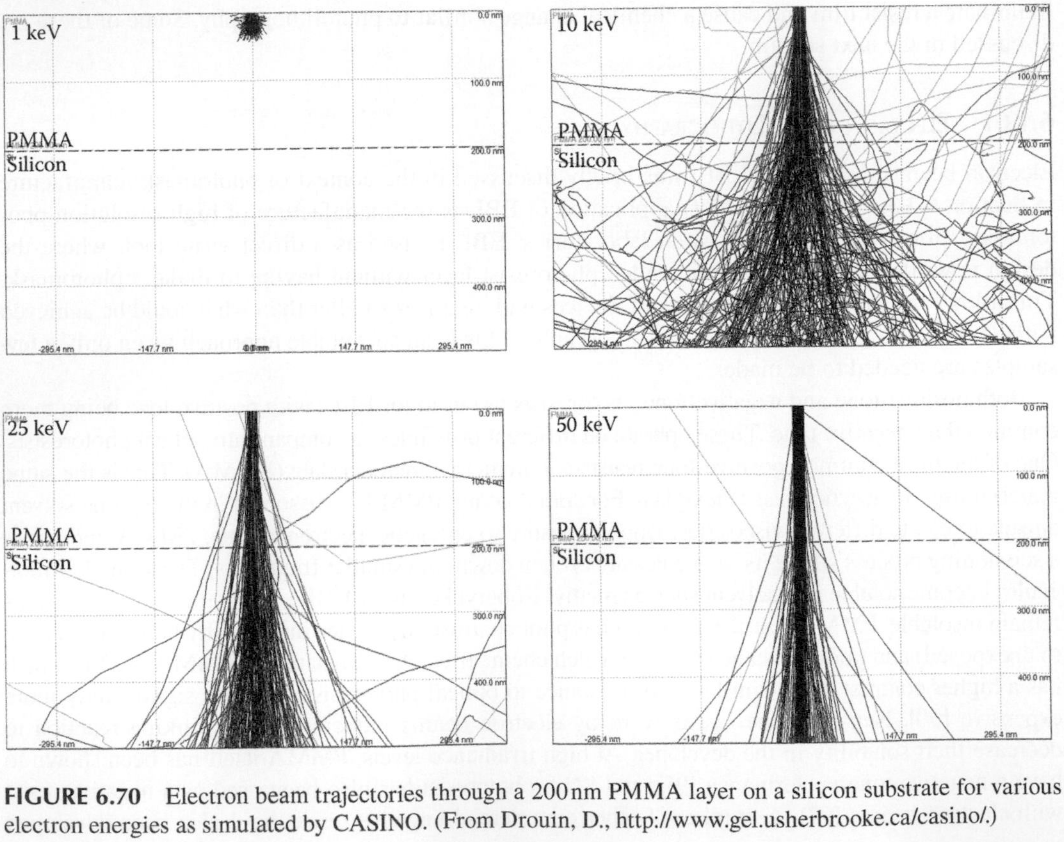

FIGURE 6.70 Electron beam trajectories through a 200 nm PMMA layer on a silicon substrate for various electron energies as simulated by CASINO. (From Drouin, D., http://www.gel.usherbrooke.ca/casino/.)

However, from the results in Figure 6.70, we can see that to achieve a reasonable sidewall angle, the electron energy has to be kept low, which results in poor spatial resolution. As a result, it is common to use a bilayer process to produce a favorable lift-off profile, similar to that shown in Figure 6.59.

The electron beam writing tool consists of an electron source, high-voltage accelerator, focusing optics, a number of beam apertures, and scanning coils all enclosed in a vacuum chamber, as shown in Figure 6.71. The beam is electronically deflected by the coils to scan the field and write the pattern. The largest writing field is determined by this scan range and is typically on the order of 100–500 μm. To write larger areas, the substrate has to be mechanically translated after each field has been completed. This can lead to what is known as stitching errors since mechanical translation may not perfectly align two adjacent writing fields.

The dose in EBL is measured in surface charge density, with 100–1000 μC/cm² being representative of the doses required to fully expose a resist. The current is in the range of 1–10 nA. Assuming 500 μC/cm² and 5 nA, the time required to write a 1 in. × 1 in. area will be $(500\times10^{-6})\times(2.54^2/(5\times10^{-9}))=179$ h. Of course, in reality, one does not have to fully expose the entire surface. The density of the patterns will determine the effective area that has to be exposed. In this example, if the pattern density is 25%, the design can be potentially written in 45 h. Therefore, the design and the resist polarity play a major role in the write times. For the same design, if an opposite resist polarity is used, the write time will increase by a factor of three because the pattern density will become 75%. The current can also be increased to improve the write times, but higher currents will result in a larger spot size and a poorer resolution.

Modern EBL systems improve the write times by dynamically adjusting the beam current based on the resolution of the local features being written. Also, while older systems used to perform a raster scan, modern systems use a vector scan to reposition the beam quickly, which enables a more intelligently driven scan patterns. Nevertheless, it should be clear that EBL throughput is significant lower than optical projection lithography systems. To write an entire 12 in. diameter wafer, assuming 25% pattern density, would require more than 210 days! This number still does not include the mechanical movement of substrate. Meanwhile, throughputs of DUV steppers are in the range of

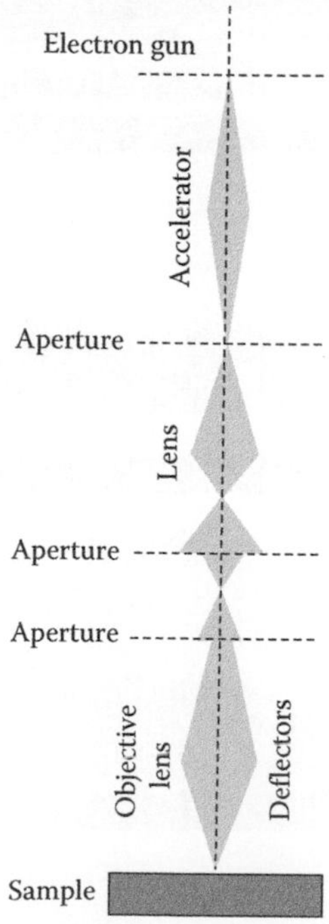

FIGURE 6.71 Vacuum column of an electron beam writing tool.

100 wafers per hour. This is a vast difference in scale. As a result, EBL is currently not considered a competition to commercial optical lithography systems. Its main application will most likely remain in photomask production and in research laboratories. Nevertheless, many techniques to improve the throughput of EBL are still under investigation, including the use of massively parallel multiple beams [94] that have shown significant potential.

6.13.2 Nanoimprint Lithography

Nanoimprint lithography (NIL) is conceptually very simple. It involves making a mold (or stamp) with surface relief features and physically compressing it into a resist film to leave its imprint, as shown in Figure 6.72. In that sense, it is very similar to contact photolithography except physical pressure is used to create the patterns instead of optical illumination. Its main advantage is that it does not require any optics or photomasks. But it does require a stamp, the equivalent of a photomask. It was first demonstrated as a viable lithographic approach in 1995 [95] and has since been developed and commercialized [96].

Despite the term "nano" in the NIL, it is quite difficult to reliably and consistently reproduce nanoscale features. Physical limitations such as substrate flatness, contamination of the stamp due to contact with the resist, and trapped air bubbles often limit what can be realistically achieved in the laboratory. Nevertheless, the absence of resolution-limiting phenomena such as diffraction or scattering has made NIL a potential candidate as a next-generation lithography tool.

The stamp is usually made using a high-resolution writing process such as EBL, just like photomasks are made for optical lithography. Whereas only a very thin film of metal such as chromium is need in photomasks, the depth of the relief structures in an NIL stamp will be dictated by the ultimate imprinted depth required in the resist film. In that sense, NIL is a 1:1 pattern transfer process both laterally and vertically (in contact photolithography, the pattern transfer was 1:1 in the lateral direction only). Therefore, NIL stamps generally require deeper structures to be etched. Silicon is

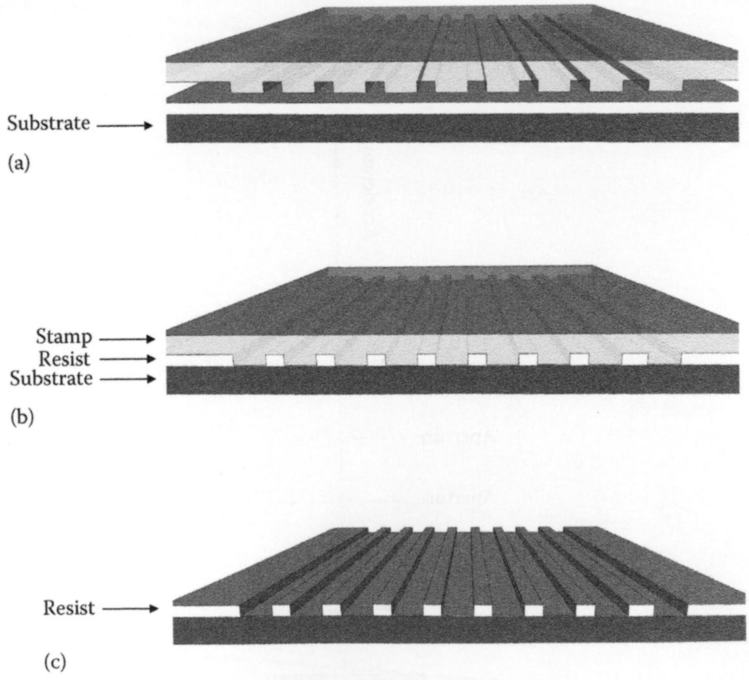

FIGURE 6.72 Simplified depiction of nanoimprint lithography. (a) The stamp is aligned and ready to imprint, (b) stamp is compressed into the resist layer, and (c) imprint is left in the resist after the stamp is removed.

the best stamp material due to the maturity of the plasma etch processes, but optical transparency is sometimes required in the stamp for alignment purposes as well as UV illumination for hardening the resist. In such cases, alternate materials such as quartz or sapphire are used.

One of the attractive features of NIL is that it can be used on flexible substrates. It is also possible to make the stamp from flexible materials such as Teflon [97]. Cylindrically shaped stamps, or roller stamps, can also be used to allow mass reproduction on continuously moving sheets [98,99]. These aspects have made NIL an attractive tool for large-area organic electronics, photonics, and solar cells.

There are several variations of the NIL process. Applying heat while the stamp is sandwiched with the resist layer will soften the resist film and allow it to flow around the features to fill the entire space between the stamp and the substrate. This is known as thermocompression NIL, or hot embossing. UV illumination can also be used during compression to cross-link and harden the resist layer, assuming a UV curable resist is used. This will require the stamp to be transparent, and a material such as quartz has to be used in such cases. Furthermore, due to the hard physical contact and the use of elevated temperatures, the thermal expansion coefficient of the stamp has to be matched with the substrate to preserve the alignment.

NIL is currently used in many research laboratories, since it is fairly inexpensive to acquire and operate. On a manufacturing scale, it is used in some photovoltaic and display applications where the dimensional tolerance and yield are less stringent. However, these are critical requirements in semiconductor electronics, and whether or not NIL will succeed DUV projection steppers (or replace EUV lithography) is still under debate. However, there are definite cost advantages compared to EUV, and Canon, who is one of the three major lithography tool manufacturers (along with ASML and Nikon), is currently developing a nanoimprint tool to meet this market need [100].

PROBLEMS

6.1 The specific gravity of the photoresist is 1.08. If the film thickness after a 30 s spin at 2000 RPM is 1.3 μm, calculate its viscosity. Then, plot the film thickness as a function of time up to 60 s. Also, plot the film thickness (at 30 s) as a function of RPM.

6.2 Examine the origins of the equation $D_{min} = 2\sqrt{\lambda z}$ (Equation 6.19) for the resolution limit of contact mask lithography.

6.3 Derive Equations 6.26 and 6.27 for the half-pitch and depth-of-focus of a projection lithography system, and summarize the assumptions made in those equations.

6.4 Using the Lloyd's mirror configuration show in Figure 6.64 for interference lithography, show that both beams incident on the substrate have equal intensities, and equal and opposite angles of incidence.

6.5 Using the Dill parameter model developed in this chapter, and using the parameters given in Tables 6.1 and 6.2, calculate the differences in the required exposure dose between a bare silicon wafer and a silicon wafer that is coated with an aluminum film.

LABORATORY EXERCISES

6.1 For an available photoresist, spin coat the film at various RPM values. Then, after a soft bake, expose with a generic photomask and use a stylus profiler to measure the film thicknesses. Based on this, calculate the viscosity of the film and determine the effect of the soft bake.

6.2 If access to a photomask with submicrometer features is available, using a contact mask aligner, conduct a series of exposures with different photoresist thicknesses and exposure doses and evaluate the results in terms of the spatial frequencies reproduced on the photoresist compared to what is on the photomask.

6.3 For a generic photoresist available in the laboratory, spin coat them on microscope glass slides. Using a mask aligner under flood exposure mode, with the lamp intensity turned down, place the glass slides on top of a UV detector and record the optical intensity on the UV detector as a function of time. Also record the optical intensity without the glass slide. From these data, calculate the Dill parameters A, B, and C for this photoresist and determine the optimum exposure times.

REFERENCES

1. H. J. Levinson, *Principles of Lithography*, 3rd edn., SPIE Press, Bellingham, WA, 2010.
2. C. Mack, *Fundamental Principles of Optical Lithography: The Science of Microfabrication*, John Wiley & Sons, Chichester, U.K., 2011.
3. C. A. Mack, *Field Guide to Optical Lithography*, SPIE Press, Bellingham, WA, 2006.
4. U. Okoroanyanwu, *Chemistry and Lithography*, SPIE Press, Bellingham, WA, 2010.
5. K. A. Reinhardt and W. Kern (Eds.), *Handbook of Silicon Wafer Cleaning Technology*, 2nd edn., William Andrew, New York, 2008.
6. W. Kern and D. A. Puotinen, Cleaning solutions based on hydrogen peroxide for use in silicon semiconductor technology, *RCA Rev.* 31 (1970) 187–206.
7. W. Kern, The evolution of silicon wafer cleaning technology, *J. Electrochem. Soc.*, 137(6) (June 1990) 1887.
8. T. Ohmi, Total room temperature wet cleaning for Si substrate surface, *J. Electrochem. Soc.*, 143(9) (1996) 2957–2964.
9. W. A. Cady and M. Varadarajan, RCA clean replacement, *J. Electrochem. Soc.*, 143(6) (1996) 2064–2067.
10. G. A. C. M. Spierings, Wet chemical etching of silicate glasses in hydrofluoric acid based solutions, *J. Mater. Sci.*, 28 (1993) 6261–6273.
11. W. Petasch, B. Kegel, H. Schmid, K. Lendenmann, and H. U. Keller, Low-pressure plasma cleaning: A process for precision cleaning applications, *Surf. Coat. Technol.*, 97(1–3) (December 1997) 176–181.
12. W. Kim, T.-H. Kim, J. Choi, and H.-Y. Kim, Mechanism of particle removal by megasonic waves, *Appl. Phys. Lett.*, 94 (2009) 081908.
13. R. J. Good, Contact angle, wetting, and adhesion: A critical review, *J. Adhes. Sci. Technol.*, 6(12) 1992 1269–1302.
14. N. P. Pham, J. N. Burghartz, and P. M. Sarro, Spray coating of photoresist for pattern transfer on high topography surfaces, *J. Micromech. Microeng.*, 15 (2005) 691.
15. L. Yu, Y. Y. Lee, F. E. H. Tay, and C. Iliescu, Spray coating of photoresist for 3D microstructures with different geometries, *J. Phys. Conf. Ser.*, 34 (2006) 937.
16. A. G. Emslie, F. T. Bonner, and L. G. Peck, Flow of a viscous liquid on a rotating disk, *J. Appl. Phys.*, 29 (1958) 858.
17. D. E. Bornside, C. W. Macosko, and L. E. Scriven, Spin coating: One-dimensional model, *J. Appl. Phys.*, 66 (1989) 5185.
18. L. W. Schwartz and R. Valery Roy, Theoretical and numerical results for spin coating of viscous liquids, *Phys. Fluids*, 16 (2004) 569.
19. D. Meyerhofer, Characteristics of resist films produced by spinning, *J. Appl. Phys.*, 49 (1978) 3993.
20. W. W. Flack, D. S. Soong, A. T. Bell, and D. W. Hess, A mathematical model for spin coating of polymer resists, *J. Appl. Phys.*, 56 (1984) 1199.
21. C. Rensch, S. Hell, M. v. Schickfus, and S. Hunklinger, Laser scanner for direct writing lithography, *Appl. Opt.*, 28(17) (1989) 3754–3758.
22. K. K. B. Hon, L. Lib, I. M. Hutchings, Direct writing technology—Advances and developments, *CIRP Ann. Manuf. Technol.*, 57(2) (2008) 601–620.
23. K.-D. Schock, F. E. Prins, S. Strähle, and D. P. Kern, Resist processes for low-energy electron-beam lithography, *J. Vac. Sci. Technol. B*, 15 (1997) 2323.
24. A. N. Broers, A. C. F. Hoole, and M. Ryan, Electron beam lithography—Resolution limits, *Microelectron. Eng.*, 32 (1996) 131–142.
25. R. Voelkel, U. Vogler, A. Bich, P. Pernet, K. J. Weible, M. Hornung, R. Zoberbier, E. Cullmann, L. Stuerzebecher, T. Harzendorf, and U. D. Zeitner, Advanced mask aligner lithography: New illumination system, *Opt. Exp.*, 18(20) (2010) 20968–20978.
26. B. Meliorisz, S. Partel, T. Schnattinger, T. Fühner, A. Erdmann, P. Hudek, Investigation of high-resolution contact printing, *Microelectron. Eng.*, 85(5–6) (May–June 2008) 744–748.
27. R. Gordon, C. A. Mack, Lithography simulation employing rigorous solutions to Maxwell's equations, *Proceedings of the SPIE, Optical Microlithography XI*, Vol. 3334, Santa Clara, CA, June 29, 1998, p. 176.

28. C. A. Mack, Lithographic simulation: A review, *Proceedings of the SPIE, Lithographic and Micromachining Techniques for Optical Component Fabrication*, Vol. 4440, Santa Clara, CA, November 9, 2001, p. 59.
29. C. A. Mack, Thirty years of lithography simulation, *Proceedings of the SPIE, Optical Microlithography XVIII*, Vol. 5754, San Jose, CA, May 12, 2005, p. 1.
30. M. Rothschild, M. W. Horn, C. L. Keast, R. R. Kunz, V. Liberman, S. C. Palmateer, S. P. Doran et al., Photolithography at 193 nm, *L. Lab. J.*, 10(1) (1997) 19–34.
31. Th. Zell, Present and future of 193 nm lithography, *Microelectron. Eng.*, 83 (2006) 624–633.
32. T. M. Bloomstein, M. Rothschild, R. R. Kunz, D. E. Hardy, R. B. Goodman, and S. T. Palmacci, Critical issues in 157 nm lithography, *J. Vac. Sci. Technol. B*, 16 (1998) 3154.
33. B. Fay, Advanced optical lithography development, from UV to EUV, *Microelectron. Eng.*, 61–62 (2002) 11–24.
34. M. McCallum, G. Fuller, S. Owa, From hyper NA to low NA, *Microelectron. Eng.*, 83 (2006) 667–671.
35. M. McCallum, M. Kameyama, S. Owa, Practical development and implementation of 193 nm immersion lithography, *Microelectron. Eng.*, 83 (2006) 640–642.
36. W. Hinsberg, G. M. Wallraff, C. E. Larson, B. W. Davis, V. Deline, S. Raoux, D. Miller et al., Liquid immersion lithography: Evaluation of resist issues, *Proceedings of the SPIE, Advances in Resist Technology and Processing XXI*, Vol. 5376, Santa Clara, CA, May 14, 2004, p. 21.
37. M. Rothschild, T. M. Bloomstein, R. R. Kunz, V. Liberman, M. Switkes, S. T. Palmacci, J. H. C. Sedlacek, D. Hardy, and A. Grenville, Liquid immersion lithography: Why, how, and when?, *J. Vac. Sci. Technol. B*, 22 (2004) 2877.
38. J. H. Burnett, S. G. Kaplan, E. L. Shirley, P. J. Tompkins, High-index optical materials for 193 nm immersion lithography, *Proceedings of the SPIE, Optical Microlithography XVIII*, Vol. 5754, San Jose, CA, May 12, 2005, p. 611.
39. D. W. Kim, J.-E. Lee, and H.-K. Oh, Heat conduction to photoresist on top of wafer during post exposure bake process: I. Numerical approach, *Jpn. J. Appl. Phys.*, 47(11) (2008) 8338–8348.
40. H.-K. Oh, D. W. Kim, and J.-E. Lee, Heat conduction to photoresist on top of wafer during post exposure bake process: II. Application, *Jpn. J. Appl. Phys.*, 47(11) (2008) 8349–8353.
41. J. A. Bruce, S. R. Dupuis, R. Gleason, and H. Linde, Effect of humidity on photoresist performance, *J. Electrochem. Soc.*, 144(9) (September 1997) 3169.
42. Lithography trouble shooter: Questions and answers around the most common problems in microstructuring, MicroChemicals GmbH, 2012.
43. R. DellaGuardia, W.-S. Huang, K. Rex Chen, and D. Kang, Correlations between dissolution data and lithography of various resists, *Proceedings of the SPIE, Advances in Resist Technology and Processing XVI*, Vol. 3678, San Jose, CA, June 11, 1999, p. 316.
44. F. H. Dill, W. P. Hornberger, P. P. Hauge, and J. M. Shaw, Characterization of positive photoresist, *IEEE Trans. Electron Devices*, 22(7) (July 1975) 445–452.
45. H. Kirchauer and S. Selberherr, Rigorous three-dimensional photoresist exposure and development simulation over nonplanar topography, *IEEE Trans. Comput. Aided Des. Integr. Circuits Syst.*, 16(12) (December 1997) 1431–1438.
46. A. Erdmann and W. Henke, Simulation of light propagation in optical linear and nonlinear resist layers by finite difference beam propagation and other methods, *J. Vac. Sci. Technol. B*, 14(6) (November/December 1996) 3734–3737.
47. X. Shao, J. Meador, S. Deshpande, R. Puligadda, K. Mizusawa, and S. Arase, *J. Photopolym. Sci. Technol.*, 14(3) (2001) 481–488.
48. C. J. Neef, V. Krishnamurthy, M. I. Nagatkina, E. Bryant, M. Windsor, and C. Nesbit, New BARC materials for the 65-nm node in 193-nm lithography, *Proceedings of the SPIE, Advances in Resist Technology and Processing XXI*, Vol. 5376, Santa Clara, CA, May 14, 2004, p. 684.
49. T. Katayama, H. Motobayashi, W.-B. Kang, M. A. Toukhy, J. E. Oberlander, S. S. Ding, and M. Neisser, Developable bottom antireflective coatings for 248-nm and 193-nm lithography, *Proceedings of the SPIE, Optical Microlithography XVII*, Vol. 5377, Santa Clara, CA, May 28, 2004, p. 968.
50. J. M. Shaw, J. D. Gelorme, N. C. LaBianca, W. E. Conley, and S. J. Holmes, Negative photoresists for optical lithography, *IBM J. Res. Dev.*, 41(1/2) (1997) 81.
51. G. L. Wolk and D. H. Ziger, Kinetic study of NH_3-catalyzed image reversal in positive photoresist, *Ind. Eng. Chem. Res.*, 30(7) (1991) 1461–1468.
52. M. L. Long and J. Newman, Image reversal techniques with standard positive photoresist, *Proceedings of the SPIE, Advances in Resist Technology I*, Vol. 0469, Santa Clara, CA, May 21, 1984.

53. H. Yanazawa, Adhesion model and experimental verification for polymer—SiO_2 system, *Colloids Surf.*, 9(2) (January 1984) 133–145.
54. J. N. Helbert and N. Saha, Application of silanes for promoting resist patterning layer adhesion in semiconductor manufacturing, *J. Adhes. Sci. Technol.*, 5(10) (1991) 905–925.
55. C.-M. Dai and D. H. Lee, Studies on the adhesion contact angle of various substrates and their photoresist profiles, *Proceedings of the SPIE, Advances in Resist Technology and Processing XII*, Vol. 2438, Santa Clara, CA, June 9, 1995, p. 709.
56. Yield Engineering Systems, Inc., Livermore, CA. http://www.yieldengineering.com.
57. S. Kishimura, Y. Kimura, J. Sakai, K. Tsujita, and Y. Matsui, Improvement of dry etching resistance of resists by deep UV cure, *Jpn. J. Appl. Phys.*, 38(1A) (1999) 250–255.
58. G. Sengo, H. A. G. M. van Wolferen, and A. Driessen, Optimized deep UV curing process for metal-free dry-etching of critical integrated optical devices, *J. Electrochem. Soc.*, 158(10) (2011) H1084–H1089.
59. K. Y. Lee, N. LaBianca, S. A. Rishton, S. Zolgharnain, J. D. Gelorme, Micromachining applications of a high resolution ultrathick photoresist, *J. Vac. Sci. Technol. B*, 13 (1995) 3012.
60. H. Lorenz, M. Despont, N. Fahrni, N. LaBianca, P. Renaud, and P. Vettiger, SU-8: A low-cost negative resist for MEMS, *J. Micromech. Microeng.*, 7(3) (1997) 121–124.
61. A. del Campo and C. Greiner, SU-8: A photoresist for high-aspect-ratio and 3D submicron lithography, *J. Micromech. Microeng.*, 17 (2007) R81–R95.
62. MicroChem Corp., Westborough, MA. http://www.microchem.com/Prod-SU8_KMPR.htm.
63. V. M. Blanco Carballo, J. Melai, C. Salm, and J. Schmitz, Moisture resistance of SU-8 and KMPR as structural material, *Microelectron. Eng.*, 86 (2009) 765–768.
64. C. K. Chung and Y. Z. Hong, Surface modification of SU8 photoresist for shrinkage improvement in a monolithic MEMS microstructure, *J. Micromech. Microeng.*, 17 (2007) 207–212.
65. H. Lee, K. Lee, B. Ahn, J. Xu, L. Xu, and K. W. Oh, A new fabrication process for uniform SU-8 thick photoresist structures by simultaneously removing edge bead and air bubbles, *J. Micromech. Microeng.*, 21 (2011) 125006.
66. P. Walker and W. H. Tarn (Eds.), *CRC Handbook of Metal Etchants*, CRC Press, Boca Raton, FL, 1991.
67. K. R. Williams, K. Gupta, and M. Wasilik, Etch rates for micromachining processing—Part II, *J. Microelectromech. Syst.*, 12(6) (December 2003) 761–778.
68. Transene Company, Inc., Danvers, MA. http://transene.com/.
69. Y. Chena, K. Peng, and Z. Cui, A lift-off process for high resolution patterns using PMMA/LOR resist stack, *Microelectron. Eng.*, 73–74 (June 2004) 278–281.
70. P. Wrschka, J. Hernandez, G. S. Oehrlein, and J. King, Chemical mechanical planarization of copper damascene structures, *J. Electrochem. Soc.*, 147(2) (2000) 706–712.
71. S. R. J. Brueck, Optical and interferometric lithography—Nanotechnology enablers, *Proc. IEEE*, 93(10) (October 2005) 1704–1721.
72. C. Lu and R. H. Lipson, Interference lithography: A powerful tool for fabricating periodic structures, *Laser Photon. Rev.*, 4(4) (June 2010) 568–580.
73. T. M. Bloomstein, M. F. Marchant, S. Deneault, D. E. Hardy, and M. Rothschild, 22-nm immersion interference lithography, *Opt. Express*, 14(14) (2006) 6434–6443.
74. A. K. Raub and S. R. J. Brueck, Deep UV immersion interferometric lithography, *Proceedings of the SPIE, Optical Microlithography XVI*, Vol. 5040, Santa Clara, CA, June 25, 2003, p. 667.
75. B. J. Lin, The ending of optical lithography and the prospects of its successors, *Microelectron. Eng.*, 83 (2006) 604–613.
76. A. K.-K. Wong, *Resolution Enhancement Techniques in Optical Lithography*, SPIE Press, Bellingham, WA, 2001.
77. M. D. Levenson, N. S. Viswanathan, R. A. Simpson, Improving resolution in photolithography with a phase-shifting mask, *IEEE Trans. Electron Devices*, 29(12) (December 1982) 1828–1836.
78. R. Jonckheere, A. Wong, A. Yen, K. Ronse, and L. Van den hove, Optical proximity correction: Mask pattern-generation challenges, *Microelectron. Eng.*, 30 (1996) 115–118.
79. A. J. Hazelton, S. Wakamoto, S. Hirukawa, M. McCallum, N. Magome, J. Ishikawa, C. Lapeyre, I. Guilmeau, S. Barnola, and S. Gaugiran, Double-patterning requirements for optical lithography and prospects for optical extension without double patterning, *J. Micro/Nanolith. MEMS MOEMS.*, 8(1) (January 06, 2009) 011003.
80. P. Xu, Y. Chen, Y. Chen, L. Miao, S. Sun, S.-W. Kim, A. Berger et al., Sidewall spacer quadruple patterning for 15 nm half-pitch, *Proceedings of the SPIE, Optical Microlithography XXIV*, Vol. 7973, San Jose, CA, March 22, 2011, p. 79731Q.

81. S.-J. Jeong, J. Y. Kim, B. H. Kim, H.-S. Moon, and S. O. Kim, Directed self-assembly of block copolymers for next generation nanolithography, *Mater. Today*, 16(12) (December 2013) 468–476.
82. S. B. Darling, Directing the self-assembly of block copolymers, *Prog. Polym. Sci.*, 32 (2007) 1152–1204.
83. C.-C. Liu, P. F. Nealey, A. K. Raub, P. J. Hakeem, S. R. J. Brueck, E. Han, and P. Gopalan, Integration of block copolymer directed assembly with 193 immersion lithography, *J. Vac. Sci. Technol. B*, 28 (2010) C6B30.
84. M. Luo and T. H. Epps, Directed block copolymer thin film self-assembly: Emerging trends in nanopattern fabrication, *Macromolecules*, 46(19) (2013) 7567–7579.
85. C. Wagner and N. Harned, Lithography gets extreme, *Nat. Photon.*, 4 (January 2010) 809–811.
86. B. Wu and A. Kumar, Extreme ultraviolet lithography: A review, *J. Vac. Sci. Technol. B*, 25 (2007) 1743.
87. V. Bakshi, *EUV Lithography*, SPIE Press, Bellingham, WA, 2008.
88. M. Singh and J. J. M. Braat, Improved theoretical reflectivities of extreme ultraviolet mirrors, *Proc. SPIE*, 3997 (*2000*) 412–419.
89. https://www.asml.com.
90. K. Koshelev, M. Ali Mohammad, T. Fito, K. L. Westra, S. K. Dew, and M. Stepanova, Comparison between ZEP and PMMA resists for nanoscale electron beam lithography experimentally and by numerical modeling, *J. Vac. Sci. Technol. B*, 29 (2011) 06F306.
91. I. Zailer, J. E. F. Frost, V. Chabasseur-Molyneux, C. J. B. Ford, and M. Pepper, Crosslinked PMMA as a high-resolution negative resist for electron beam lithography and applications for physics of low-dimensional structures, *Semicond. Sci. Technol.*, 11 (1996) 1235–1238.
92. B. Bilenberg, S. Jacobsen, M. S. Schmidt, L. H. D. Skjolding, P. Shib, P. Bøggild, J. O. Tegenfeldt, and A. Kristensen, High resolution 100 kV electron beam lithography in SU-8, *Microelectron. Eng.*, 83(4–9) (April–September 2006) 1609–1612.
93. D. Drouin, Universite de Sherbrooke, Sherbrooke, Québec, Canada. http://www.gel.usherbrooke.ca/casino/.
94. G. de Boer, M. P. Dansberg, R. Jager, J. J. M. Peijster, E. Slot, S. W. H. K. Steenbrink, and M. J. Wieland, MAPPER: Progress toward a high-volume manufacturing system, *Proceedings of the SPIE, Alternative Lithographic Technologies V*, Vol. 8680, March 26, 2013, p. 86800O.
95. S. Y. Chou, P. R. Krauss, and P. J. Renstrom, Imprint of sub-25 nm vias and trenches in polymers, *Appl. Phys. Lett.*, 67 (1995) 3114.
96. H. Schift, Nanoimprint lithography: An old story in modern times? A review, *J. Vac. Sci. Technol. B*, 26(2) (March/April 2008) 458–480.
97. L. Jay Guo, Nanoimprint lithography: Methods and material requirements, *Adv. Mater.*, 19 (2007) 495–513.
98. H. Tan, A. Gilbertson, and S. Y. Chou, Roller nanoimprint lithography, *J. Vac. Sci. Technol. B*, 16 (1998) 3926.
99. H. J. Lim, K.-B. Choi, G. H. Kim, S. Y. Park, J. H. Ryu, and J. J. Lee, Roller nanoimprint lithography for flexible electronic devices of a sub-micron scale, *Microelectron. Eng.*, 88 (2011) 2017–2020.
100. H. Takeishi and S. V. Sreenivasan, Nanoimprint system development and status for high volume semiconductor manufacturing, *Proceedings of the SPIE, Alternative Lithographic Technologies VII*, Vol. 9423, Monterey, CA, March 19, 2015, p. 94230C.

7 Wet Chemical and Plasma Etching

Etching is the process of removing portions of a thin film (or substrate) by chemical or physical processes. Wet chemical and plasma etching are two of the most common methods used in research and manufacturing. As the name implies, wet chemical etching is done in a liquid bath. Plasma etching, also referred to as "dry etching," is done in a vacuum chamber with reactive gases and is based on chemical and physical action on the substrate.

7.1 WET CHEMICAL ETCHING

7.1.1 Basic Principles

Wet chemical etching is conceptually very simple and does not require a significant infrastructure. The process basically consists of immersing a photoresist-patterned substrate in a chemical bath to allow the liquid chemicals to work their way through the openings in the photoresist pattern and dissolve the underlying film. It is a highly scalable process—a large number of wafers can be loaded into a carrier and immersed in the wet bath at the same time. The main drawback of wet etching is its isotropic nature, especially on amorphous or polycrystalline films. The isotropic nature results in an etch rate that is equal along all directions. This produces an undercut below the photoresist pattern, as illustrated in Figure 7.1. As discussed in Chapter 6, isotropic etching results in a significant narrowing of the etched line and a loss of resolution, especially when the thickness of the film being etched is on the same order of magnitude as the lithographically patterned dimensions. Wet etching was widely used in electronic circuit manufacturing until the 1980s until the device geometries became too small for this process. Nevertheless, it is still used in research laboratories when the device dimensions are large and also in the manufacturing process for micro-electro-mechanical systems (MEMS).

The chemicals used in a wet etching process must satisfy three primary requirements:

- It must etch the material of interest at reasonable rates (not too slow or not too fast, resulting in etch times on the order of minutes)
- It must not significantly attack the photoresist
- It must not etch the underlying substrate or other materials on the substrate

Etch selectivity is defined as the ratio between the etch rate of the target material and the etch rate of the masking material (normally photoresist):

$$\text{Selectivity} = \frac{\text{Etch rate of the film}}{\text{Etch rate of the photoresist (or etch mask)}} \tag{7.1}$$

A large etch selectivity is the desired goal. One big advantage of wet chemical etching (compared to plasma etching) is the exceptionally high etch selectivity. The nature of chemical reactions is such that they have vastly varying reaction rates from one material to another. However, chemical action is not the only form of mask erosion. Mask deterioration can also occur through nonchemical actions such as mass diffusion or diffusion through pinhole defects. A photoresist mask, for example, will never be totally free of defects. The evaporation of solvents from the photoresist

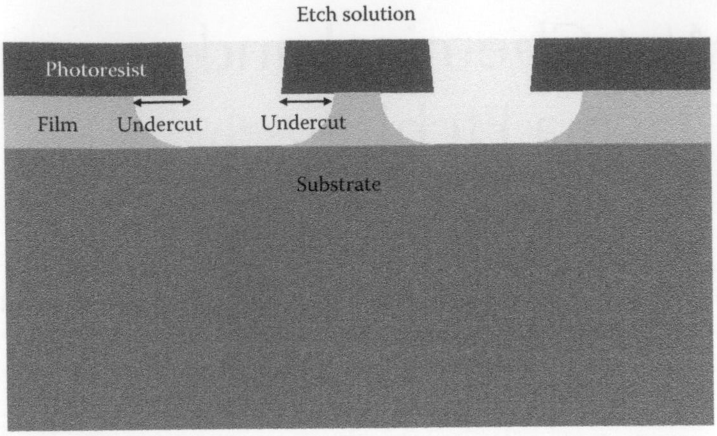

FIGURE 7.1 Wet etch profiles through a photoresist mask.

will often leave small pores through which the etchants can diffuse. Aggressive hard baking of the photoresist after the photolithography step can also produce stress cracks through which the etch chemicals can diffuse.

Another advantage of wet chemical etching (compared to plasma etching) is the low damage, since there are no ions or energetic particles involved. Since the etching process is strictly a surface action, the atoms immediately below the surface will remain damage free. In plasma etching, the ions penetrate a finite depth below the surface and can cause some surface and subsurface damage.

The overall etch rate in a wet chemical process is limited by one of the following rate-limiting steps:

- Transport of reactant species from the bulk solution to the surface
- Reaction at the surface
- Transport of end products from the surface to the bulk solution

Which step limits the overall reaction rate greatly depends on the solution temperature and viscosity. The surface reaction rate J_s (in $cm^{-2} \cdot s^{-1}$) is exponentially dependent on the temperature, as

$$J_s = kC_s e^{-\frac{E_A}{kT}} = k_s C_s \tag{7.2}$$

where

k_s is the rate constant
C_s is the volumetric concentration of the reactants at the surface
E_A is the activation energy
T is the absolute temperature
k is the Boltzmann constant

The transport rate of the reactants from bulk to the surface J_{bs} has a form

$$J_{bs} = h_g (C_b - C_s) \tag{7.3}$$

where

C_b is the concentration of the reactants in the bulk fluid
h_g is a diffusion coefficient that depends on the viscosity and boundary layer thickness

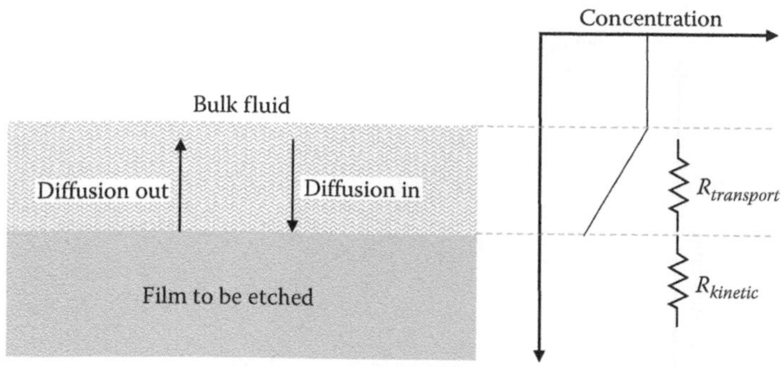

FIGURE 7.2 Flow of reactants into and away from the surface during wet chemical etching.

At higher solution temperatures and lower viscosities, the surface reaction in Equation 7.2 will be the rate-limiting step, and at lower temperatures and higher viscosities, the transport rate in Equation 7.3 will be the rate-limiting step. Under steady state, $J_s = J_{bs}$, from which we can get

$$J = J_s = J_{bs} = \frac{C_b}{\frac{1}{k_s} + \frac{1}{h_g}}. \tag{7.4}$$

The reactant concentration falls across the boundary layer from the bulk value C_b to the surface value C_s, which is then consumed by the reaction at the surface. These can be modeled as a kinetic resistance and a transport resistance acting in series as shown in Figure 7.2 [1], where $1/k_s$ is the kinetic resistance and $1/h_g$ is the transport resistance. The larger of the two resistances will be the limiting factor of the overall etch rate.

In any etch process, the uniformity of the etch across the wafer and from one wafer to another is an important consideration. The boundary layer thickness and the diffusion rate will vary depending on the flow characteristics and viscosity of the bulk fluid. These are difficult parameters to control in an etch bath. The temperature uniformity is much easier to maintain, especially on substrates such as silicon where thermal conductivity is high. Therefore, an etch process that is limited by surface reaction rate will yield a greater uniformity compared to one that is limited by the transport rate. This means, a chemical bath with a relatively high concentration of reactant species at a low bath temperature is preferred compared to a weak solution at high temperatures. However, from Equation 7.2, the temperature has to be sufficiently high to achieve a reasonable etch rate, but it should not be too high to make it a transport-limited reaction.

The rate-limiting steps can be evaluated by plotting the etch rate as a function of temperature on a log(rate) vs. $1/T$ scale, as shown in Figure 7.3. The straight-line portion is characteristic of the surface-limited reaction rate in Equation 7.2. The change in slope indicates a transition to the transport-limited range.

When photoresist is used as the etch mask, the etchant can also diffuse laterally, penetrating the photoresist/film interface. Depending on the film material and etchant, this is sometimes the most dominant failure mechanism during etching. The extent of this effect largely depends on the surface energy of the underlying film. The surface energy determines the adhesion of the photoresist to the film and the adhesion of the etchant to the film. The competition between these two effects will dictate if the photoresist will remain attached to the film or if it will allow the etchant to penetrate the interface. Once the etchant penetrates the interface, it can travel long distances due to the capillary effect and peel the entire photoresist off the film surface, as shown in Figure 7.4. For this reason, the surface is often coated (or primed) with a monolayer of organosilane such as hexamethyldisilazane (HMDS). This coating is used to adjust the surface energy to reach the best balance between the photoresist's adhesion and the etchant's adhesion to the film. The HMDS treatment was discussed

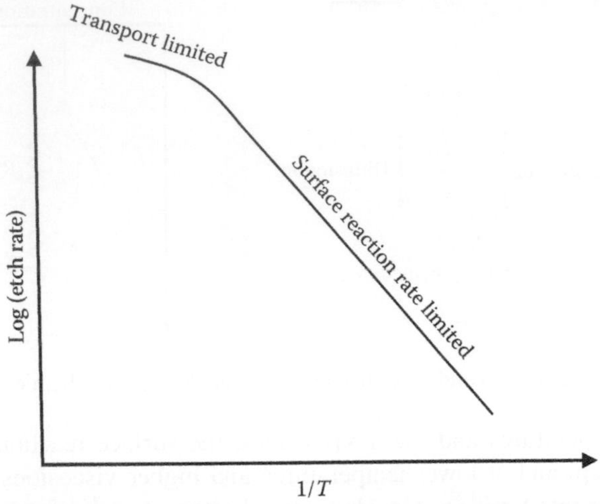

FIGURE 7.3 Plot of etch rate vs. temperature for determining the boundaries of surface reaction limit and transport limit.

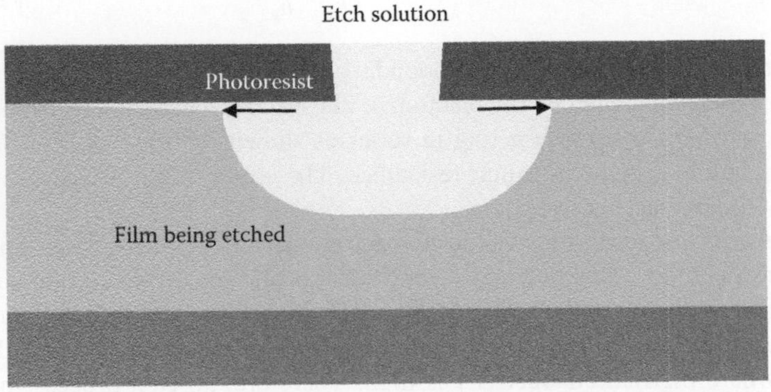

FIGURE 7.4 Penetration of the etchant between the photoresist and the film by capillary action.

in Chapter 6 in the context of aqueous development of the photoresist, but it is also important for the survivability of the photoresist in an aqueous wet etchant.

A large number of etch chemicals have been developed for different metals and dielectrics assuming silicon as the substrate and photoresists as the etch mask. These can be commercially bought in premixed containers for relatively low cost [2]. However, it should be noted that these etch chemicals may not necessarily work with other substrates, such as III–V semiconductors, unless they have been tested and verified to be chemically compatible [3].

In the following section, a few common wet chemical etch processes are discussed. The details of etching for a large number of materials, substrates, and photoresists have been well documented in the *CRC Handbook* [4] and the *IEEE paper* by Kurt Williams et al. [5].

7.1.2 Wet Chemical Etch of Selected Materials

7.1.2.1 Silicon Dioxide Etch

SiO_2 is one of the most widely used thin film in device fabrication processes. It is most commonly etched with hydrofluoric (HF) acid. Upon reaction with HF, SiO_2 forms a water-soluble by-product SiF_6^2 [6–9]. Full-strength HF (49%) etches SiO_2 too rapidly to have accurate control of etch depth. It also attacks photoresist by penetrating through microscopic cracks. Therefore, HF is normally used diluted, in the range of 1%–10%. However, a simple dilution with water (referred to as diluted HF—DHF) will not produce repeatable results because the concentration of fluorine ions will deplete more rapidly than hydrogen ions. This will produce a declining etch rate depending on the volume of the etchant and the volume of the SiO_2 being etched. The standard technique is to buffer the diluted HF with its conjugate base, such as NH_4F. This is referred to as buffered oxide etch (BOE) or buffered HF. The ammonium fluoride acts as a reservoir that supplies fluorine ions as they become depleted. Commercially produced BOE solutions also contain other agents and surfactants to improve their stability and surface wetting properties. The etch rate is ideally isotropic, although some variability between vertical depth and horizontal distance can arise due to ion transport bottlenecks in the solution, especially through very small structures. The biggest advantage of BOE is that it does not attack the photoresist or silicon. Once the SiO_2 film is removed, the etch process comes to a virtual stop, assuming the material under the SiO_2 is silicon. In addition to patterned etching, BOE is also used to clean bare silicon wafers by removing the native oxide.

BOE is produced and sold in various concentrations, such as 5:1 and 50:1. The 5:1 refers to 5 parts of 40% NH_4F mixed with 1 part of 49% HF. A 5:1 BOE will etch thermally grown silicon dioxide at approximately 15 Å/s and plasma-enhanced chemical vapor deposition (PECVD) silicon dioxide at 50 Å/s.

Even though BOE does not chemically etch the photoresist, it can diffuse through the resist film easily due to the extremely small size of the HF molecules, as illustrated in Figure 7.5. Once the HF reaches the photoresist/SiO_2 interface, it will start reacting with the SiO_2 and lift the resist film entirely from the substrate. Furthermore, if the surface is not adequately treated with an antiwetting agent, such as HMDS, the HF can also diffuse laterally through capillary action by penetrating the SiO_2/photoresist interface. As a result, even though the selectivity of this etch is large, it is not trivial to conduct prolonged etching of SiO_2 with BOE to achieve deep profiles [10].

A common difficulty with any etching is to determine when to stop the etch process. A significantly under-etched case is easy to identify by the coloration as well as the nonuniformities across the etched areas. However, a very thin layer of oxide will be difficult to distinguish from the bare surface by appearance alone. Even though the vertical etch rate comes to a stop when the oxide is fully consumed, it can still continue laterally. Over-etching can cause pattern widening and loss of resolution. There are a few practical techniques for identifying when an etch is complete. First, bare silicon surface is hydrophobic. Even the presence of a thin SiO_2 film will make it hydrophilic. This can be observed by lifting the wafers off the BOE solution and observing how the water drains. If SiO_2 is still present, the etchant (which is mostly water) will cling to the surface. Once the SiO_2 film is completely removed, the water will bead-up in the etched

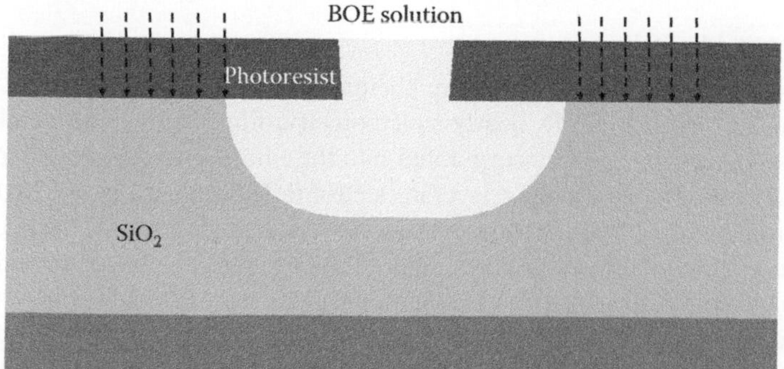

FIGURE 7.5 Mask erosion due to diffusion of BOE through the photoresist.

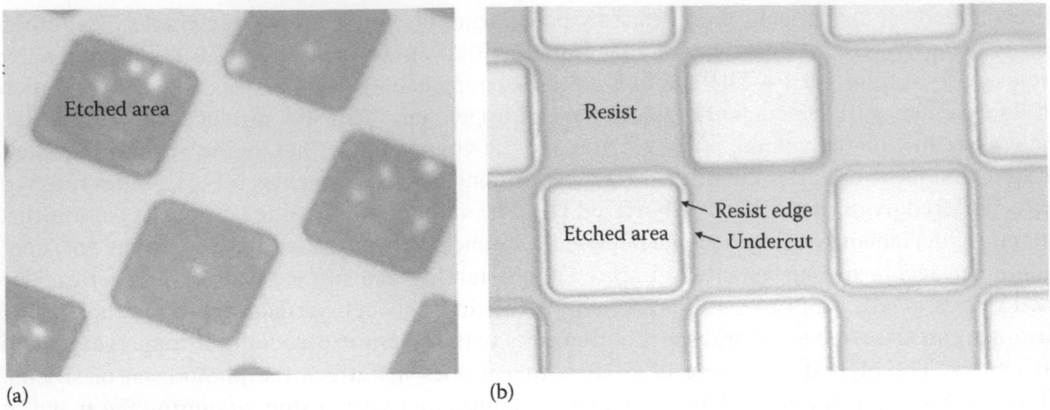

FIGURE 7.6 Optical microscope inspection of (a) under-etched, and (b) over-etched cases with BOE and a photoresist mask.

areas and roll off. This is a clear indication that the film has been fully etched. However, this only works if the wafer contains at least some areas that are large enough to observe with the naked eye. When the photoresist patterns are very small, it becomes difficult to identify the wetting characteristics. Novolak resins are also hydrophobic, so the wafer may not produce a noticeable wetting difference. The second technique is to look for an undercut using a microscope. For this, the wafer needs to be removed and dried periodically from the etchant and inspected under a microscope. When the film starts to undercut the photoresist film, two parallel lines will be observed at every photoresist edge (because the photoresist is partially transparent). This is shown in Figure 7.6. The outer edge is the photoresist edge, and the receding edge is the SiO_2 film. As long as the distance between the two edges is small, the etch can be terminated at an optimal point without compromising the resolution too much. Nevertheless, due to the resolution limit of optical microscopes the undercut becomes difficult to identify when the line widths are smaller than about 1 μm.

HF etching is also used in the release process in MEMS devices. Suspended devices have to be initially supported by an underlying film. SiO_2 is the most commonly used support film. After all the layers have been fabricated, a final etching is performed to dissolve away all the interior SiO_2 layers. This is known as the HF release step [11]. Since the HF has to be able to reach all the interior layers, the design of the suspended layer has to include access holes. In many MEMS devices, these access holes can be easily identified by the presence of a perforated structure. The holes in the perforations are designed to allow equal access of the HF to penetrate under the film to release the supporting SiO_2.

7.1.2.2 Silicon Nitride Etch

Silicon nitride is usually grown by low-pressure chemical vapor deposition (LPCVD) or PECVD, as discussed in Chapter 3. With LPCVD, nearly stoichiometric films (Si_3N_4) can be obtained. PECVD nitride tends to have excess hydrogen incorporated into the film due to the lower deposition temperatures. Silicon nitride is used as a passivation layer in electronic devices and also as a high refractive index film (index of approx. 2.0) in optical coatings and in optical waveguides. Another application of LPCVD silicon nitride is as a more robust etch mask compared to photoresist or silicon dioxide. For potassium hydroxide (KOH) etching of silicon, LPCVD silicon nitride is the only known material that can withstand extended etch times at high temperatures. Silicon nitride can be etched with phosphoric acid, but it requires elevated temperatures in the range of 150°C which makes photoresist unsuitable as an etch mask. In most cases, LPCVD-grown silicon nitride is etched in a fluorine plasma. PECVD-grown silicon nitride is not stoichiometric and can be etched to some degree with BOE with a photoresist mask.

7.1.2.3 Silicon Etch

Silicon can be etched in a solution of hydrofluoric (HF), nitric (HNO_3), acetic (CH_3COOH) acids, also known as the HNA etch. One of the characteristic of this etch is that it is isotropic, even when the silicon is crystalline. Photoresist can survive in this etchant as long as the HNA is highly diluted. Another etchant for silicon is KOH. This has a unique property of etching silicon at vastly different rates depending on the crystal orientation. The etching of silicon with KOH is discussed separately in Section 7.3.1. Other etchants that behave similar to KOH are tetramethyl ammonium hydroxide (TMAH) and ethylene diamine pyrocatechol (EDP) [12].

7.1.2.4 Aluminum Etch

Aluminum etchants contain nitric acid (HNO_3), phosphoric acid (H_3PO_4), and acetic acid (CH_3COOH). The nitric acid oxidizes the aluminum and the phosphoric acid etches the aluminum oxide, and the acetic acid is used for improved surface wetting. The etch rate is about 100 Å/s at 50°C. The reaction is exothermic, so agitation is necessary to maintain a uniform temperature and to prevent areas of high etch rates due to local heating.

7.1.2.5 Copper Etch

Copper can be etched with ferric chloride ($FeCl_3$) or with ammonium persulfate ($(NH_4)_2S_2O_8$), both of which are also used in etching copper traces on printed circuit boards (PCB). Etch rate for $FeCl_3$ is in the range of 0.5 µm/s at 50°C, which is too fast for most precise applications so it is mostly used in etching PCBs. Ammonium persulfate etches at a more reasonable rate of 100 Å/s at 50°C.

7.1.2.6 Titanium Etch

Titanium is etched with a mixture of hydrogen peroxide (H_2O_2) and HF acid. The H_2O_2 oxidizes the titanium, and the HF removes the oxide. Etch rates are in the range of 10–100 Å/s at 50°C depending on the concentration of the oxidizing agent and the HF.

7.1.2.7 Gold Etch

A mixture of hydrochloric (HCl) and nitric (HNO_3) acids, known as aqua regia, rapidly etches gold and other noble metals, but due to its strong oxidative property it will also attack photoresist. Instead, a mixture of potassium iodide (KI) and iodine (I_2) can be used. Its etch rate is slower, but it can be used with photoresist. Etch rate of aqua regia is about 100 Å/s at 30°C and the etch rate of $KI + I_2$ is about 25 Å/s at 30°C.

7.1.2.8 Silver Etch

Silver will also etch in the mixture of KI and I_2, just like gold. It can also be etched in a mixture of sodium hydroxide (NH_4OH) and hydrogen peroxide (H_2O_2).

7.1.3 Orientation-Dependent Wet Etching of Silicon

7.1.3.1 (100) Silicon Etch with KOH

Crystalline silicon has different chemical properties along different crystal axes. A (100) wafer surface will etch different than a (110) or (111) [13]. This effect is most widely exploited with KOH etching of silicon. KOH comes as solid pellets which can be easily dissolved in water. It will etch {100} and {110} surfaces significantly faster than {111}. The {111} planes can therefore be considered as etch-stop planes [14]. As described in Chapter 5, a (100) wafer surface will have {111} planes that are inclined at 54.7° to the surface. When a rectangular opening is etched, the resulting hole will have perfectly tapered sidewalls at 54.7°, and the etch will come to a stop when all the {111} planes meet at a point (in the case of a square opening) or along a line (in the case of a rectangular opening). This is known as orientation-dependent etching and is used extensively in MEMS device fabrication to make deep trenches and through-wafer holes. An example is shown in Figure 7.7.

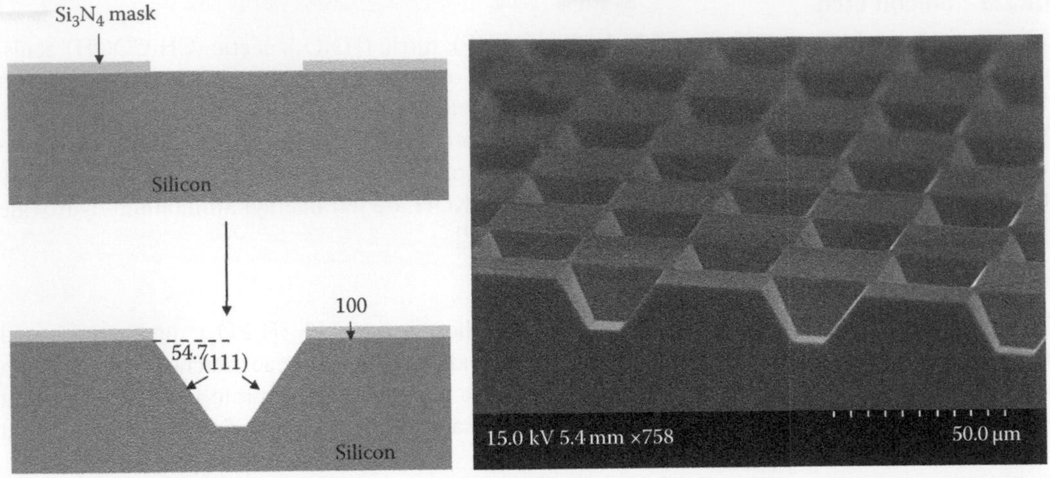

FIGURE 7.7 Etching of (100) silicon surface with KOH.

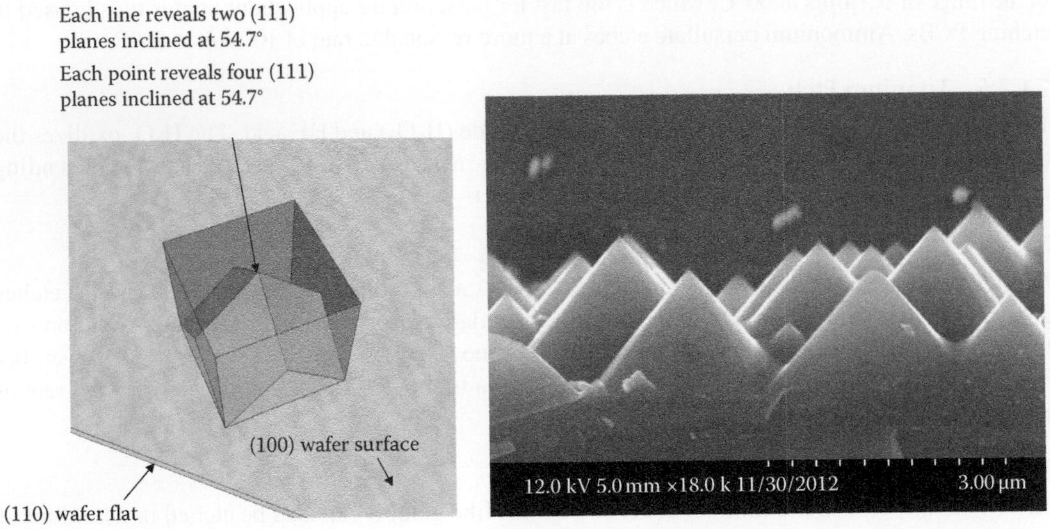

FIGURE 7.8 Pyramidal-shaped etched structures in KOH due to micro masking.

A single point mask, on the other hand, will create a pyramidal shape with four walls tapering outward at 54.7°, as shown in Figure 7.8. This could be an intentional mask or a contamination or micro masking from etch by-products that act as a temporary etch mask. It is not uncommon to see these pyramidal shapes shown on the bottom (100) planes of the etched structure [15].

As in any surface reaction rate-limited wet chemical etch, the etch rate increases exponentially with temperature as shown in Equation 7.1. The etch rates with KOH are known to reach a peak value at 20%–30% concentration and then decline both above and below these concentrations. Figure 7.9 shows the etch rates as a function of temperature for a 20% KOH concentration by weight, as an Arrhenius plot. It is commonplace to also add isopropyl alcohol (IPA) to the KOH solution to improve the anisotropy and roughness of the resulting etched faces [16,17]. For example, at 80°C the etch rate is 1.5 µm/min for the {100}-planes and 2.0 µm/min for the {110}-planes. With 25% IPA added to the KOH solution, the etch rates drop to 1.0 µm/min for the {100}-planes but

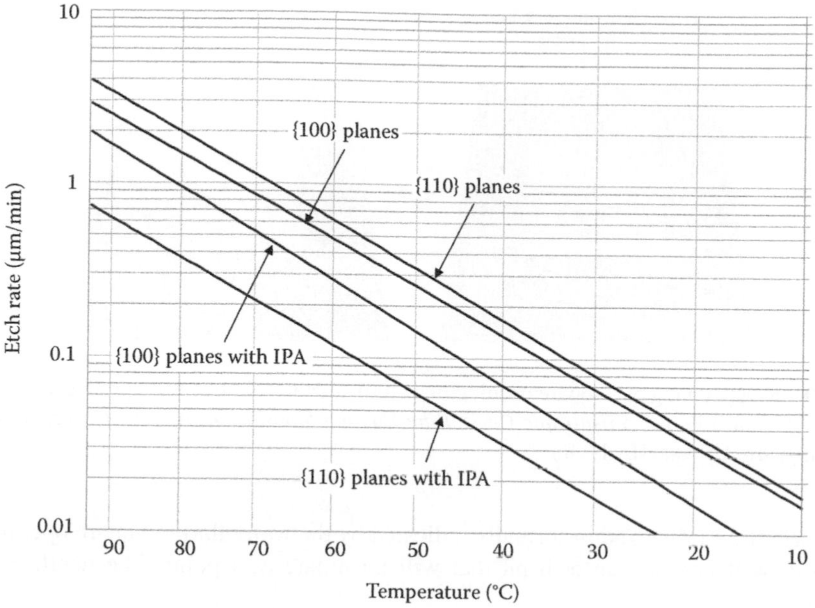

FIGURE 7.9 KOH etch rates at 20% solution concentration.

the {110} etch drops even more significantly to 0.4 µm/min, resulting in a marked improvement in the preferential etch for {100} compared to {110} [13]. The {111}-planes are not shown on this plot because their etch rate is about 400 times smaller.

The etch rates of silicon with KOH are several orders of magnitude greater than the etch rates discussed earlier in the context of thin films. Therefore, KOH is mostly used for etching deep troughs in the silicon substrate and rarely for etching thin films. It is possible, for example, to etch a through hole in a 400 µm thick silicon wafer in about 5 hours.

There are some additional etching considerations that are unique to KOH. The etchant is very strongly alkaline and will destroy most metals and photoresists [18,19]. Therefore, photoresist is an unsuitable masking film. Some metals like gold and chromium as well as dielectrics like Si_3N_4 are suitable etch masks because they do not get etched in KOH, but the films have to be free of pin holes when etching for long durations. KOH can easily diffuse through these pin holes and reach the underlying silicon. Evaporated films are generally unsuitable because they will have a large density of pin holes due to the low energy of the incident atoms during deposition. Sputtered films are slightly better but may still be unsuitable for prolonged etching with KOH. The most commonly used masking film is LPCVD silicon nitride. The high-temperature LPCVD growth can produce nearly pin hole-free films well suited for this application. Therefore, the photolithography has to involve a two-step process to do a mask conversion. The LPCVD silicon nitride film is first deposited on bare silicon wafers, and a photoresist mask is used to etch the nitride film, generally with a plasma since there aren't many wet chemicals that will etch silicon nitrde while leaving the photoresist intact. Furthermore, LPCVD is a high-temperature process (>800°C); therefore, the silicon wafers have to be free of any sensitive devices or metals. PECVD nitride can be used if there are devices on the silicon that are sensitive to temperature, but PECVD films are slightly inferior to LPCVD and may not hold up to long durations in the KOH bath. However, both are viable etch masks, and the choice ultimately depends on the intended application and process design [20].

In order to produce a square or rectangular etch on a (100) wafer, the mask opening has to be parallel to the lines where the {111} planes intersect the (100) surface (which is also where the {110}

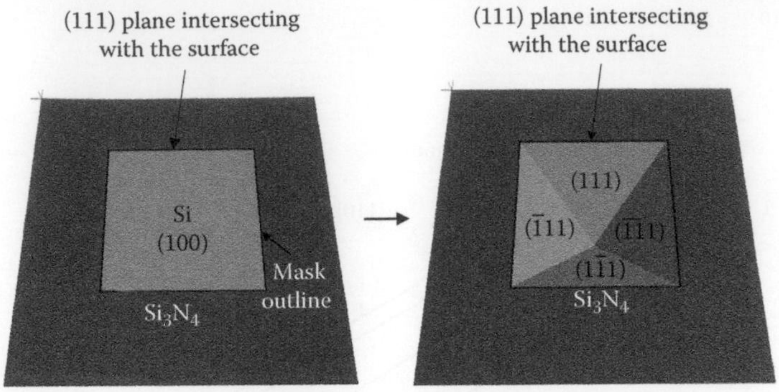

FIGURE 7.10 KOH etching of a square opening in a (100) silicon. (Figures were produced using the ACES software Zhu, Z. and Liu, C., *Anisotropic Crystalline Etching Simulation (ACES)*, University of Illinois at Urbana-Champaign, Urbana, IL, 1998.)

planes intersect the surface and is normally indicated by the wafer flat). A square opening as shown in Figure 7.10 will produce an etch pit that will terminate at a point. The depth can be easily calculated as

$$D = \frac{W \tan(54.7)}{2} \tag{7.5}$$

where
W is the width of the square opening
D is the depth (also see Figure 7.14)

For example, a 30 μm square opening will etch down to a depth of 21 μm. Continued etching after reaching the maximum depth will very slowly expand the top opening diameter because the etch rate along the {111} planes, although small, is not exactly zero.

Another aspect of orientation-dependent KOH etching is that if the mask opening is not perfectly aligned to the {111} planes that intersect with the (100) surface, due to a rotational error in photolithography, the etched hole will expand until it becomes bounded by {111} planes on all sides. This will result in an opening that is slightly larger than the original by

$$W' = W(\sin\theta + \cos\theta) \tag{7.6}$$

where
θ is the rotational error
W is the width of the original opening

This effect is illustrated in Figure 7.11.

Along the same line of argument, any arbitrarily shaped mask opening will also expand outward until the entire opening is bounded by rectangular {111} planes that intersect the (100) surface. This is essentially an undercut of the etch mask and can be exploited to make suspended structures like cantilevers, as illustrated in Figure 7.12. In this example, the mass of silicon under the cantilever is progressively removed until the etched structure is bounded by only {111} planes. Silicon nitride cantilevers are routinely made using this technique. The biggest challenge with this method is the stress in the silicon nitride film. Any intrinsic stress in the film will force the cantilever to curl up or down when it becomes fully released from underneath. Stress-free or stress-compensated silicon nitride films are essential for this process to work.

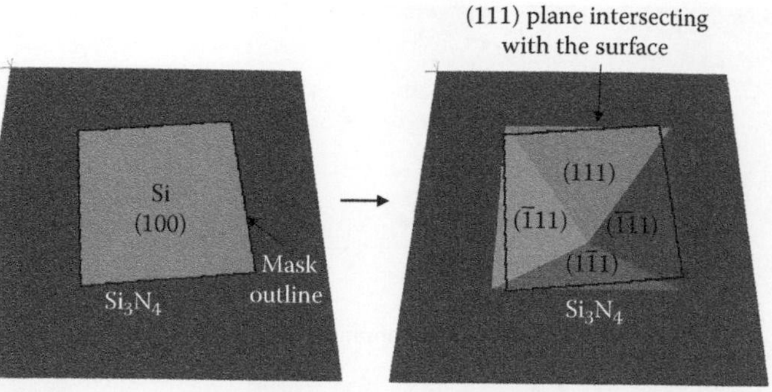

FIGURE 7.11 Expansion of the etched hole when the mask is rotated so that its edges are not parallel to the (110) surfaces. (Figures were produced using the ACES software Zhu, Z. and Liu, C., *Anisotropic Crystalline Etching Simulation (ACES)*, University of Illinois at Urbana-Champaign, Urbana, IL, 1998.)

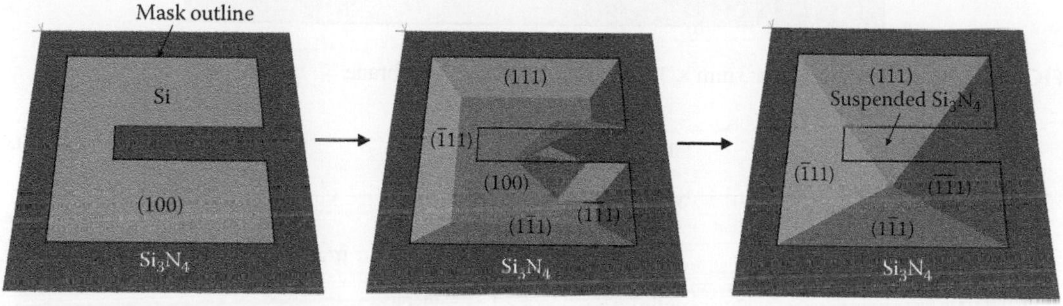

FIGURE 7.12 Suspended silicon nitride structures by KOH etching. (Figures were produced using the ACES software Zhu, Z. and Liu, C., *Anisotropic Crystalline Etching Simulation (ACES)*, University of Illinois at Urbana-Champaign, Urbana, IL, 1998.)

Assuming the silicon wafer is coated on both sides with LPCVD silicon nitride, the wafer can be etched all the way through to produce a suspended silicon nitride membrane on the opposite side. This process is used in pressure sensors, strain gauges, and MEMS microphones. The membrane can be as thin as 200 nm and still be robust enough to remain intact after the etch [22,23]. An example of a large 3 mm × 3 mm × 200 nm LPCVD silicon nitride membrane is shown in Figure 7.13. Such thin membranes will deflect under a tiny pressure difference, which can then be detected with a piezo-resistive or piezo-electric films fabricated on the membrane surface.

If the desired opening size on the bottom of the wafer is W', the required top opening can be calculated as

$$W = \frac{2D}{\tan(54.7)} + W' \tag{7.7}$$

where D is the wafer thickness, as shown in Figure 7.14. For example, if the wafer thickness is 400 μm, and the desired bottom opening is 250 μm, the top opening has to be 816 μm. As this example illustrates, one disadvantage of this process is that it requires the top opening to be significantly larger than the bottom side membrane opening.

7.1.3.2 (110) Silicon Etch with KOH

KOH etching of (110) wafers can produce vertical sidewalls along certain directions because some of the {111} planes intersect the surface at right angles [24]. However, the hole will be bounded by

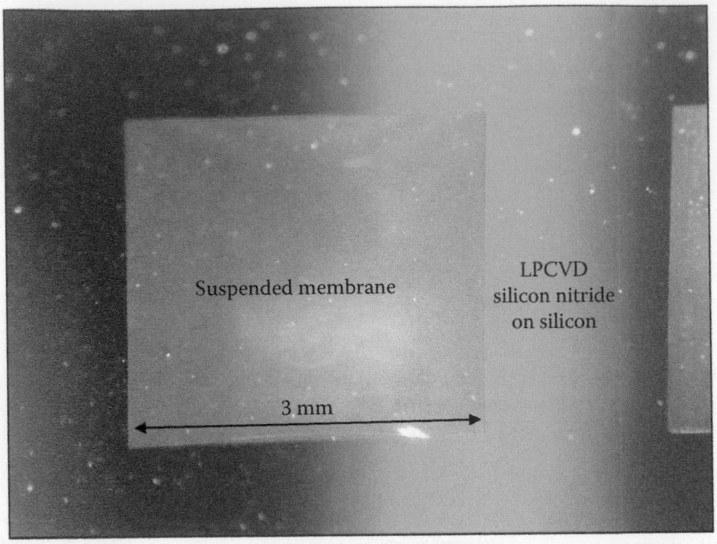

FIGURE 7.13 Photograph of a 3 mm × 3 mm Si_3N_4 suspended membrane.

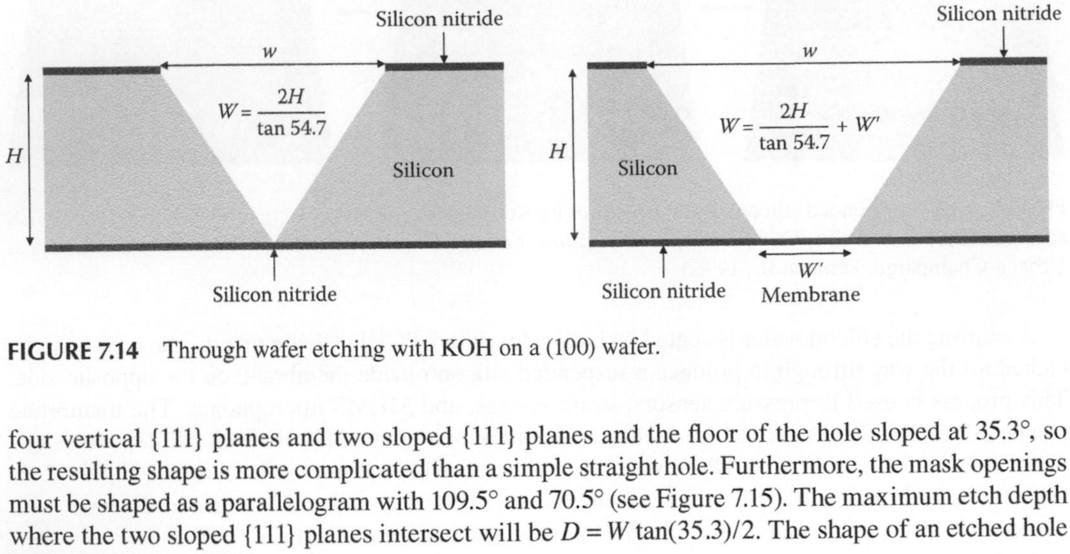

FIGURE 7.14 Through wafer etching with KOH on a (100) wafer.

four vertical {111} planes and two sloped {111} planes and the floor of the hole sloped at 35.3°, so the resulting shape is more complicated than a simple straight hole. Furthermore, the mask openings must be shaped as a parallelogram with 109.5° and 70.5° (see Figure 7.15). The maximum etch depth where the two sloped {111} planes intersect will be $D = W \tan(35.3)/2$. The shape of an etched hole is shown in Figure 7.16.

7.1.3.3 Other Etchants for Orientation-Dependent Etching of Silicon

KOH is incompatible with complementary metal oxide semiconductor (CMOS) device fabrication because the potassium ions can diffuse into the MOS gate dielectric and cause leakages. Therefore, clean rooms where CMOS devices are made will generally prohibit the use of KOH. Furthermore, KOH will attack SiO_2 at a slow rate of about 3 nm/min, which makes SiO_2 unsuitable as an etch mask for prolonged etching. A CMOS-compatible alternative is tetramethyl ammonium hydroxide (TMAH), which is the same ingredient in metal-ion-free photoresist developers, except the developers are made at a much smaller concentration. TMAH behaves very similar to KOH, but the etch ratio between the {100} and the {111} is only about 30 (compared to 200 for KOH) [25]. Additionally, unlike KOH, TMAH does not appreciably attack SiO_2. Nevertheless, KOH is significantly cheaper than TMAH, so it is more widely used where contamination of CMOS is not a concern. Another alternative is EDP, which reportedly produces smoother sidewalls compared to TMAH, but it has

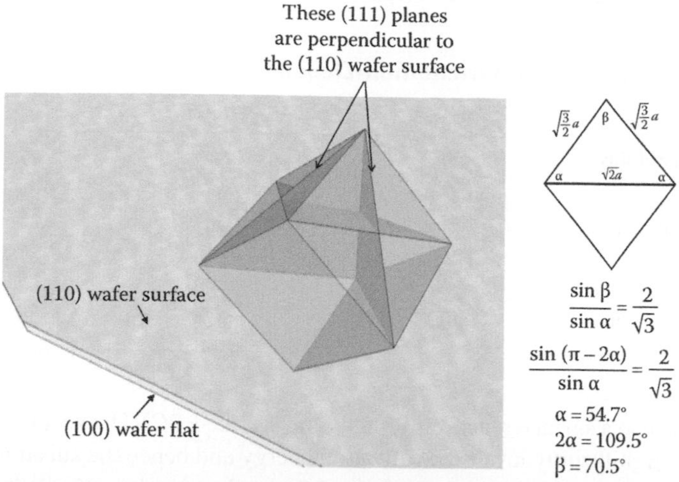

FIGURE 7.15 Intersection of the {111} planes on a (110) wafer surface.

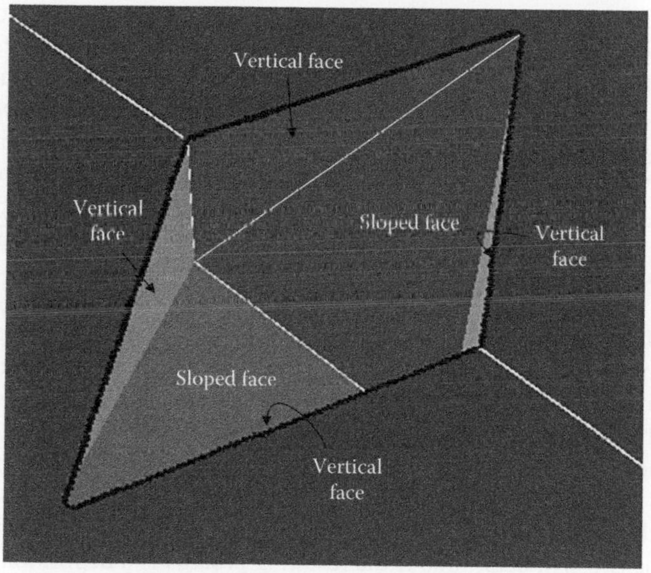

FIGURE 7.16 Shape of an etched hole with four vertical walls and two sloped floors on a (110) wafer. (Figures were produced using the ACES software Zhu, Z. and Liu, C., *Anisotropic Crystalline Etching Simulation* (*ACES*), University of Illinois at Urbana-Champaign, Urbana, IL, 1998.)

also been shown to be carcinogenic so its use should generally be limited. A summary of the three major orientation-dependent etchants used with silicon is shown in Table 7.1.

7.2 PLASMA ETCHING

Plasma etching is best understood from the perspective of chemical vapor deposition (CVD). Referring to our earlier discussion on CVD in Chapter 3:

- LPCVD was a surface reaction-limited process. As a result, the deposition rate was independent of the gas flow characteristics and was primarily a function of the substrate surface temperature. The main drawback of LPCVD was the requirement for high substrate temperatures.

TABLE 7.1
Comparison of Different Orientation-Dependent Etchants for Silicon

	KOH	TMAH	EDP
CMOS Compatibility	**No**	**Yes**	**Yes**
Etch rates	1 μm/min (80°C)	0.5 μm/min (80°C)	0.3 μm/min (80°C)
(100)/(111) selectivity	200	30	20
Hazards	Corrosive	Corrosive	Carcinogenic
(100)/SiO_2 selectivity	300	2000	30,000
(100)/Si_3N_4 selectivity	High	High	High

- PECVD was also a surface reaction-limited process like LPCVD, except the plasma made it possible to significantly lower the activation energy and hence the substrate temperature. The decomposition of the reactant species and the ion bombardment were the main reasons for this reduction in temperature.

Now consider a CVD process where the film growth rate is negative. This could occur, for example, if the gases react with the substrate to produce a volatile by-product instead of a nonvolatile thin film. This is in fact the chemical vapor etching (CVE) process. A low-pressure chemical vapor etching (LPCVE) process can be considered a corollary to the LPCVD deposition process. All aspects of LPCVE will be identical to LPCVD, except the substrate is etched rather than deposited upon. LPCVE would be a surface reaction-limited process just like LPCVD. Since the role of diffusion is minimal, the etch rate will be independent of the gas flow characteristics and, more importantly, almost independent of the etch geometries. However, just like LPCVD, LPCVE would require high substrate temperatures. This makes it unsuitable for etching with photoresist or other delicate masking material.

Now consider plasma enhanced chemical vapor etching (PECVE). By introducing a plasma at the substrate surface, the reaction temperature can be significantly reduced, just like in PECVD. This is in fact the most commonly used configuration in plasma etching systems. Despite the similarity to PECVD, this process is almost never referred to as PECVE. Instead, it is most commonly known as reactive ion etching (RIE) or simply as "plasma etching." It is also sometimes referred to as "dry etching" to make a distinction with wet chemical etching.

In any chemically reactive process, both deposition and etching take place simultaneously. The balance between the two will determine if the resulting process is a deposition process or an etch process. If the rate of production of nonvolatiles exceeds that of volatile products, the process can be considered a deposition process, and if volatile products exceed nonvolatile products it can be considered an etch process. This is depicted in Figure 7.17. Plasma etching is the preferred etching technique in nearly all of today's nanofabrication processes. The basic principles and practice of plasma etching will be covered in this chapter. For more specific details on this topic, the reader is referred to the references listed at the end of the chapter [26–30].

7.2.1 Basic Construction of a Plasma Etcher

The basic plasma etching system is almost identical to a PECVD system. It consists of a vacuum chamber with a controlled gas in-flow rate and a throttled pumping rate to control the chamber pressure. The cathode electrode is energized by an RF source (typically 13.56 MHz) and the substrate to be etched is placed on this electrode, as illustrated in Figure 7.18. The gases decompose in the plasma and generate ions and highly reactive free radical species. This allows the substrate temperature to be significantly lowered compared to LPCVE. Etching can be done at room temperature or even

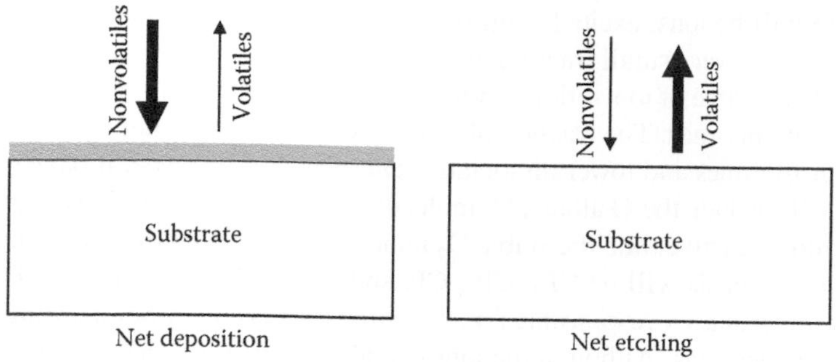

FIGURE 7.17 Chemical vapor process with net deposition and net etching.

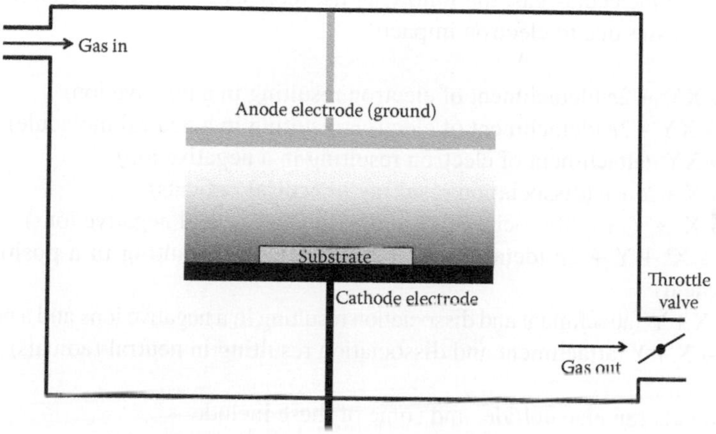

FIGURE 7.18 Basic configuration of a parallel plate plasma etcher.

below room temperatures. However, because etch rates have an exponential temperature dependence, the substrate temperature has to be maintained very stable and very uniform.

Similar to LPCVD, the surface reaction rate can be written as

$$J_s = k_s P_s \tag{7.8}$$

where

$$k_s = Ae^{-\left(\frac{E_a}{kT}\right)} \tag{7.9}$$

where

E_a is the activation energy of the rate-limiting step
P_s is the partial pressure of the reactant species on the substrate surface

Therefore, using this simplistic model and excluding all other factors that influence etch rate, we can see that increasing the pressure and temperature will increase the etch rate.

7.2.2 Free Radicals and Ions in a Plasma and Their Roles

As discussed in Chapter 2, the number density of gas molecules is $N = 9.6 \times 10^{18} \frac{P}{T}\ \text{cm}^{-3}$ where P is in Torr and T is in Kelvin. For example, at 200 mT and 30°C there will be 6.4×10^{15} molecules/cm^3.

In a plasma, there will be ions, excited neutral species (free radicals), and unexcited feed gas molecules. In general, only a very small fraction of the gas will be free radicals and an even smaller fraction will be ions. The degree of ionization can range from 10^{-5} in conventional parallel plate plasmas to 10^{-2} in high density plasmas. The fraction of free radicals is much higher, in the range of 1%–10%, due to their longer lifetimes and lower dissociative energies. For instance, if oxygen is the feed gas, the free radicals will include the O atoms, O_3 molecules, and other metastable oxygen species, all of which are much more reactive than the stable O_2 molecule. The ions will be O_2^+ and O^-. In the case of CF_4 gas, the free radicals will be CF_3, CF_2, CF, and F, and the ions will be CF_3^+, CF_3^-, and F^-. The positive ions are primarily responsible for sustaining the plasma by bombarding the cathode and releasing secondary electrons. Although the ions can also be reactive, they are typically far smaller in number than the free radicals and are not considered to be the primary source of etching.

The collision of electrons with the gas species can produce a number of different outcomes. Assuming XY is the molecular gas, the following list represents some of the common dissociative and ionization processes due to electron impact:

- $e + XY \rightarrow XY^+ + 2e$ (detachment of electron resulting in a positive ion)
- $e + XY^- \rightarrow XY + 2e$ (detachment of electron resulting in a neutral molecule)
- $e + XY \rightarrow XY^-$ (attachment of electron resulting in a negative ion)
- $e + XY \rightarrow X + Y + e$ (dissociation resulting in neutral radicals)
- $e + XY \rightarrow X^+ + Y^- + e$ (dissociation resulting in positive and negative ions)
- $e + XY \rightarrow X^+ + Y + 2e$ (detachment and dissociation resulting in a positive ion and a neutral radical)
- $e + XY \rightarrow X + Y^-$ (attachment and dissociation resulting in a negative ions and a neutral radical)
- $e + XY^+ \rightarrow X + Y$ (attachment and dissociation resulting in neutral radicals)

The ions and radicals can also collide, and some of these include

- $X^+ + Y^- \rightarrow XY$ (recombination resulting in a neutral molecule)
- $X^+ + Y^- \rightarrow X + Y$ (recombination resulting in neutral radicals)
- $X + Y^- \rightarrow XY + e$ (recombination and detachment resulting in a neutral molecule)
- $X + Y \rightarrow XY$ (recombination resulting in a neutral molecule)
- $X + Y \rightarrow XY^+ + e$ (recombination and detachment resulting in a positive ion)

The free radicals are neutral species and are highly reactive, which are denoted by X and Y in these equations. They are the main ingredients in a chemical etch process. In the case of CF_4 gas, the F radical is the most reactive species responsible for etching silicon. Being neutral species, they are not driven along any particular direction by the electric fields in the plasma. Their transport is governed primarily by diffusion. In the absence of any other effects like ion bombardment, this will result in an isotropic etch profile, just like in wet chemical etching. From Equations 7.8 and 7.9, we can see that temperature has a profound effect on the etch rate due to the Arrhenius relationship $k_s = Ae^{-(E_a/kT)}$ (this ignores other rate-limiting steps in the etch process which are discussed next). Increasing the free radical density will also result in an increased etch rate. This can be accomplished by increasing the pressure or by increasing the RF discharge power, or both.

However, it is not possible to create free radicals in a plasma without also creating charged species. Charged species are required for sustaining the plasma. Positive ions strike the cathode to release secondary electrons that sustain the plasma. But they can also release neutral atoms from the cathode in the form of sputtering. If the substrate is placed on the cathode electrode, this will result in sputter etching of the substrate, and the etch rate would depend on the sputter yield of the substrate. In the case of an inert gas like argon, this will result in a purely mechanical removal of the substrate, but in the case of a gas like CF_4, the ions can be partially reactive. Nevertheless, in a typical plasma, the reaction rate due to the free radical density far exceeds that of the ion density, so for all practical

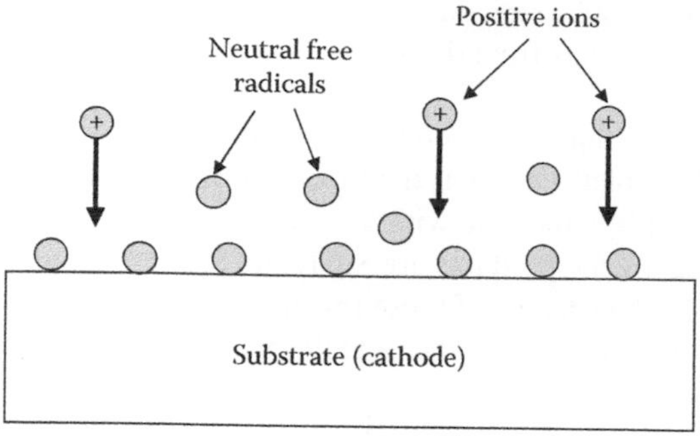

FIGURE 7.19 Illustration of the role of free radicals (chemical action) and positive ions (sputter action) in a plasma etch process.

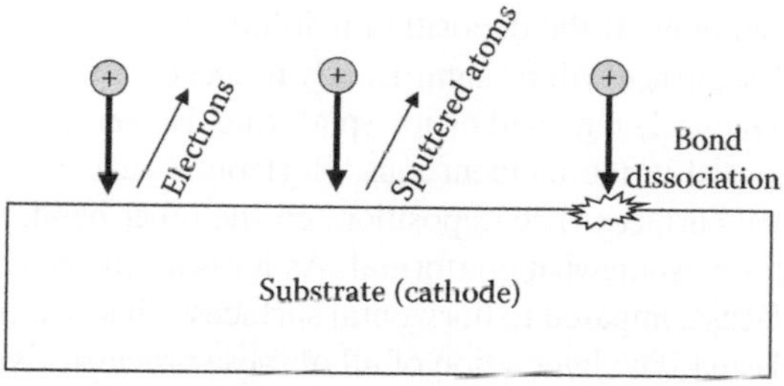

FIGURE 7.20 The three main roles of positive ions in a plasma etch process.

purposes the role of ions can be considered as primarily contributing to a nonchemical mechanical action only. For this reason, the term "Reactive Ion Etching (RIE)" is incorrect [28]. A more apt term is "ion-assisted reactive etching." Alternatively, an even better descriptor is PECVE, which draws from the similarity with PECVD. Nevertheless, RIE is the most widely used term in plasma etching-Figure 7.19 illustrates the role of free radicals and positive ions in a plasma etch process.

The overall etch rate is not a simple sum of sputter rate and chemical etch rate. In addition to sputtering, the ion bombardment can also break or loosen the substrate bonds and increase their reactivity with the free radicals in the plasma. The resulting increase in etch rate will be significantly larger than the sum of the two effects alone. Ion bombardment effectively results in a reduction of the activation energy E_a, which from Equation 7.9 we can see that it will result in an exponential increase in the etch rate. Furthermore, due to the impingement direction of the ions, this reduction in activation energy occurs only on horizontal surfaces of the substrate. The vertical surfaces like sidewalls will not experience this increase in etch rate. This is one of the several reasons for the anisotropic etch profiles in a plasma reactor.

In summary, ion bombardment in plasma etching results in the following primary effects (see Figure 7.20):

- Sustaining the plasma by releasing electrons from the cathode
- Purely mechanical sputter etch
- Bond dissociation resulting in a reduced activation energy E_a on horizontal surfaces

The etch process itself consists of a number of steps that occur in a sequential fashion. The overall etch rate will be limited by the slowest step in the sequence. The exact sequence can vary depending on the specific reaction, but in general the process will involve at least some of the following steps:

1. Dissociation of substrate surface bonds by the incident ions at a rate of R_{ion}
2. Adsorption of free radicals on the substrate surface at a rate of R_{ads}

3. Chemical reaction of the substrate atoms with the radicals to produce by-products R_{ch}
4. Desorption of the by-products from the surface R_{des}

In addition, any or all of these steps can be enhanced by the rate of ion bombardment. For example, if the by-products are highly volatile species, their desorption rate will be large and mostly independent of ion bombardment. This is the case with silicon etching where the by-products are SiF_2 and SiF_4. With chlorine etching, the by-products are $SiCl_2$ and $SiCl_4$. These are less volatile and require ion bombardment to desorb. The details of these reactions are discussed later in this chapter.

Using the sequence of these four steps, the overall reaction rate can be written as

$$\frac{1}{R} = \frac{1}{R_{ion}} + \frac{1}{R_{ads}} + \frac{1}{R_{ch}} + \frac{1}{R_{des}}. \tag{7.10}$$

The smallest of these rates will dominate the overall etch rate R. Along with this etching process, a deposition process can also occur in parallel. The rate and type of the deposition will depend on the plasma gas and process conditions. If the deposition is inhibitory (i.e., if it forms a protective film), step 1 and step 2 in the etch sequence will be significantly reduced or even come to a halt. The protective film will also be simultaneously removed by ion sputtering. However, this removal can only occur along surfaces that are exposed to the incident ions. Horizontal surfaces will experience a greater sputter removal than vertical surfaces. The deposition, on the other hand, will have similar characteristics to PECVD and will be somewhat conformal. As a result, the etch rate will be significantly reduced along vertical surfaces compared to horizontal surfaces. This is another reason for the anisotropic etch property in a plasma. The interaction of all of these processes is illustrated in Figure 7.21.

The arrival energy of the ions on the substrate will be determined by the DC bias and the gas pressure. Higher DC bias will result in a higher bombardment energy, and higher pressure at a given DC bias will result in a lower energy due to the increased collisions of the ions with the neutral gas molecules. The three main parameters of interest in a plasma etching are as follows:

1. Free radical density (which controls the chemical etch rate)
2. Current density (which directly translates to the ion flux density at the substrate)
3. DC bias (which indicates the bombardment energy of the ions)

In general, increasing the RF discharge power and/or the pressure will increase the ion density and the free radical density. Increasing the pressure while holding the discharge power constant will result in a lowering of the DC bias. This is because the higher ion density will lead to a higher ion current toward the cathode, and hence a lower RF amplitude voltage will be required to maintain

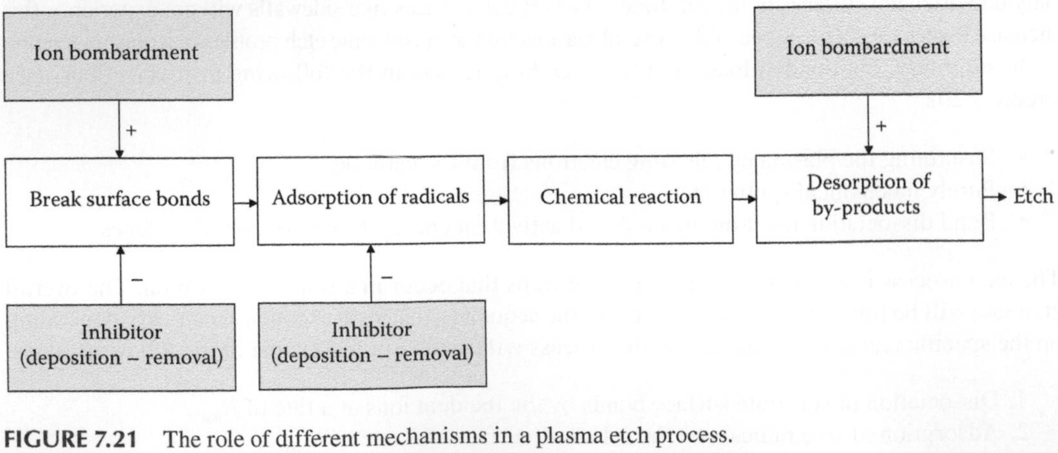

FIGURE 7.21 The role of different mechanisms in a plasma etch process.

the same power dissipation. Increasing the RF discharge power while holding pressure constant will result in an increase in DC bias. As we can see, all of these are interrelated parameters and it is not possible to exercise independent control over the ion density, free radical density, and DC bias. This is one of the drawbacks of the parallel plate plasma etching configuration. The ICP–RIE configuration described next provides a greater degree of control to overcome at least some of these parameters.

7.2.3 Inductively Coupled Plasma Etching

Whereas in a parallel plate capacitive configuration the plasma is excited via a time-varying electric field, in an inductively coupled plasma (ICP) the energy is supplied via a magnetic field produced from a time-varying current. In the most typical ICP configuration, a coil is wound around the outside of the vacuum chamber and is excited by a RF current. The plasma is generated and sustained by the continuous ionization of the gas by the time-varying magnetic field. This field will accelerate the electrons in a circular path, but the ions will remain stationary since there will be no net static electric field in any direction [31,32]. As a result, even though there will be plenty of ions present in the plasma (which will be nearly equal to the number of electrons), none of the ions will bombard the substrate. Increasing the RF excitation will increase the plasma density, and unlike in a parallel plate configuration, the ion energy incident on the substrate will not increase. Furthermore, since the entire excitation source is located outside the vacuum chamber, the system is less prone to contamination from the electrodes and leads. For these reasons, the ICP configuration is widely used in atomic emission spectroscopy and mass spectroscopy, in addition to plasma etching [33].

One could also combine the ICP with a capacitive excitation by providing a separate RF source to the wafer electrode, using the chamber wall as the grounded anode. This will induce a negative DC bias on the electrode and attract the ions toward the wafer causing ion bombardment. This second RF power is referred to as the bias power or RIE power. The DC bias can be increased or decreased by adjusting the bias power, without significantly affecting the plasma density or the ion density. This effectively allows one to decouple the plasma density from the ion bombardment energy, giving more flexibility to select between a purely chemical etch to a purely mechanical sputtering etch to anywhere in between. Such a configuration is also referred to as an ICP–RIE system. An example of an ICP–RIE is illustrated in Figure 7.22.

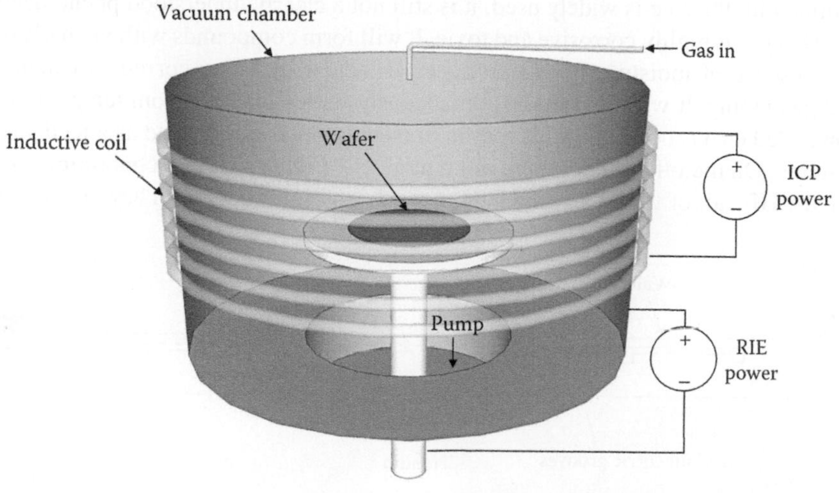

FIGURE 7.22 ICP–RIE plasma configuration.

7.2.4 Substrate Temperature

Due to the exponential dependence of etch rates on temperature, as shown in Equation 7.9, the wafer temperature needs to be maintained constant and uniform to achieve a consistent and reproducible etch rate. This is not trivial due to the heating effects from the plasma and the etch reaction itself. Heating or cooling with electric elements will generally have a slow time constant, and it is also difficult to provide good thermal contact with the wafer due to the vacuum conditions and the uneven backside surface of most wafers. Placing such elements inside the chamber may also introduce unwanted contaminations.

A commonly used approach is to float the wafer on a cushion of temperature-controlled pressurized helium gas, as shown in Figure 7.23. This provides good thermal contact to the wafer without any heat transfer compounds or pastes. The wafer typically has a small thermal mass, so it is able to reach the temperature of the helium much more quickly than if it were thermally bonded to something with a larger thermal mass. Due to the continuous flow of helium it is also possible to induce quick changes in the wafer temperature even during the etch process. Furthermore, helium is inert and light, so any small leaks into the chamber will not significantly alter the reaction chemistry or cause additional sputtering. Nevertheless, the wafer needs to be held down by some means, such as clamps or electrostatic forces to prevent excessive helium from escaping into the chamber. Known as back-side helium cooling, this approach is widely used in etch reactors to maintain a constant temperature [34].

7.2.5 Silicon Etching

For silicon, the main requirement is to create silicon compounds that are volatile at the working substrate temperature. The end products must have a high vapor pressure so they can thermally evaporate or they must have a high enough sputter yield to be removed by ion sputtering. Examples include silicon tetra fluoride (SiF_4), silicon tetra chloride ($SiCl_4$), and silane (SiH_4). Of these, etching silicon with fluorine to produce SiF_4 is the most widely used method. SiF_4 is a gas at room temperature with a melting temperature of −90°C. If silicon can be combined with fluorine (F) free radicals, then it can be converted to the gaseous SiF_4 and be pumped out of the chamber resulting in the clean removal of the silicon atoms. This is the basic idea behind all types of plasma etching. However, the reaction is rarely a straightforward single-step reaction. For example, silicon can react with fluorine radicals to produce SiF_2, which can then desorb from the surface by ion bombardment, and then subsequently recombine with fluorine to form SiF_4. When a number of parallel reaction pathways exist, the one with the lowest activation energy will dominate the overall mechanism. Even though silicon etching with fluorine is widely used, it is still not a clearly understood phenomenon.

Fluorine (F_2) gas is highly corrosive and toxic. It will form compounds with virtually everything, and in the presence of moisture it will produce HF acid which can corrode vacuum chambers, pumps, and gas piping. It will also react spontaneously with silicon at room temperature resulting in an uncontrolled etch process. For this reason, molecular F_2 is rarely used as a feed gas in plasma etching systems. On the other hand, there are a number of other fluorine-containing gases such as SF_6, CF_4, and CHF_3, all of which are much more stable and safer than F_2. They are also fairly inert.

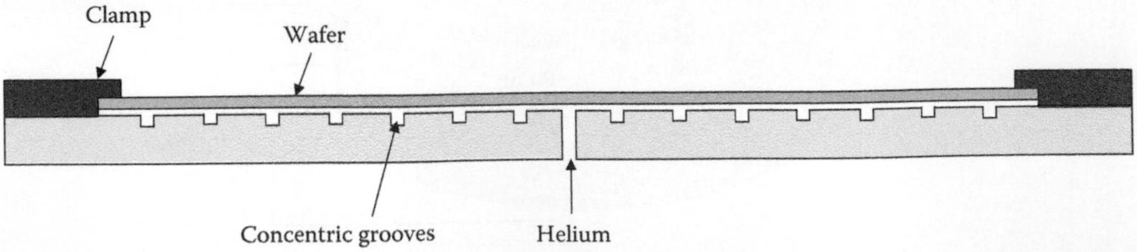

FIGURE 7.23 Helium backside cooling.

Yet, once excited in the plasma, they can be used to generate copious amounts of atomic F free radicals. Therefore, we can use plasma activation as a means to use safer and stable gases in place of the more reactive F_2 gas. SF_6 and CF_4 are two of the most commonly used etch gases for silicon [35].

There is, however, an important consideration with these gases. They contain at least one other element in addition to the desired fluorine atoms. This can significantly alter the way the reaction proceeds. The sulfur in SF_6 and the carbon in CF_4 can produce nonvolatile by-products during the reaction. Carbon and fluorine can produce solid fluorocarbons which are nonvolatile and inert to further chemical reactions. Any accumulation of these materials on the substrate will passivate the substrate and prevent further etching from taking place. This will result in a film growth rather than etching. Therefore, it might initially appear that these gases are unsuitable for etching silicon compared to pure F_2. This is where the ions in the plasma play a critical role. These ions can be utilized to remove the nonvolatile deposition via the sputtering action. Once the nonvolatile films are sputtered away, new silicon will be exposed for continued reaction and etching. In addition, these nonvolatile deposition actually improves the overall etch quality. When straight F_2 is used as a feed gas, the aggressive nature of fluorine will produce pitting and surface roughness. The nonvolatile fluorocarbon compounds from CF_4 have a moderating effect on this and generally produce smoother surfaces and sidewalls.

7.2.5.1 SF_6 Plasma for Etching Silicon

SF_6 (sulfur hexafluoride) will decompose in the plasma to produce F radicals and a number of S–F negative and positive ions, as well as sulfur atoms. SF_6 and other neutral SF_x species are dissociated by electron impact such as

$$e + SF_x \rightarrow SF_{x-1} + F + e. \tag{7.11}$$

The F radicals then react with Si to form SiF_4:

$$Si + 4F \rightarrow SiF_4. \tag{7.12}$$

One of the possible by-products of the reaction in Equation 7.11 is sulfur. In most SF_6 etching, this can be visually seen on the wafer as a milky white film. However, sulfur is not a hard passivating film, so it does not prevent the etch from continuing.

SF_6 is a highly electronegative gas. This means it has a very high tendency to attract electrons and produce negatively charged ions. In an SF_6 plasma, the number of negative ions can easily exceed the number of electrons. These negative ions are generated by electron impact ionization and are lost due to recombination in the volume of the plasma. Due to their negative charge, they are repelled from the cathode, but they also do not normally reach the anode region. Therefore, these negative ions play a minimal role in the etch process. Nevertheless, they significantly affect the electrical characteristics of the plasma. The plasma can be considered to be approximately charge neutral:

$$n^+ - n^- - e = 0 \tag{7.13}$$

where

n^+ is the number of positive ions
n^- is the number of negative ion
e is the number of electrons

In an electropositive gas, such as argon, the only ionized species are positive, which results in an equal number of electrons and positive ions:

$$e = n^+. \tag{7.14}$$

In an electronegative gas such as SF_6, the presence of negative ions will result in a smaller number of electrons compared to the positive ions:

$$e = n^+ - n^- \tag{7.15}$$

$$e < n^+. \tag{7.16}$$

Following the discussion on how a DC self-bias is created at the cathode from an RF excitation, the lower electron density compared to the positive ion density will result in a lower DC bias. Therefore, for the same discharge power, the ion bombardment with SF_6 will be significantly lower compared to an electropositive gas such as Ar. Furthermore, the dissociation of SF_6 has fairly low threshold energies, some of which are shown here [36]:

$$SF_6 \rightarrow SF_5 + F\ (\Delta G = +340\ \text{kJ/mol})$$

$$SF_5 \rightarrow SF_4 + F\ (\Delta G = +179\ \text{kJ/mol})$$

$$SF_4 \rightarrow SF_3 + F\ (\Delta G = +296\ \text{kJ/mol})$$

$$SF_3 \rightarrow SF_2 + F\ (\Delta G = +245\ \text{kJ/mol})$$

$$SF_2 \rightarrow SF + F\ (\Delta G = +351\ \text{kJ/mol})$$

$$SF \rightarrow S + F\ (\Delta G = +76\ \text{kJ/mol})$$

The low dissociation energies combined with the low DC bias results in an etch process that tends to be mostly driven by chemical action than by mechanical action. This will lead to a partly isotropic etch profile, even in an RIE-only configuration. Furthermore, due to the lack of passivation, the etch tends to produce rough surfaces as shown in Figure 7.24, which is similar to pure F_2 gas [37].

When performed in an ICP-only configuration, due to the lack of any ion bombardment as well as the higher free radical density, the etch profile will become much more isotropic and highly

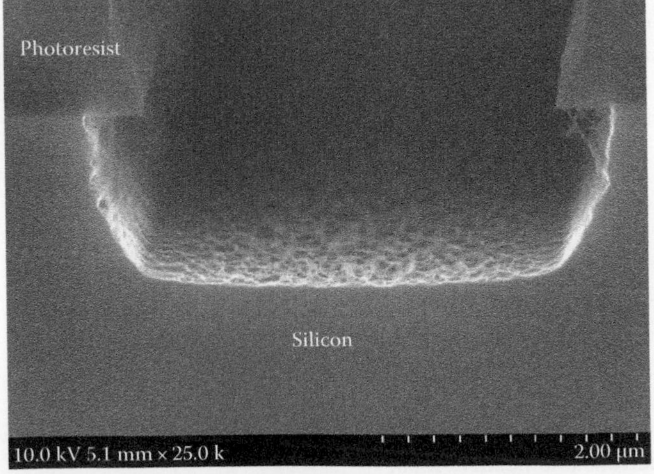

FIGURE 7.24 Silicon etch profile with an SF_6 plasma, at a pressure of 20 mT and a discharge power of 2.2 W/cm² (100 W total) and a DC bias of −200 V. Etch time was 5 min.

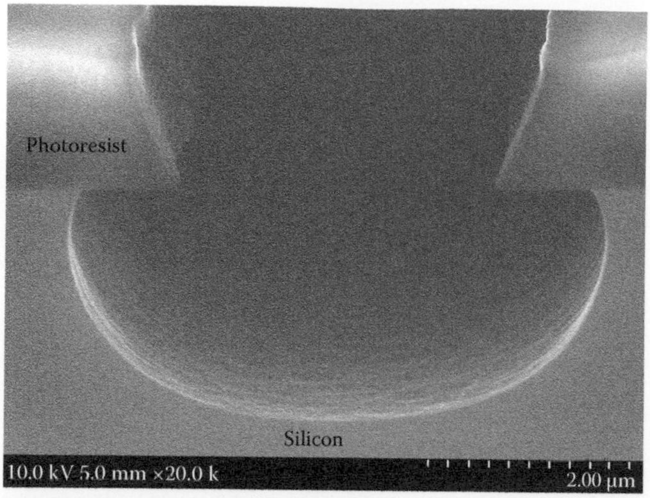

FIGURE 7.25 Silicon etch profile with an SF_6 plasma, at a pressure of 20 mT and an ICP discharge power of 800 W. Etch time was 2 min.

selective and will be very similar to an isotropic wet chemical etch. An example of an etched structure with SF_6 ICP discharge is shown in Figure 7.25. The isotropic nature of the etch also removes some of the surface roughness that can be seen with the RIE-only etch.

A small amount of oxygen or argon gas can be added to SF_6 to reduce the electronegativity (to reduce the number of negative ions) and hence increase the electron concentration and the DC bias [38]. The greater electron concentration will increase the dissociation of the feed gas and produce more reactive F radicals, and the larger DC bias will make the etch profile more directional. Oxygen can also scavenge sulfur to form SO_2 and other volatile oxides, effectively freeing up more fluorine radicals in the plasma. Typical values for oxygen or argon is in the range of 5%–10%. One trade-off of adding oxygen is the increased erosion of the photoresist due to oxidation of its organic components.

The sulfur also produces interesting effects on the photoresist surface. Sulfur can make certain polymers crosslink, which is known as the vulcanization process. This can make the photoresist more durable and less susceptible to erosion in the plasma. However, this crosslinking only occurs within a narrow process window, so the plasma parameters have to be selected with special attention to maximize the vulcanization process while also maintaining the other desirable properties such as etch rates and anisotropy. Optimizing all of these simultaneously is not a trivial process since there are no simple equations or physical models. Typically, this is done through process iterations, refinements, and characterizations.

7.2.5.2 CF_4 Plasma for Etching Silicon

CF_4 (tetrafluoromethane) works similar to SF_6, except it is less electronegative, resulting in a higher DC bias, and its dissociation energies are also generally higher. Some of the dissociation energies are as follows [36]:

$$CF_4 \rightarrow CF_3 + F\left(\Delta G = +494 \text{ kJ/mol}\right)$$

$$CF_3 \rightarrow CF_2 + F\left(\Delta G = +327 \text{ kJ/mol}\right)$$

$$CF_2 \rightarrow CF + F\left(\Delta G = +477 \text{ kJ/mol}\right)$$

$$CF \rightarrow C + F\left(\Delta G = -161 \text{ kJ/mol}\right)$$

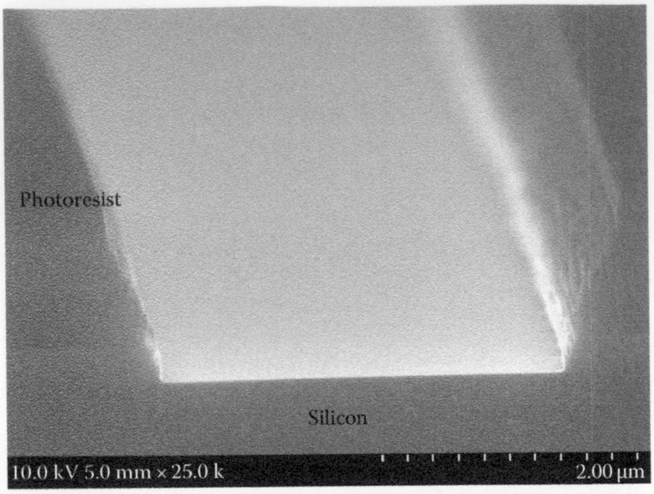

FIGURE 7.26 Silicon etch profile with CF_4 plasma, at a pressure of 20 mT and a discharge power of 2.2 W/cm^2 and a DC bias of −320 V. Etch time was 10 min.

Comparing these numbers with SF_6, we can see that SF_6 will dissociate more readily than CF_4. So for the same pressure and discharge power, CF_4 will produce fewer free radicals, resulting in a lower etch rate.

Additionally, some of the CF compounds can produce long chain fluoropolymers. Being non-volatile, these would accumulate on the surfaces as a thin film deposition. These fluoropolymers are known for their chemical and physical resistance. They belong to the same family as polytetrafluoroethylene (PTFE), also known under the trade name Teflon. They act as very effective passivating films and can significantly reduce or even stop the etching from proceeding. Therefore, a sufficiently high DC bias is necessary to maintain a reasonable etch rate. Nevertheless, there are several beneficial aspects to etching with CF_4. The etch profiles tend to be anisotropic, and the moderating effect of the fluorocarbon films results in smoother surfaces. Comparing Figure 7.26 with Figure 7.24, we can see the difference in surface roughness and the anisotropic etch characteristic between SF_6 and CF_4.

A major drawback with CF_4 is the increased erosion rate of the photoresist. Depending on the sidewall angle of the photoresist this can lead to a sloped sidewall in the etched structure.

As with SF_6, a small amount of oxygen can be used with CF_4 to increase the density of fluorine free-radicals. by scavenging carbon atoms to form CO_2 and other volatile oxides, although this comes at the expense of significantly reduced selectivity due to the increased rate of photoresist erosion [39].

7.2.5.3 Mixed Gas Fluorine Plasmas for Etching Silicon

Most etch processes are rarely performed with a single gas. It is common practice to optimize the etch recipe by mixing different gases that perform different roles. For example, Ar, SF_6, and C_4F_8 could be combined to optimize the process for high anisotropy with high selectivity. SF_6 decomposes easily in the plasma and will be the primary driver of the etch chemistry. Fluorocarbon gases tend to form fluoropolymer passivation. A higher carbon-to-fluorine ratio will result in a more significant accumulation of fluoropolymer deposition. Compared to CF_4, C_4F_8 (octafluorocyclobutane) has a very large carbon-to-fluorine ratio and is commonly used to increase the rate of fluoropolymer deposition. Also, while the etch rate increases with increasing temperature, the fluoropolymer deposition rate increases at lower temperatures. Therefore, temperature can be used to tune the balance between the deposition rate and the etch rate. Argon is a good sputtering gas as well as a dilutant. Therefore, by combining all of these three gases one could synthesize a recipe for producing smooth vertical sidewalls with a very large etch selectivity. The ICP and RIE power, as well as the pressure can also be individually adjusted to control the plasma density and DC bias. This is a very large

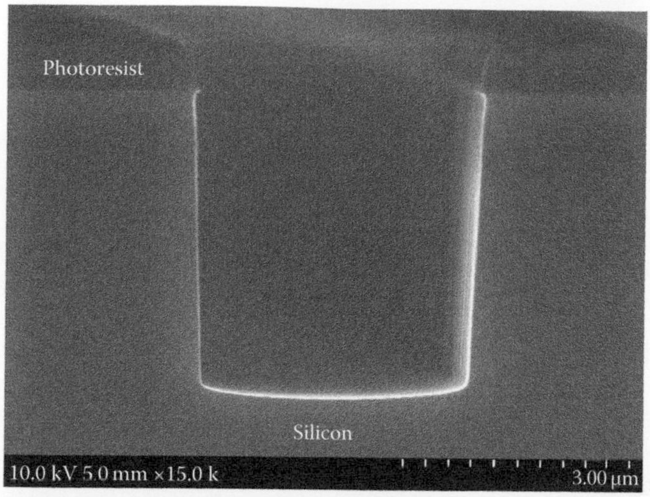

FIGURE 7.27 Etch profile from a SDRIE, using SF_6 = 13 sccm, C_4F_8 = 27 sccm, Ar = 10 sccm, pressure = 20 mT, temperature = 10°C, ICP = 800 W, RIE = 9 W. Etch time was 30 min.

parameter space and it is virtually impossible to examine all possible parametric combinations. Most of the optimized etch recipes are arrived at by using a design of experiment approach that focuses on a relatively small parameter space with certain assumed values.

Figure 7.27 shows the etch profile from a mixed gas etch, known as the single-step deep-RIE (SDRIE) [40]. The gas flows for this process were SF_6 = 13 sccm, C_4F_8 = 27 sccm, Ar = 10 sccm, at a pressure of 20 mT with a substrate temperature of 10°C, and an ICP power of 800 W and RIE power of 9 W. We can see that the sidewalls are nearly vertical and the bottom surface is rectangular and smooth. This arises due to the combined effect of all three gases acting on the silicon. The etch selectivity with photoresist, although not shown in this figure, was estimated to be greater than 8.

The SDRIE is actually a variant of a more popular process known as the Bosch etch process. This was invented by the Robert Bosch company and has been commercialized in many etch reactors to create deep vertical structures in silicon [41]. It is widely used in the fabrication of MEMS and through silicon vias (TSVs).

The originally patented Bosch process consists of two steps: (1) an isotropic deposition of fluorocarbon polymers; (2) directional ion-assisted etch with argon ions and fluorine radicals (from SF_6). During the first step, a carbon-rich gas such as CHF_3 or C_4F_8 is used with an ICP or microwave-coupled plasma. This is an isotropic process, and it produces a thin fluorocarbon passivation on all surfaces of the etched structure. The second step is executed with an RIE excitation. The ions bombard the surface and clears away the passivation film on horizontal surfaces while the fluorine radicals perform a shallow semi-isotropic etch in the cleared area. Then the first step is repeated to passivate the freshly etched silicon surface. The two-step etch process is repeatedly cycled a number of times to etch through a silicon substrate to great depths, as shown in Figure 7.28. Each step only lasts a few seconds. The first step typically produces a deposition of about 50 nm thick and the second step etches to a depth of about 1–2 μm. As a result of the repeated cycling, a characteristic sidewall ripple emerges as shown in Figure 7.28. Many process modifications have subsequently been developed to reduce or eliminate this sidewall ripple.

7.2.5.4 Cl_2 Plasma for Etching Silicon

Silicon can also be etched with Cl_2 plasmas to produce a number of different $SiCl_x$ compounds including $SiCl_4$ and $SiCl_2$. Unlike F_2, Cl_2 does not spontaneously react with silicon and is also less reactive than F_2 (but still very toxic and dangerous). $SiCl_4$ is a volatile liquid with a vapor pressure of 300 Torr. Hence, under vacuum conditions it will mostly be in vapor form.

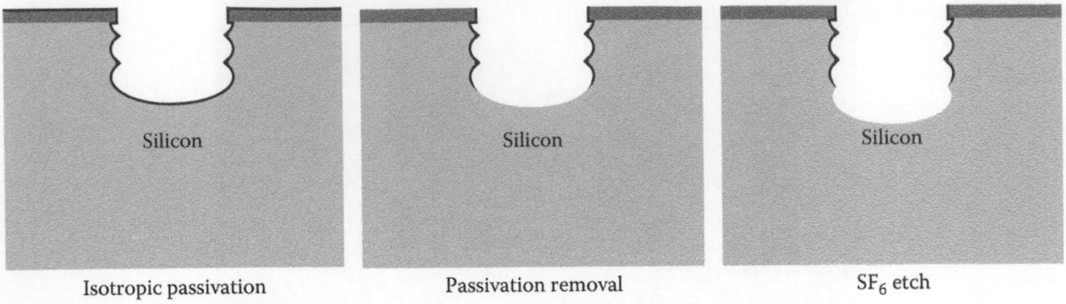

FIGURE 7.28 Deposition and etch cycle of the Bosch etch process.

7.2.6 Photoresist Erosion in a Plasma Etch

Although photoresists are the most commonly used etch mask in plasma etching, their etch properties are not clearly understood [42–44]. This is partly because not all photoresists are exactly the same. Even within the same photoresist family, process conditions such as hard baking and UV curing will make a significant difference in their etch rates.

To act as an effective etch mask, the photoresist needs to survive as long as possible in a plasma etch process. It is interesting to note that the original plasma etching systems were developed for exactly the opposite purpose—to more effectively strip photoresists from wafers compared to using liquid hydrocarbons such as acetone [45]. Oxygen plasma can selectively etch organic photoresists without affecting any of the inorganic films and substrates. In this process, also known as plasma ashing, oxygen radicals react with the carbon and hydrogen atoms in the photoresist to produce volatile CO_2 and H_2O.

In a fluorine plasma, photoresists undergo a number of different mechanisms, including ion-assisted chemical sputtering, direct ion sputtering, and implantation. Ion-assisted chemical sputtering occurs when free radicals in the plasma react with the photoresist to form a compound, which is then removed by ion sputtering [46]. This is believed to be the dominant mechanism in novolak-based photoresists in a fluorine plasma. The photoresist does not spontaneously etch with fluorine radicals. Instead, fluorine radicals adsorb onto the surface of the photoresist, react with the carbon-containing organic molecules to produce CF_x species, which are then sputtered away by ions. One reason for this mechanism could be that photoresists are too soft to be directly sputtered by ions. Depending on the energy, ions will get implanted rather than induce sputtering. By first forming a denser CF_x species, the threshold for implantation is increased, so sputter removal becomes more likely.

Direct ion sputtering, by Ar^+ ions or other species, can also slowly etch the photoresist [47]. But depending on the ion energy it can also cause implantation rather than sputter removal [48]. Sputter yields also have an angular dependence, reaching a peak value when the incident ions are at 60°–80° from the surface normal, as discussed in Chapter 3. This can lead to tilted facets developing at the corners of photoresist lines, known as photoresist faceting. These will eventually get transferred into the underlying substrate if the etch is continued until the entire photoresist is exhausted. This is illustrated in Figure 7.29.

Even if the etch process is inherently perfectly vertical, any sidewall angle on the photoresist shape will still lead to a sloped sidewall in the etched profile. Positive tone photoresists have a natural sidewall that tapers away from the structure. This effect becomes most noticeable when the selectivity is low. For example, an etched structure in a CF_4 plasma is shown in Figure 7.30. CF_4 tends to erode the photoresist at fairly fast rates, resulting in low selectivity values on the order of 1.0. Assuming the etch rate only exists along the vertical axis (a perfectly anisotropic etch), we can

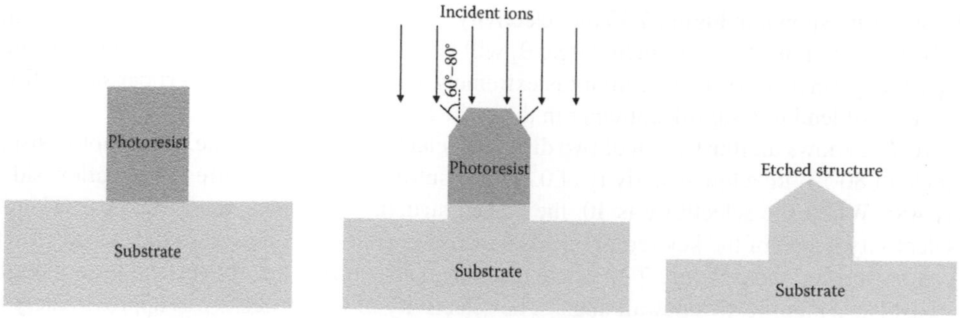

FIGURE 7.29 Photoresist faceting due to angle-dependent sputter yield.

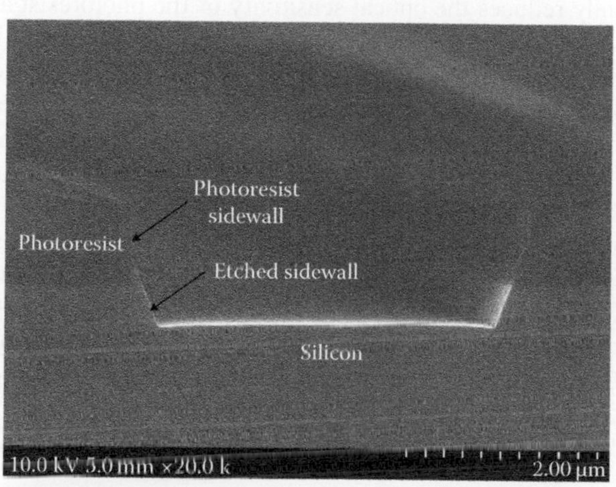

FIGURE 7.30 Silicon etch profile with a CF_4 plasma at a pressure of 20 mT and an ICP discharge power of 800 W and a RIE discharge power of 10 W.

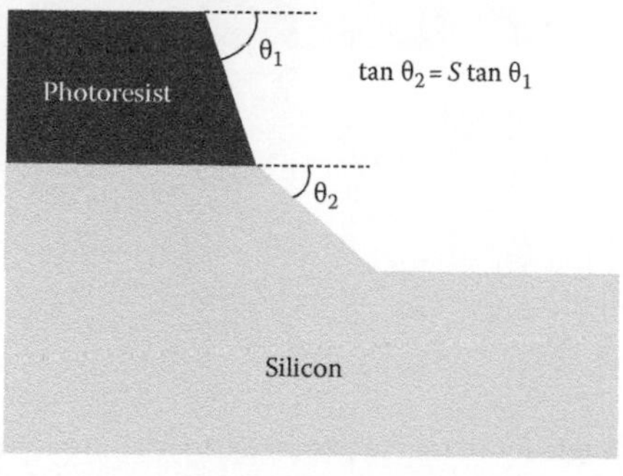

FIGURE 7.31 Relationship between sidewall angle and the etch selectivity.

show from simple geometry that the relationship between the photoresist sidewall angle θ_1 and the etched sidewall angle θ_2 is

$$\tan\theta_2 = S\tan\theta_1 \tag{7.17}$$

where S is the selectivity defined in Equation 7.1, that is, S is the etch rate of the material being etched divided by the etch rate of the photoresist. This relationship is illustrated in Figure 7.31.

For the structure shown in Figure 7.30 the selectivity is nearly 1.0, so the angles θ_1 and θ_2 are almost equal. If the selectivity S is extremely large, θ_2 will be close to 90° even if the photoresist sidewall is not perfectly vertical. If the selectivity is extremely small, even a nearly vertical sidewall in the photoresist will lead to a significant taper in the etched structure.

Figure 7.32 shows an illustration of two different selectivity values for the same photoresist sidewall angle of 80°. With a low selectivity of 0.2, the resulting etched structure has a shallow sidewall of only 48°. When the selectivity is 10, the etched structure becomes nearly vertical. Therefore, high selectivity is one of the key requirements to achieve a vertical sidewall in the etched structure. This can be verified from Figure 7.27 where the etched structure is nearly vertical even though the photoresist has a significant sidewall angle. The selectivity of that process was approximately 8.

High selectivity is most effectively accomplished by optimizing the plasma chemistry and process conditions. Hard baking is the next most effective technique to increase the etch selectivity. Since baking significantly reduces the optical sensitivity of the photoresist and increases the dark erosion rate, this step has to be done after all of the photolithography steps have been completed. Hard baking is performed on a hot plate or an oven at a temperature below the glass transition temperature, T_g, of the photoresist. The goal is to drive off all of the solvents and harden the photoresist, and also cause some crosslinking in the resin. One drawback of hard baking is that it makes the remaining photoresist difficult to remove afterward. It may have to be etched in an oxygen plasma or an aggressive wet chemical etch such as Piranha.

The glass transition temperatures of most photoresists are in the range of 120°C–130°C. Hard baking at temperatures above the glass transition will cause significant thermal reflow and distortion of the patterns. An example is shown in Figure 7.33. This reflow is not a sharp transition but occurs gradually over a temperature window of about 10°C. Although reflow is generally undesired, it can also serve as a useful technique for smoothening and removing some of the surface roughness in photoresist surfaces, albeit at the loss of some resolution. It is also used for making microlenses because curved surface can be created this way [49].

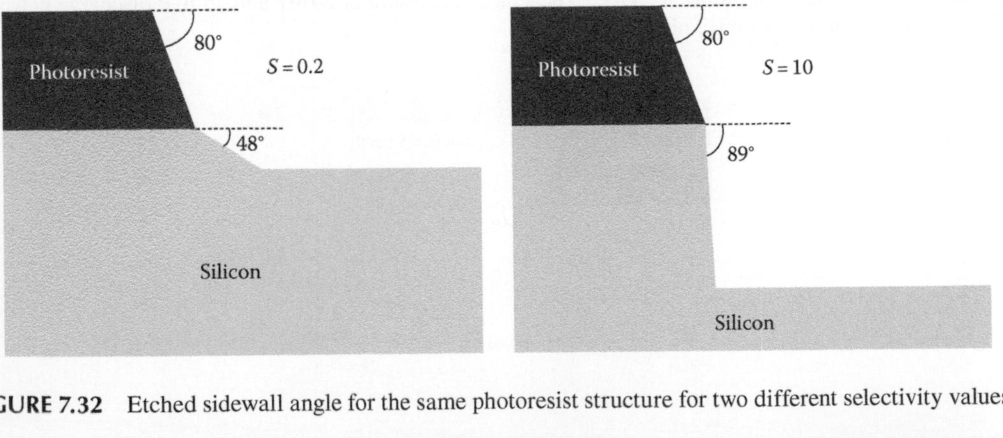

FIGURE 7.32 Etched sidewall angle for the same photoresist structure for two different selectivity values.

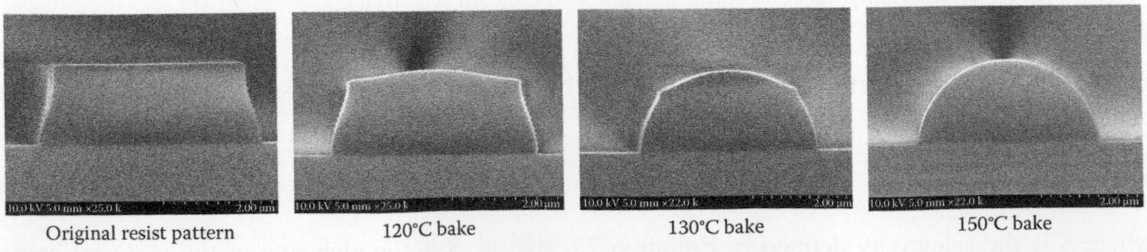

FIGURE 7.33 Photoresist pattern as a function of hard bake temperatures.

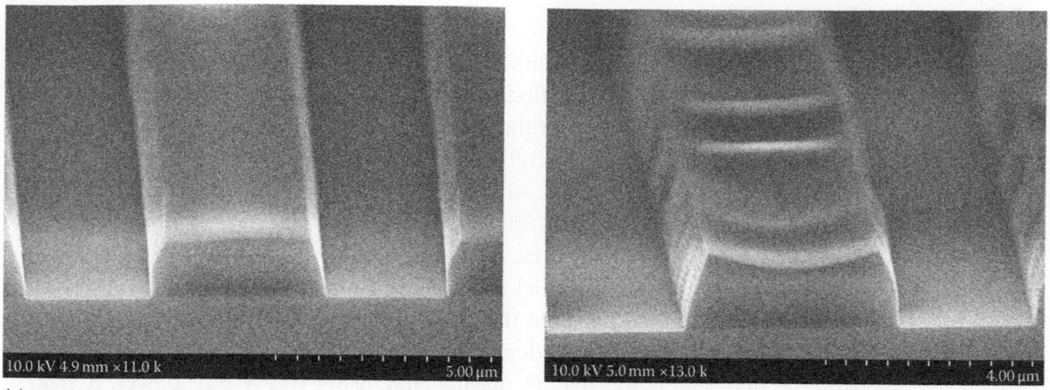

(a) (b)

FIGURE 7.34 Wrinkling in the photoresist skin due to baking at 250°C. (a) DUV exposure was done under ambient air; (b) DUV exposure was done under N_2.

Another technique used to crosslink and harden the photoresist is by exposing it to deep UV radiation. This can be done with a simple low-pressure mercury vapor lamp (254 nm wavelength emission). Even positive photoresists will become crosslinked when exposed to a sufficiently high energy dose [50–52]. This allows the photoresist to be subsequently baked at very high temperatures, as high as 250°C, without causing significant thermal reflow. At such high temperatures, the photoresist becomes fully polymerized. It is speculated that the deep UV exposure creates a thin hardened crust around the photoresist profile, preventing the interior from reflowing.

However, baking at high temperatures with a hardened crust can also create wrinkles in the photoresist skin, as shown in Figure 7.34. This wrinkling has been noted to occur only when the DUV exposure is conducted with a nitrogen purge. It does not occur if the deep UV exposure is done under ambient air. However, the deep UV light will generate ozone from the oxygen molecules in air, which can react with the photoresist and create an oxygen-rich crust. When etched in a fluorine plasma, this results in an isotropic etch profile of the photoresist, presumably because the oxygen-rich film etches spontaneously with fluorine. Therefore, the nitrogen purge appears to be important although that leads to wrinkling in the photoresist.

PROBLEMS

7.1 Consider a 50 nm thick metal film patterned with a photoresist line that is 250 nm wide and 250 nm thick. Using the photoresist as the etch mask, the metal is etched in a wet chemical until all of the exposed metal is removed. Calculate the width of the etched metal at the top and bottom of the metal trace. Assume the etch selectivity is infinite.

7.2 Consider the same metal film patterned with the photoresist line as in problem 1, but this time it is etched with a plasma. The selectivity of the etch is 0.8. Calculate the thickness of the remaining photoresist when the etch just comes to completion.

7.3 A through-hole is being etched in a 150 μm thick (100) silicon using KOH at a rate of 1 μm/min at 80°C. The etch mask is SiO_2 and the selectivity is 300. Calculate the minimum thickness of SiO_2 needed to perform this etch, and the size of the mask opening such that the through-hole at the bottom of the wafer is 50 μm × 50 μm.

7.4 Derive equation (7.6), which describes the enlargement of the hole in a KOH etch when the mask is rotated by an angle θ.

7.5 Derive equation (7.17), which relates the sidewall angle of the photoresist to the sidewall angle of the etched structure in a plasma etch process.

LABORATORY EXERCISES

7.1 Prepare four (100) wafers for etching in a KOH bath. Then deposit 500 nm of silicon dioxide by thermal evaporation on the first wafer, by sputtering on the second wafer, by PECVD on the third wafer and by wet thermal oxidation on the fourth wafer. Pattern all four oxides with the same patterns, etch the oxides with BOE, and conduct the KOH etch using 25% concentration and 80°C until the entire mask disappears. Calculate the selectivity and evaluate the etch quality in all four cases.

7.2 Etch selectivity can be calculated by performing three measurements using a stylus profiler: (1) measure the photoresist thickness, (2) perform the etch and then measure the total step height difference between the photoresist and the etched areas, (3) strip the photoresist and measure the etched depth in the substrate. Using a standard photoresist, measure the etch selectivity in a SF_6 plasma etch, using 100 W RIE power. Repeat the exercise after hard baking the photoresist. Evaluate the sidewall properties under a scanning electron microscope.

7.3 Using the same measurement technique as the previous exercise, conduct a short etch and a long etch, and determine if the etch rate is linear with time. If it is not linear, explain the possible reasons for this. Then determine the maximum depth can be etched with this photoresist.

REFERENCES

1. M. S. Kulkarni and H. F. Erk, Acid-based etching of silicon wafers: Mass-transfer and kinetic effects, *J. Electrochem. Soc.*, 147(1) (2000) 176–188.
2. Transene Company, Inc., Danvers, MA. http://transene.com/.
3. A. R. Clawson, Guide to references on III-V semiconductor chemical etching, *Mater. Sci. Eng.*, 31 (2001) 1–438.
4. P. Walker and W. H. Tarn, *CRC Handbook of Metal Etchants*, CRC Press, Boca Raton, FL, 1991.
5. K. R. Williams, K. Gupta, and M. Wasilik, Etch rates for micromachining processing—Part II, *J. Microelectromech. Syst.*, 12(6) (December 2003) 761.
6. S. Verhaverbeke, I. Teerlinck, C. Vinckier, G. Stevens, R. Cartuyvels, and M. M. Heyns, The etching mechanisms of SiO_2 in hydrofluoric acid, *J. Electrochem. Soc.*, 141(10) (1994) 2852–2857.
7. J. S. Judge, A study of the dissolution of SiO_2 in acidic fluoride solutions, *J. Electrochem. Soc.*, 118(11) (1971) 1772–1775.
8. H. Nielsen and D. Hackleman, Some illumination on the mechanism of SiO etching in HF solutions, *J. Electrochem. Soc.*, 130(3) (1983) 708–712.
9. G. A. C. M. Spierings, Wet chemical etching of silicate glasses in hydrofluoric acid based solutions, *J. Mater. Sci.*, 28(23) (December 1993) 6261–6273.
10. D. C. S. Bien, P. V. Rainey, S. J. N. Mitchell, and H. S. Gamble, Characterization of masking materials for deep glass micromachining, *J. Micromech. Microeng.*, 13 (2003) S34–S40.
11. A. Witvrouw, B. D. Bois, P. De Moor, A. Verbist, C. A. Van Hoof, H. Bender, and C. Baert, Comparison between wet HF etching and vapor HF etching for sacrificial oxide removal, *Proceedings of the SPIE, Micromachining and Microfabrication Process Technology VI*, Vol. 4174, Santa Clara, CA, August 25, 2000, p. 130
12. W. Lang, Silicon microstructuring technology, *Mater. Sci. Eng.*, R17 (1996) I-55.
13. H. Seidel, L. Csepregi, A. Heuberger, and H. Baumgärtel, Anisotropic etching of crystalline silicon in alkaline solutions, *J. Electrochem. Soc.*, 137(11) (1990) 3612–3626.
14. A. J. Nijdam, J. van Suchtelen, J. W. Berenschot, J. G. E. Gardeniers, and M. Elwenspoek, Etching of silicon in alkaline solutions: A critical look at the {111} minimum, *J. Cryst. Growth*, 198/199 (1999) 430–434.
15. H. Schröder, E. Obermeier, and A. Steckenborn, Micropyramidal hillocks on KOH etched {100} silicon surfaces: Formation, prevention and removal, *J. Micromech. Microeng.*, 9 (1999) 139–145.
16. I. Zubel and M. Kramkowska, The effect of isopropyl alcohol on etching rate and roughness of (100) Si surface etched in KOH and TMAH solutions, *Sensors Actuators A Phys.*, 93(2) (September 30, 2001) 138–147.

17. I. Zubel and M. Kramkowska, Etch rates and morphology of silicon (hkl) surfaces etched in KOH and KOH saturated with isopropanol solutions, *Sensors Actuators A*, 115 (2004) 549–556.
18. D. S. Bodas, S. K. Mahapatra, and S. A. Gangal, Comparative study of spin coated and sputtered PMMA as an etch mask material for silicon micromachining, *Sensors Actuators A*, 120 (2005) 582–588.
19. G. Canavese, S. L. Marasso, M. Quaglio, M. Cocuzza, C. Ricciardi, and C. F. Pirri, Polymeric mask protection for alternative KOH silicon wet etching, *J. Micromech. Microeng.*, 17 (2007) 1387–1393.
20. A. Stoffel, A. Kovács, W. Kronast, and B. Müller, LPCVD against PECVD for micromechanical applications, *J. Micromech. Microeng.*, 6 (1996) 1–13.
21. Z. Zhu and C. Liu, *Anisotropic Crystalline Etching Simulation (ACES)*, University of Illinois at Urbana-Champaign, Urbana, IL, 1998.
22. A. Torkkeli, O. Rusanen, J. Saarilahti, H. Seppä, H. Sipola, and J. Hietanen, Capacitive microphone with low-stress polysilicon membrane and high-stress polysilicon backplate, *Sensors Actuators A Phys.*, 85(1–3) (August 25, 2000) 116–123.
23. P. Ekkels, R. W. Tjerkstra, G. J. M. Krijnen, J. W. Berenschot, J. Brugger, and M. C. Elwenspoek, Fabrication of functional structures on thin silicon nitride membranes, *Microelectron. Eng.*, 67–68 (June 2003) 422–429; *Proceedings of the 28th International Conference on Micro- and Nano-Engineering*, Lugano, Switzerland.
24. A. Hölke and H. Thurman Henderson, Ultra-deep anisotropic etching of (110) silicon, *J. Micromech. Microeng.*, 9 (1999) 51–57.
25. M. Shikida, K. Sato, K. Tokoro, and D. Uchikawa, Differences in anisotropic etching properties of KOH and TMAH solutions, *Sensors Actuators*, 8 (2000) 179–188.
26. H. Jansen, H. Gardeniers, M. de Boer, M. Elwenspoek, and J. Fluitman, A survey on the reactive ion etching of silicon in microtechnology, *J. Micromech. Microeng.*, 6 (1996) 14–28.
27. A. Fridman, *Plasma Chemistry*, Cambridge University Press, Cambridge, U.K., 2008.
28. D. M. Manos and D. L. Flamm, *Plasma Etching: An Introduction*, Academic Press Inc., Boston, MA, 1989.
29. K. Nojiri, *Dry Etching Technology for Semiconductors*, Springer International Publishing, Cham, Switzerland, 2012.
30. V. M. Donnelly and A. Kornblit, Plasma etching: Yesterday, today, and tomorrow, *J. Vac. Sci. Technol. A*, 31(5) (September/October 2013) 050825.
31. J. Hopwood, Review of inductively coupled plasmas for plasma processing, *Plasma Sources Sci. Technol.*, 1 (1992) 109–116.
32. J. H. Keller, Inductive plasmas for plasma processing, *Plasma Sources Sci. Technol.*, 5 (1996) 166–172.
33. A. Montaser, *Inductively Coupled Plasma Mass Spectrometry*, Wiley-VCH Inc., New York, 1998.
34. D. R. Wright, D. C. Hartman, U. C. Sridharan, M. Kent, T. Jasinski, and S. Kang, Low temperature etch chuck: Modeling and experimental results of heat transfer and wafer temperature, *J. Vac. Sci. Technol. A*, 10 (1992) 1065.
35. R. Legtenberg, H. Jansen, M. de Boer, and M. Elwenspoek, Anisotropic reactive ion etching of silicon using $SF_6/O_2/CHF_3$ gas mixtures, *J. Electrochem. Soc.*, 142(6) (June 1995) 2020.
36. C. W. Bale, E. Bélisle, P. Chartrand, S. A. Decterov, G. Eriksson, K. Hack, I. H. Jung, Y. B. Kang, J. Melançon, A. D. Pelton, C. Robelin, and S. Petersen, FactSage thermochemical software and databases: Recent developments, *Calphad*, 33 (2009) 295–311. www.factsage.com.
37. K. P. Larsen, D. Hjorth Petersen, and O. Hansen, Study of the roughness in a photoresist masked, isotropic, SF6-based ICP silicon etch, *J. Electrochem. Soc.*, 153(12) (2006) G1051–G1058.
38. S. Rauf, P. L. G. Ventzek, I. C. Abraham, G. A. Hebner, and J. R. Woodworth, Charged species dynamics in an inductively coupled Ar/SF6 plasma discharge, *J. Appl. Phys.*, 92 (2002) 6998.
39. I. C. Plumb and K. R. Ryan, A model of the chemical processes occurring in $CF4/O_2$ discharges used in plasma etching, *Plasma Chem. Plasma Process.*, 6(3) (September 1986) 205–230.
40. Y.-J. Hung, S.-L. Lee, B. J. Thibeault, and L. A. Coldren, Fabrication of highly ordered silicon nanowire arrays with controllable sidewall profiles for achieving low-surface reflection, *IEEE J. Sel. Top. Quant. Electron.*, 17(4) (July/August 2011) 869–877.
41. F. Laermer and A. Schilp, Method of anisotropically etching silicon, United States Patent No. 5,501,893, March 26, 1996.
42. D. Zhang, S. Rauf, and T. Sparks, Modeling of photoresist erosion in plasma etching processes, *IEEE Trans. Plasma Sci.*, 30(1) (February 2002) 114–115.
43. R. Paul Bray and R. Russell Rhinehart, A simplified model for the etch rate of novolac-based photoresist, *Plasma Chem. Plasma Process.*, 21(1) (2001) 149–161.
44. M. F. Doemling, N. R. Rueger, G. S. Oehrlein, and J. M. Cook, Photoresist erosion studied in an inductively coupled plasma reactor employing CHF_3, *J. Vac. Sci. Technol. B*, 16 (1998) 1998.

45. S. M. Irving, A plasma oxidation process for removing photoresist films, *Solid State Technol.*, 14(6) (1971) 47.
46. F. Greer, J. W. Coburn, and D. B. Graves, Vacuum beam studies of photoresist etching kinetics, *J. Vac. Sci. Technol. A*, 18(5) (September/October 2000) 2288.
47. T. Takeuchi, C. Corbella, S. Grosse-Kreul, A. von Keudell, K. Ishikawa, H. Kondo, K. Takeda, M. Sekine, and M. Hori, Development of the sputtering yields of ArF photoresist after the onset of argon ion bombardment, *J. Appl. Phys.*, 113, (2013) 014306.
48. R. B. Guimarães, L. Amaral, M. Behar, P. F. P. Fichtner, F. C. Zawislak, and D. Fink, Implanted boron depth profiles in the AZ111 photoresist, *J. Appl. Phys.*, 63 (1988) 2083.
49. F. T. O'Neilla and J. T. Sheridan, Photoresist reflow method of microlens production Part I: Background and experiments, *Optik—Int. J. Light Electron Opt.*, 113(9) (2002) 391–404.
50. S. Kishimura, Y. Kimura, J. Sakai, K. Tsujita, and Y. Matsui Improvement of dry etching resistance of resists by deep UV cure, *Jpn. J. Appl. Phys.*, 38 (1999) 250–255.
51. G. Sengo, H. A. G. M. van Wolferen, and A. Driessen, Optimized deep UV curing process for metal-free dry-etching of critical integrated optical devices, *J. Electrochem. Soc.*, 158(10) (2011) H1084–H1089.
52. E. B. Vázsonyi, S. Holly, and Z. Vértesy, Characterization of UV hardening process, *Microelectron. Eng.*, 5(1–4) (December 1986) 341–347.

8 Doping, Surface Modifications, and Metal Contacts

In this chapter, we cover several processes related to the introduction of controlled impurities in semiconductors (doping), implantation and related process such as exfoliation, and metal–semiconductor electrical contacts. One common theme that runs across all of these processes is elevated temperature. Although many of the previously described processes also involved the use of elevated temperatures, such as chemical vapor deposition (CVD), evaporation, and even photoresist processing, the temperatures used here can be much higher. They can range from 400°C for contact metal annealing to as high as 1300°C for thermal diffusion and oxidation of silicon. These processes generally fall outside the scope of thin films and etching, but are crucial in most device fabrication process. The examination of these processes is the subject of this chapter.

8.1 THERMAL BUDGET

Due to the vast differences in the temperatures used for device fabrication, the overall fabrication process has to be designed such that high temperature processes occur earlier in the sequence to prevent the later steps from adversely affecting the already completed steps. The term "thermal budget" is often used to describe the maximum envelope of temperatures and durations that are acceptable during a fabrication process. In order to reduce the impact of a specific process on the overall thermal budget, the process temperature can be reduced or its duration can be reduced. However, this is not a simple product of the temperature and time. For example, the diffusion distance of dopant species increases exponentially with temperature but it increases only as the square root of time. Therefore, temperature will have a far greater impact on the thermal budget than durations. Other processes may have a hard temperature limit regardless of duration. For instance, aluminum melts at 660°C. It also forms an Al–Si eutectic with a melting temperature of 577°C. Therefore, all process steps that are performed after aluminum has been deposited should stay well below these temperatures. Substrate type also places a limit on the process temperature. The melting temperature of silicon is fairly high at 1414°C. Compound semiconductors like GaAs, InP, and InSb have significantly lower melting temperatures, 1238°C, 1062°C, and 527°C respectively. Even at temperatures well below their melting temperatures, the substrates may warp or deform, or one atomic species may out-diffuse more than the other. Gallium and indium are known to out-diffuse more than the arsenic, phosphorous or antimony so techniques such as encapsulation or the introduction of a gallium or indium vapor in the background will be necessary to prevent compositional changes to the substrate surface.

In a complementary metal oxide semiconductor (CMOS) circuit fabrication process, high temperatures are required for creating the doped areas of the transistors. This phase of the manufacturing process is referred to as the front-end-of-line (FEOL), and all the materials used have to be compatible with high temperatures. Since metals tend to form eutectic alloys and compounds with silicon at relatively low temperatures, they are not used during the FEOL process. The common materials used during FEOL are dielectrics and polysilicon. The second phase of the process, known as the back-end-of-line (BEOL), utilizes metals for interconnecting the transistors to create the functional circuitry. BEOL processes are done at significant lower temperatures to prevent the device junctions from diffusing as well as the metals from contaminating the devices. The thermal budget of the manufacturing process will also place a limit on the maximum temperatures that the finished chip can be exposed to. This includes brief exposures to high temperatures during soldering as well as degradation due to elevated temperatures during prolonged operation.

8.2 DOPING BY THERMAL DIFFUSION

In thermal diffusion, the wafers are placed in a hot furnace along with the dopant sources to allow the impurity atoms to diffuse through the substrate. The process consists of two main steps—deposition of the dopant atoms and thermal drive-in. Although the dopant atoms can be deposited on the substrate surface directly from elemental sources, the most common method is to use precursors which thermally decompose or chemically react with the substrate to produce the dopant atoms. The precursors could be delivered by gas, or by evaporating a liquid or solid. Using precursors may also leave behind a solid film on the surface. During the drive-in step, the dopant atoms are thermally diffused through the substrate to create the desired doping profile. The deposition and drive-in steps can be done separately, or together in a single process.

The diffusive transport that occurs during the drive-in will be isotropic, that is, it will diffuse vertically and laterally at equal rates. The doped profile will be similar to the etched profile from wet chemical etching. If the device geometry to be doped is very small compared to the desired depth, this can result in significant lateral diffusion and merging of adjacent doped areas, as shown in Figure 8.1. This is the main drawback of thermal diffusion. As a result, doping by thermal diffusion has now been mostly replaced by ion implantation, especially in critical CMOS processes. Nevertheless, for larger geometries, such as discrete devices, photodetectors, and solar cells, thermal diffusion still remains attractive because it allows for batch processing of a large number of wafers simultaneously.

During doping by thermal diffusion, the wafers are stacked vertically in a wafer carrier and placed in a quartz tube, and heated using infrared lamps or resistive heaters from outside the tube. The setup is very similar to low pressure chemical vapor deposition (LPCVD) except the temperatures here are higher. In cases where the diffusion temperatures are close to the softening temperature of the wafers, the wafers may have to be placed horizontally in the furnace to prevent warpage. Such cases are encountered with III–V semiconductors with low melting temperatures, such as InSb. A vacuum environment is preferred to reduce background contaminations, but the high temperatures often make it difficult to maintain a vacuum in a quartz tube without compromising the structural integrity of the tube. So, near-atmospheric pressures can also be used as long as there is a flow of a purge gas such as nitrogen or argon at sufficiently high rates to maintain the purity inside the chamber. With a gas precursor source, a reaction-rate-limited condition has to be maintained to ensure uniformity, so the partial pressure of the species needs to be kept high, and the gas flow patterns need to be uniform.

8.2.1 Vapor, Liquid, and Solid Dopant Sources

For silicon, the commonly used dopant species are boron (for p-type) or phosphorous (for n-type), but antimony and arsenic are also used as n-type dopants. In III–V semiconductors, silicon, carbon, sulfur, selenium, or zinc are used as dopants. Examples of dopant precursor gases include diborane (B_2H_6), boron trichloride (BCl_3), boron trifluoride (BF_3), phosphine (PH_3), arsine (AsH_3), silane (SiH_4), and methane (CH_4). The gases react and decompose at the silicon surface releasing

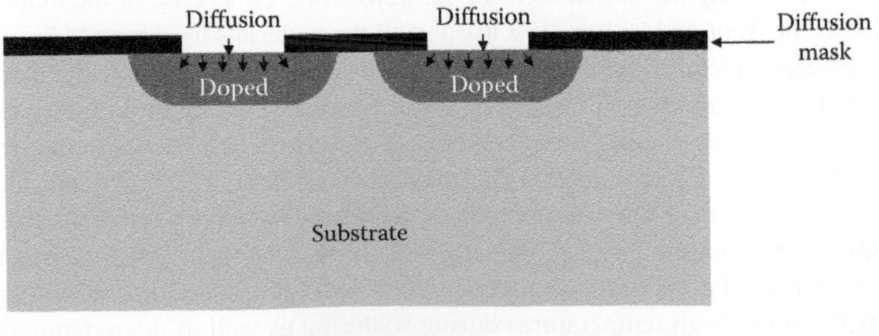

FIGURE 8.1 Isotropic doping profiles from thermal diffusion.

the dopant atoms, which subsequently diffuse into the substrate. For example, diborane decomposes by pyrolysis into BH_3 and BH_2, and then form boron and hydrogen gas [1]. Phosphine is reacted with oxygen to produce P_2O_5 deposition on the substrate, which then reacts with silicon to produce phosphorus and SiO_2. More importantly, these gases are highly toxic or highly explosive, so they require significant infrastructure for safe handling and use. The exhaust gases also have to be scrubbed or neutralized before being released into the atmosphere. Despite the hazards, vapor phase doping offers the flexibility to switch from one dopant source to another quickly during the same diffusion process. It can also result in greater uniformity and greater manufacturing throughput. An example of a quartz furnace setup using gas dopant source is shown in Figure 8.2.

Liquid source dopants include boron tribromide (BBr_3) and phosphoryl chloride ($POCl_3$) [2,3]. The liquid is heated to increase its vapor pressure, and then bubbled with a carrier gas such as Ar, argon, nitrogen or oxygen to increase the flow rate and distribution of the vapor in the chamber, as illustrated in Figure 8.3. Although these liquid sources are still hazardous, they have significantly lower vapor pressures and are more contained than the gas phase dopant. They also offer the same flexibilities and advantages of gas sources.

In a small cleanroom laboratory, solid source dopants are the most convenient to use because they require the least infrastructure support. Solid dopants for silicon include boron nitride (BN), boron oxide (B_2O_3), and phosphorous oxide (P_2O_5) [4]. These are typically manufactured as ceramic disks in the same diameter as the intended wafer size [5,6]. They also include a number of other binding agents

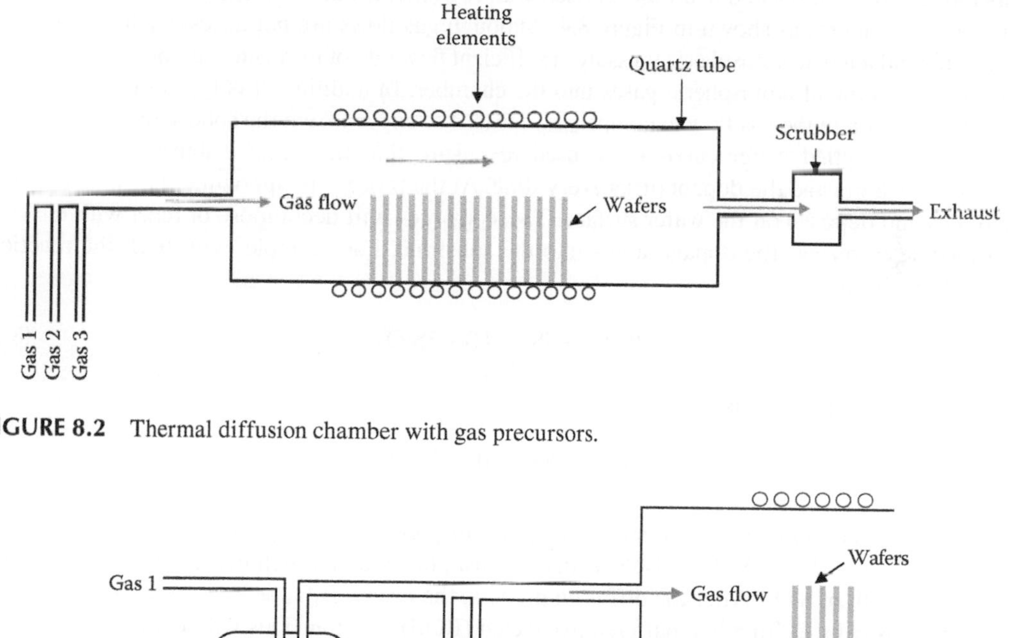

FIGURE 8.2 Thermal diffusion chamber with gas precursors.

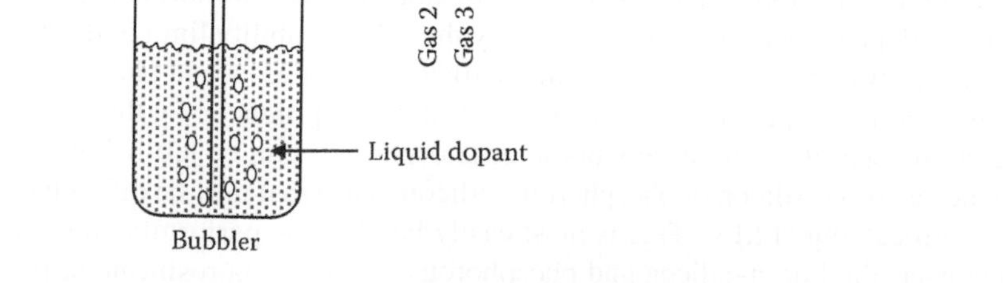

FIGURE 8.3 Thermal diffusion with liquid source dopants in a bubbler.

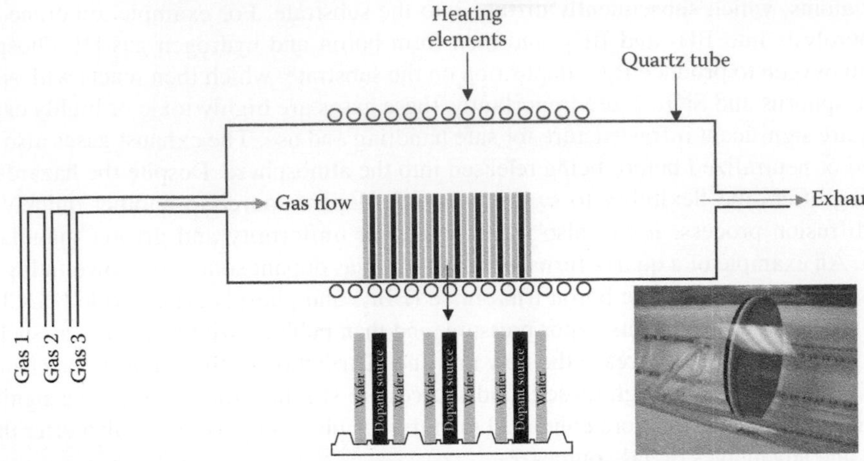

FIGURE 8.4 Thermal diffusion with solid dopants. The inset photograph shows the actual arrangement of two wafers sandwiching a dopant source.

such as silicates and alumina to give the proper evaporation rate and stability at the desired process temperature. They are stacked in a wafer carrier, with two silicon wafers on each side of a dopant disk in a repeating pattern, as shown in Figure 8.4. Although gas flows are not necessary for the doping process, if conducted at atmospheric pressure, a sufficient flow rate of nitrogen or argon will be needed to prevent backflow of atmospheric gases into the chamber. In addition, it is typical to add a small quantity of oxygen in the gas flow to compensate for loss of oxygen from the solid sources.

Specially designed wafer carriers are used to ensure that the same distance is maintained between the wafers and the dopant disks every time. At the process temperature, the dopant source evaporates and deposits on the wafer surface. These sources will decompose or react with the silicon substrate to release the dopant atoms into the substrate. For example, with B_2O_3 the reaction that produces boron is

$$2B_2O_3 + 3Si = 4B + 3SiO_2 \tag{8.1}$$

and with P_2O_5, the reaction is

$$2P_2O_5 + 5Si = 4P + 5SiO_2 \tag{8.2}$$

These reactions consume a small amount of silicon to produce the dopant atoms. A thin layer of SiO_2 is also formed. Since SiO_2 is also a diffusion mask, this reaction will eventually come to a stop due to the growth of SiO_2. After the diffusion process, the excess B_2O_3 or P_2O_5, as well as the SiO_2, are stripped away by etching in a buffered oxide etch (BOE) solution. This will leave a small etched recess in the doped areas due to the conversion of silicon into SiO_2. For example, with boron, for a total dose of 1×10^{15} cm^{-2}, which is a typical value, we can calculate the recess depth using the density value of 2.32 g/cm^3 and an atomic mass of 28 g/mol for silicon, which works out to a depth of just 3 Å.

The maximum dopant concentration is limited by the solid solubility limit of the dopant atoms in the substrate [7–11]. Boron has a solubility limit of 10^{20} cm^{-3} in silicon, and phosphorous has a limit of 10^{21} cm^{-3}. As a result, depending on the amount of dopants produced by the precursors and the length of the subsequent drive-in, excess boron or phosphorous can be left behind on the surface. This can produce a boron–silicon or phosphorous–silicon phase that may be difficult to remove during subsequent processing [12,13]. This is most easily handled by performing a high-temperature oxidation to convert the boron–silicon and phosphorous–silicon to borosilicate or phosphosilicate glass, which can then be easily etched away in BOE [14].

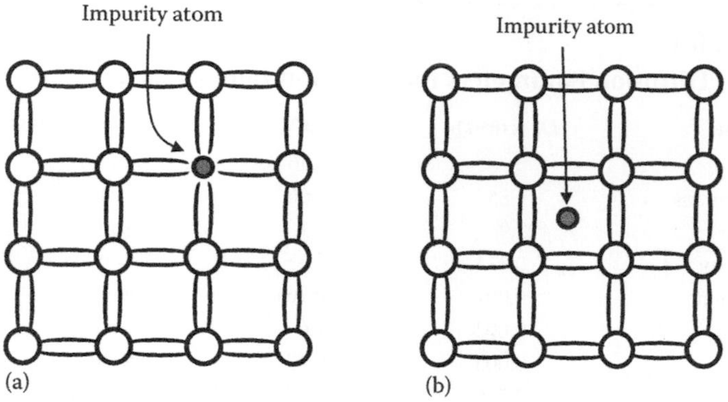

FIGURE 8.5 (a) Substitutional and (b) interstitial impurity atoms.

Once the dopant atoms are on the surface of the substrate, the drive-in mechanism depends on the type of source atom and the substrate. For semiconductor doping, vacancy-mediated substitutional diffusion is the primary mechanism. A substitutional doping occurs when impurity atoms take the place of the substrate lattice atoms. Substitutional diffusion occurs when the impurity atom hops from one vacancy site to another. In order to diffuse, the impurity atom needs to break its chemical bonds with the substrate. This is a high-energy process, so it generally requires high temperatures. The other mechanism is interstitial diffusion. An interstitial is a defect site within the substrate. When an atom occupies an interstitial site, it is only weakly bonded to the substrate atoms. It can easily break free from those bonds and hop to other defect sites. Therefore, interstitial diffusion can take place at lower temperatures compared to substitutional diffusion. These two cases are illustrated in Figure 8.5.

In both cases, the diffusion is governed by Fick's law of diffusion, which can be written as

$$\frac{\partial N}{\partial t} = \nabla(D\nabla N) \tag{8.3}$$

where

D is the diffusion coefficient
N is the dopant concentration

D is a strong function of temperature:

$$D = D_0 e^{-\frac{Ea}{kT}}. \tag{8.4}$$

In Equation 8.3, E_a is the activation energy, which will be large for substitutional diffusion, and will be smaller for interstitial diffusion. The diffusion parameters for some of the common impurities in silicon are listed in Table 8.1. In addition, D will also be a function of dopant concentration. This makes Equation 8.3 nonlinear. Nevertheless, analytical solutions for Equation 8.3 can be found by assuming D to be a constant for a given temperature. The solution is often expressed in terms of the complementary error function *erfc*(*z*) [10]. Though this is not difficult, numerical solution of Equation 8.3 is even simpler to obtain, and more importantly it can be easily adapted to account for changes in D as a function of doping concentration as well as a number of other position-dependent parameters. This numerical approach is briefly outlined in the next section. Although the derivations shown here may initially appear to be lengthy, the final Equation 8.32 is actually quite easy to solve numerically.

TABLE 8.1
List of Diffusion Parameters for Common Impurities in Silicon

Dopant	D_0 (cm²/s)	E_a (eV)	Diffusion Mechanism
Boron	0.76	3.46	Vacancy
Phosphorous	3.85	3.66	Vacancy
Arsenic	22.9	4.1	Vacancy
Antimony	0.214	3.65	Vacancy
Potassium	0.001	0.76	Interstitial
Silver	0.002	1.6	Interstitial
Copper	0.005	0.43	Interstitial
Iron	0.006	0.87	Interstitial

Source: Jones, S.W., *Diffusion in Silicon*, IC Knowledge LLC, Georgetown, MA.

8.2.2 Calculation of Diffusion Profiles

Using a one-dimensional z-axis, the term $D\nabla N$ in Equation 8.3 can be discretized into

$$D\nabla N = D\frac{\partial N}{\partial z} = D\left(z+\frac{\Delta z}{2}\right)\left[\frac{N(z+\Delta z,t)-N(z,t)}{\Delta z}\right]. \tag{8.5}$$

As a result, $\nabla(D\nabla N)$ becomes

$$\nabla(D\nabla N) = \frac{\partial}{\partial z}\left(D\frac{\partial N}{\partial z}\right)$$

$$= \frac{D\left(z+\frac{\Delta z}{2}\right)\left[\frac{N(z+\Delta z,t)-N(z,t)}{\Delta z}\right] - D\left(z-\frac{\Delta z}{2}\right)\left[\frac{N(z,t)-N(z-\Delta z,t)}{\Delta z}\right]}{\Delta z}. \tag{8.6}$$

The term $\partial N/\partial t$ in finite differences becomes

$$\frac{\partial N}{\partial t} = \frac{N(z,t+\Delta t)-N(z,t)}{\Delta t}. \tag{8.7}$$

Therefore, the discretized version of Equation 8.3 becomes

$$\frac{N(z,t+\Delta t)-N(z,t)}{\Delta t}$$

$$= \frac{D\left(z+\frac{\Delta z}{2}\right)\left[\frac{N(z+\Delta z,t)-N(z,t)}{\Delta z}\right] - D\left(z-\frac{\Delta z}{2}\right)\left[\frac{N(z,t)-N(z-\Delta z,t)}{\Delta z}\right]}{\Delta z}. \tag{8.8}$$

This is known as the explicit finite-difference scheme. Implementation of Equation 8.8 is fairly straightforward because we can rearrange it such that the unknown term $N(z,t+\Delta t)$ is on the left side and everything else (which are known) are on the right side. However, such a scheme is not always numerically stable. For stability purposes, the right-hand side should be taken as an average between t and $t+\Delta t$. This is known as the Crank–Nicholson scheme, and it results in

$$\frac{N(z,t+\Delta t)-N(z,t)}{\Delta t}$$

$$=\frac{1}{2}\left\{\frac{D\left(z+\frac{\Delta z}{2}\right)\left[\frac{N(z+\Delta z,t)-N(z,t)}{\Delta z}\right]-D\left(z-\frac{\Delta z}{2}\right)\left[\frac{N(z,t)-N(z-\Delta z,t)}{\Delta z}\right]}{\Delta z} + \frac{D\left(z+\frac{\Delta z}{2}\right)\left[\frac{N(z+\Delta z,t+\Delta t)-N(z,t+\Delta t)}{\Delta z}\right]-D\left(z-\frac{\Delta z}{2}\right)\left[\frac{N(z,t+\Delta t)-N(z-\Delta z,t+\Delta t)}{\Delta z}\right]}{\Delta z}\right\} \tag{8.9}$$

Now, the $t+\Delta t$ terms (unknowns) can be grouped on the left-hand side and the t terms (knowns) can be grouped on the right-hand side and written as

$$\begin{aligned}&A_m(z)N(z-\Delta z,t+\Delta t)+A_0(z)N(z,t+\Delta t)+A_p(z)N(z+\Delta z,t+\Delta t)\\&=B_m(z)N(z-\Delta z,t)+B_0(z)N(z,t)+B_p(z)N(z+\Delta z,t).\end{aligned} \tag{8.10}$$

Denoting $D^+=D\left(z+\frac{\Delta z}{2}\right)$ and $D^-=D\left(z-\frac{\Delta z}{2}\right)$, the coefficients $A_m(z)$, $A_0(z)$, $A_p(z)$, $B_m(z)$, $B_0(z)$, and $B_p(z)$ are

$$A_m(z)=-\left(\frac{D^-}{2}\right)\frac{\Delta t}{\Delta z^2} \tag{8.11}$$

$$A_0(z)=1+\left(\frac{D^++D^-}{2}\right)\frac{\Delta t}{\Delta z^2} \tag{8.12}$$

$$A_p(z)=-\left(\frac{D^+}{2}\right)\frac{\Delta t}{\Delta z^2} \tag{8.13}$$

$$B_m(z)=\left(\frac{D^-}{2}\right)\frac{\Delta t}{\Delta z^2} \tag{8.14}$$

$$B_0(z)=1+\left(\frac{D^++D^-}{2}\right)\frac{\Delta t}{\Delta z^2} \tag{8.15}$$

$$B_p(z)=\left(\frac{D^+}{2}\right)\frac{\Delta t}{\Delta z^2}. \tag{8.16}$$

Since everything on the right-hand side of Equation 8.10 is known, it can be written as

$$A_m(z)N(z-\Delta z,t+\Delta t)+A_0(z)N(z,t+\Delta t)+A_p(z)N(z+\Delta z,t+\Delta t)=c(z,t). \tag{8.17}$$

If there are z_p number of points along the z-axis, Equation 8.17 will result in z_p number of equations, each with three coefficients that connect three adjacent unknowns. Written in a matrix form, it becomes

$$AN(t+\Delta t)=C(t) \tag{8.18}$$

where A will be a $z_p \times z_p$ two-dimensional matrix, and N and C will be a single-column vector with z_p elements. Furthermore, since the three coefficients of each equation multiply adjacent elements of the unknown vector N, the matrix A will be tridiagonal (i.e., it will only have non-zero elements in the three central diagonal positions). Tridiagonal matrices are fairly easy to solve by lower-upper factorization (LU decomposition), forward and back substitution. However, in order to complete the solution, we would need boundary conditions at both ends of the computational window.
The first equation in the array, for $z = \Delta z$, is

$$A_m(\Delta z)N(0,t+\Delta t)+A_0(\Delta z)N(\Delta z,t+\Delta t)+A_p(\Delta z)N(2\Delta z,t+\Delta t)=c(\Delta z,t). \tag{8.19}$$

Since the dopant concentration at the surface will most often be set by the solubility limit, we can arbitrarily set this to a normalized value of 1.0 for all values of t as

$$N(0,t+\Delta t)=1.0. \tag{8.20}$$

As a consequence, the first equation becomes

$$A_0(\Delta z)N(\Delta z,t+\Delta t)+A_P(\Delta z)N(2\Delta z,t+\Delta t)=c(z,t)-A_m(\Delta z) \tag{8.21}$$

where the right-hand term $c(z, t)$ also includes the boundary term $N(0, t) = 1.0$. Therefore, the first equation can be written as

$$A_0(\Delta z)N(\Delta z,t+\Delta t)+A_p(\Delta z)N(2\Delta z,t+\Delta t)=c'(\Delta z,t) \tag{8.22}$$

where

$$c'(\Delta z,t)=c(\Delta z,t)-A_m(\Delta z). \tag{8.23}$$

At the other end of the z-axis, we allow the diffusive flux to flow out of the computation windows by setting

$$D\frac{\partial N}{\partial z}\bigg|_{Zp+\frac{\Delta z}{z}}=D\frac{\partial N}{\partial z}\bigg|_{Zp-\frac{\Delta z}{z}}. \tag{8.24}$$

Furthermore, assuming D to be constant near the boundary, this results in

$$N(z_p\Delta z,t+\Delta t)-N((z_p-1)\Delta z,t+\Delta t)=N((z_p+1)\Delta z,t+\Delta t)-N(z_p\Delta z,t+\Delta t) \tag{8.25}$$

$$N((z_p+1)\Delta z,t+\Delta t)=2N(z_p\Delta z,t+\Delta t)-N((z_p+1)\Delta z,t+\Delta t). \tag{8.26}$$

The last equation in the array is

$$\begin{aligned}&A_m(z_p\Delta z)N((z_p-1)\Delta z,t+\Delta t)+A_0(z_p\Delta z)N(z_p\Delta z,t+\Delta t)\\&+A_p(z_p\Delta z)N((z_p+1)\Delta z,t+\Delta t)=c(z_p\Delta z,t).\end{aligned} \tag{8.27}$$

Substituting for $N((z_p+1)\Delta z,t+\Delta t)$ using Equation 8.23, this becomes

$$\begin{aligned}&[A_m(z_p\Delta z)-A_p(z_p\Delta z)]N((z_p-1)\Delta z,t+\Delta t)\\&+[A_0(z_p\Delta z)+2A_p(z_p\Delta z)]N(z_p\Delta z,t+\Delta t)=c(z_p\Delta z,t)\end{aligned} \tag{8.28}$$

where the right-hand term $c(z_p\Delta z,t)$ includes the boundary term $N(z_p+\Delta z,t)$ from the previous time step t as per Equation 8.26. Equation 8.28 can now be re-expressed as

$$A'_m(z_p\Delta z)N((z_p-1)\Delta z,t+\Delta t)+A'_0(z_p\Delta z)N(z_p\Delta z,t+\Delta t)=c(z_p\Delta z,t) \tag{8.29}$$

where

$$A'_m(z_p\Delta z)=A_m(z_p\Delta z)-A_p(z_p\Delta z) \tag{8.30}$$

$$A'_0(z_p\Delta z)=A_0(z_p\Delta z)+2A_p(z_p\Delta z). \tag{8.31}$$

This results in a matrix of the form

$$\begin{bmatrix} A_0(\Delta z) & A_p(\Delta z) & 0 & 0 & 0 & \cdot & \cdot & \cdot & & \cdot \\ 0 & A_m(2\Delta z) & A_0(2\Delta z) & A_p(2\Delta z) & \cdot & \cdot & \cdot & \cdot & & \cdot \\ 0 & 0 & \cdot & \cdot & \cdot & \cdot & \cdot & \cdot & & \cdot \\ \cdot & \cdot & \cdot & \cdot & \cdot & \cdot & \cdot & \cdot & & \cdot \\ & & & & & & & & & \\ \cdot & \cdot & \cdot & 0 & 0 & 0 & & & & \\ \cdot & \cdot & \cdot & \cdot & \cdot & 0 & 0 & A'_m(z_p\Delta z) & & A'_0(z_p\Delta z) \end{bmatrix}$$

$$\times\begin{bmatrix} N(\Delta z,t+\Delta t) \\ N(2\Delta z,t+\Delta t) \\ N(3\Delta z,t+\Delta t) \\ N(4\Delta z,t+\Delta t) \\ \vdots \\ N((z_p-1)\Delta zt+\Delta t) \\ N(z_p\Delta z,t+\Delta t) \end{bmatrix} = \begin{bmatrix} c'(\Delta z,t) \\ c(2\Delta z,t) \\ c(3\Delta z,t) \\ c(4\Delta z,t) \\ \vdots \\ c((z_p-1)\Delta zt) \\ c(z_p\Delta z,t) \end{bmatrix} \tag{8.32}$$

where all $3Z_p-2$ elements of the tridiagonal matrix are defined. This allows this system to be solved by canned numerical subroutines, such as *tridag* [15].

The initial doping profile is assumed to be $N(0, t) = 1.0$ and zero for all other values of z. After solving Equation 8.32 once, we will get the doping profile at time step $t+\Delta t$. This needs to be continued for many time steps to predict the evolution of the doping profile as a function of time. Furthermore, Δt and Δz need to be sufficiently small to ensure that discretization errors do not creep up. An analysis can be done to optimize for the largest Δt and Δz that will still ensure acceptable accuracy, but we will not pursue those details here.

In Table 8.1, the diffusion parameters of several elements in silicon are given. The diffusion coefficient of boron in silicon has a value of $D_0 = 0.76\ \text{cm}^2/\text{s}$ with $E_a = 3.46$ eV. As a result, at 1100°C, $D=1.5\times10^{-13}\ \text{cm}^2/\text{s}$. Using $\Delta t = 1$ s and $\Delta z = 1$ Å, Figure 8.6 shows the resulting doping profile for different time steps. Assuming the surface concentration is limited by the solid solubility limit of $10^{20}\ \text{cm}^{-3}$, the profiles can be used to find the dopant concentration at any depth. For the substrate that is doped n-type with a background concentration of $10^{15}\ \text{cm}^{-3}$, the metallurgical junction will

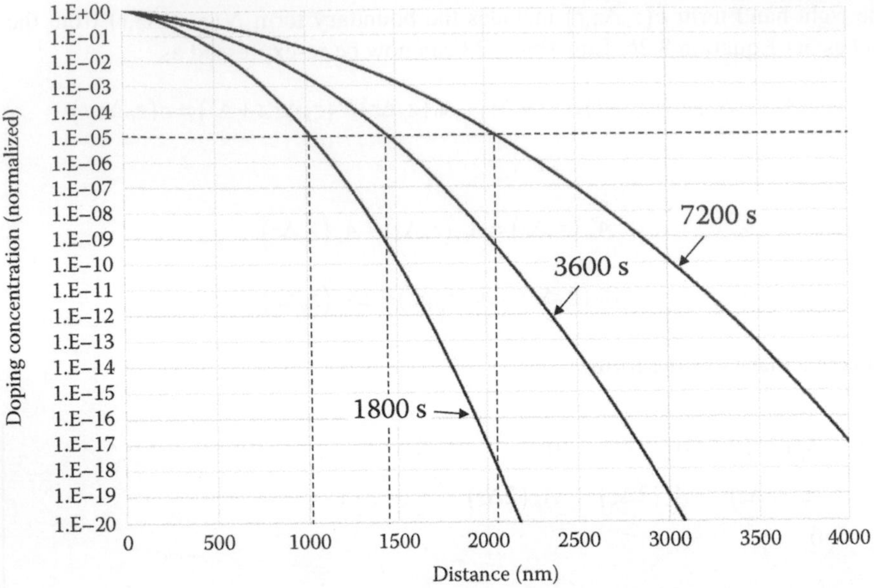

FIGURE 8.6 Diffusion profile of boron in silicon at 1100°C as a function of depth for different times.

occur at a depth of 1.03, 1.45, and 2.06 μm for the 1800 s (30 min), 3600 s (1 h), and 7200 s (2 h) diffusion respectively. These points are shown by the intersection of the dashed lines in Figure 8.6.

An important point to note is that time t only has a weak dependence on the diffusion depth. For the example shown in Figure 8.6, for a fourfold increase in diffusion time the junction depth increases by only a factor of 2 (1.03–2.06 μm). Temperature, on the other hand, has a much more dramatic effect. Increasing the temperature by a mere 50°C from 1100°C to 1150°C will increase the junction depth from 1.45 to 2.43 μm for the 1800s (30min) diffusion. At 1200°C, the junction depth becomes 3.9 μm. This is shown in Figure 8.7. This also demonstrates the importance of maintaining a consistent and accurate substrate temperature during the diffusion process. Even a 5°C difference could lead to significant difference in the junction depth.

We can also use these plots to determine the total dose of dopants delivered into the silicon substrate. This is the area under the curves. For example, the area under the 1100°C curve in Figure 8.7 is 2.63×10^{-5} cm. Assuming the same peak surface concentration of 10^{20} cm^{-3}, the total dopant dose delivered becomes 2.63×10^{15} cm^{-2}.

This diffusion model is also known as an "infinite source" model, because it assumes there is a never-ending supply of dopant atoms at the surface. This is the reason the peak concentration remains pinned at the solubility limit at the surface. As time progresses, the total dose delivered to the substrate will increase. In many applications, the surface concentration has to be reduced below the solubility limit values. This is done by performing a second drive-in step with all of the excess precursors and dopants removed from the surface. This will diffuse the dopants deeper into the substrate by reducing the surface concentration. This is known as a "constant source" model because the total dose of the dopant species in the substrate will remain constant throughout the diffusion process. We can make a small modification to the surface boundary condition in the model to account for this.

In Equation 8.20 instead of maintaining a constant value for $N(0,t+\Delta t)$, we will assume that the incoming flux of dopants is zero. This results in

$$D\left.\frac{\partial N}{\partial Z}\right|_{\frac{\Delta z}{2}} = 0 \tag{8.33}$$

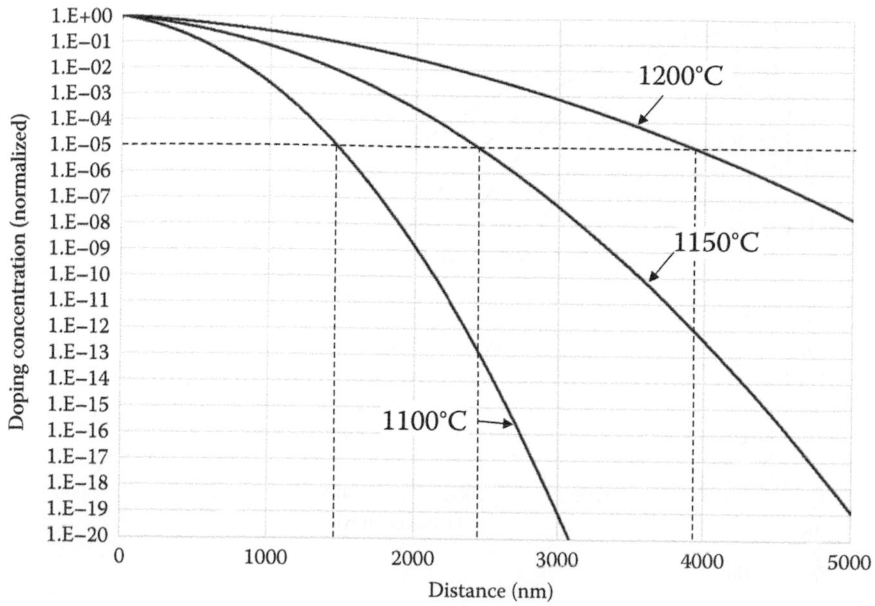

FIGURE 8.7 Diffusion profile of boron in silicon at different temperatures for 3600 s (1 h) diffusion.

which becomes

$$N(\Delta z, t+\Delta t) - N(0, t+\Delta t) = 0. \tag{8.34}$$

Returning to the first equation in the array

$$A_m(\Delta z)N(0, t+\Delta t) + A_0(\Delta z)N(\Delta z, t+\Delta t) + A_p(\Delta z)N(2\Delta z, t+\Delta t) = c(\Delta z, t) \tag{8.35}$$

it becomes

$$[A_m(\Delta z) + A_0(\Delta z)]N(\Delta z, t+\Delta t) + A_p(\Delta z)N(2\Delta z, t+\Delta t) = c(z,t). \tag{8.36}$$

where the right-hand term $c(z, t)$ also includes the boundary condition $N(0,t) = N(\Delta z,t)$. As a result, the first equation becomes

$$A_0'(\Delta z)N(\Delta z, t+\Delta t) + A_p(\Delta z)N(2\Delta z, t+\Delta t) = c(z,t) \tag{8.37}$$

where

$$A_0'(\Delta z) = A_m(\Delta z) + A_0(\Delta z). \tag{8.38}$$

Using the 1100°C, 1800 s (30 min) boron diffusion profile from Figure 8.6 as the starting point, Equation 8.32 can be repeatedly solved for additional drive-ins with the constant source boundary condition. The results are shown in Figure 8.8. Both the logarithmic and linear scales are shown for clarity. We can see that the peak surface concentration declines under the constant source boundary. It can also be easily verified during the calculation that the area under the curves remains constant for all diffusion times.

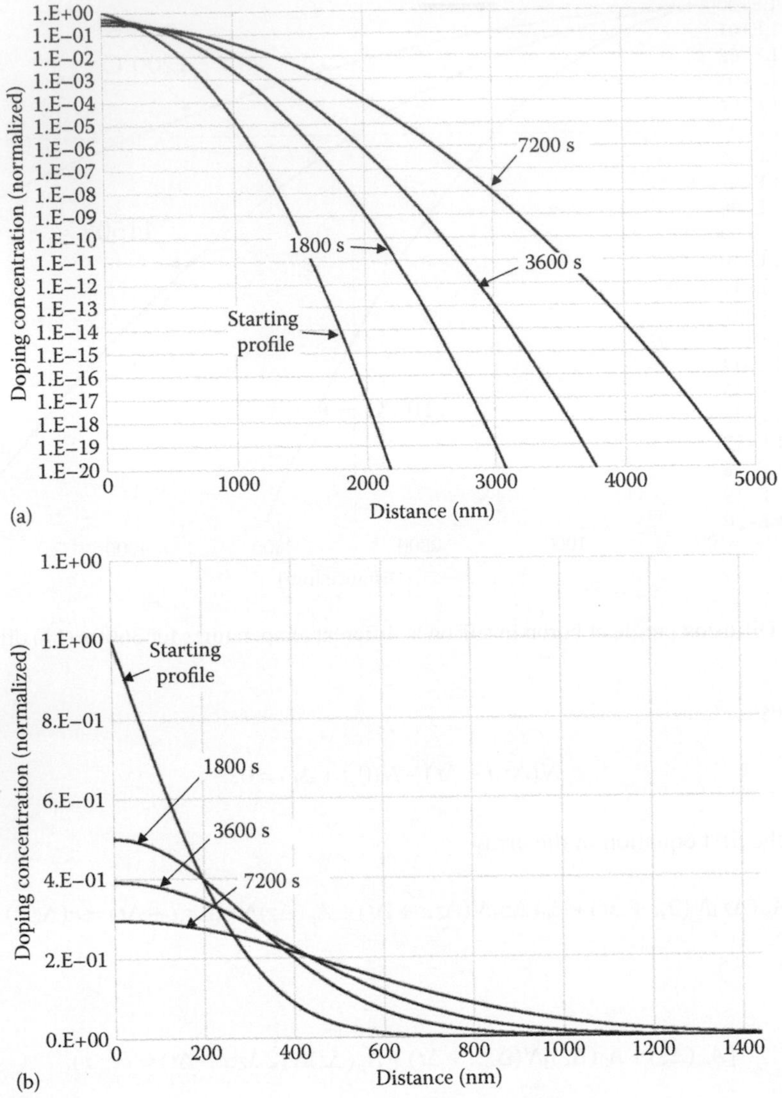

FIGURE 8.8 Additional constant source diffusion of the profile initially produced by a 1800 s infinite source diffusion under 1100°C (Figure 8.6). (a) The plot is on a logarithmic scale and (b) the plot is on a linear scale.

8.2.3 Masking for Thermal Diffusion

Selective doping on a substrate is performed by masking off the unwanted areas with a suitable material before the doping process. Due to the high temperatures used in thermal diffusion, photoresist will obviously not serve as a masking material. Metals will also be unsuitable because they can penetrate the silicon to great depths by fast interstitial diffusion. The most commonly used materials are dielectrics such as SiO_2 or Si_3N_4. These masking films must be robust and free of pinholes to prevent even microscopic amounts of dopants from penetrating through. They are typically deposited by thermal oxidation, plasma-enhanced CVD (PECVD) or low pressure CVD (LPCVD). They also have to be thick enough to prevent the dopant atoms from diffusing through. Fortunately, the diffusion coefficient of typical dopant atoms through these dielectrics is several orders of magnitude smaller than that through silicon. Some of these values are listed in Table 8.2.

TABLE 8.2
List of Diffusion Parameters for Common Impurities in SiO_2

Dopant	D_0 (cm/s)	E_a (eV)
Boron	3.1×10^{-4}	3.53
Phosphorous	0.2	4.03
Arsenic	67	4.7
Antimony	$1.3 \times 10^{+16}$	8.75

Source: Jones, S.W., *Diffusion in Silicon*, IC Knowledge LLC, Georgetown, MA.

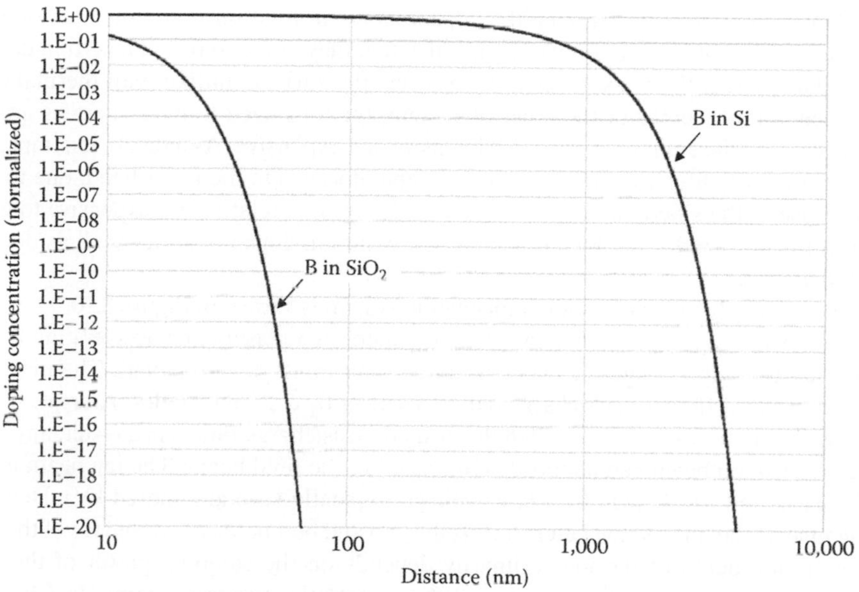

FIGURE 8.9 Comparison of the diffusion of boron in Si and SiO_2 at 1100°C.

For example, the diffusion coefficient of boron through thermally grown SiO_2 is 3.4×10^{-17} cm^2/s at 1100°C, which is about 3–4 orders of magnitude smaller than the diffusion through silicon [16]. Therefore, in principle only a few nanometers of thickness is needed to prevent diffusion. However, there are other mechanisms besides diffusion that reduces the effectiveness of these masking materials. With oxide and nitrides, the inclusion of sufficient quantities of boron or phosphorous will reduce their glass transition temperatures, and the films could potentially melt at the diffusion temperatures. This can consume a certain fraction of the masking material before the boron or phosphorous dilution drops down to a sufficiently low level to resolidify the masking film [14]. As a result, the film thickness should be chosen several times larger than the minimum required value.

Figure 8.9 shows the diffusion profiles of boron through thermally grown SiO_2 and Si at 1100°C for 7200 s. The plot is shown on a log–log scale for clarity. We can see that the penetration depth (for a 10^{-15} concentration) in SiO_2 is only 30 nm whereas it is 2060 nm in silicon. Even accounting for the formation of borosilicate glass and its consumption by melting, a mask thickness of 150 nm can be considered a safe practical value. In the case of PECVD oxide (which is inferior to thermal oxide), an even thicker value, such as 250 nm, can be used.

8.3 ION IMPLANTATION

8.3.1 Doping by Ion Implantation

As mentioned earlier, one of the major disadvantages of doping by thermal diffusion is the isotropic nature of the doped profile. As shown in Figure 8.1, when the doped areas are spaced close to each other, diffusion along the lateral direction can compromise the separation between the two adjacent regions. As the size of components has continued to shrink, this has become a serious limitation. Another major drawback is the fact that the peak concentration occurs at the surface, with a declining concentration as a function of depth. While this is desirable for making ohmic metal contacts to the semiconductor, there are applications where it is desirable to locate the peak concentration below the surface or tailor a doping profile other than the *erfc* function. Ion implantation is an alternative doping technique where most of these limitations can be overcome. Just like wet chemical etching has been replaced by plasma etching for nanoscale features, doping by thermal diffusion has also been largely replaced by ion implantation in nanoscale devices.

Ion implantation is done by ionizing the dopant atoms in a vacuum and then accelerating and driving them into the substrate at high energies such that they come to rest at a certain depth below the surface. This places the peak concentration below the surface, unlike with thermal diffusion. To achieve penetration depths of the order of 100 nm, the ions need to have energies greater than 100 keV. This makes the equipment resource intensive and expensive. As a result, ion implant equipment is not commonly found in small cleanroom laboratories. On the other hand, due to the large volume of its use in the silicon integrated circuits industry, it is relatively inexpensive to get wafers implanted from commercial vendors, especially for common dopant sources such as boron, phosphorous, and arsenic.

A representative schematic of an ion implantation system is shown in Figure 8.10. The ion source is generated by vaporing the source atoms, ionizing them in a plasma and accelerating them. They are then passed through a magnetic mass separating unit (mass spectrometer). The magnetic field causes the ions with different mass/charge ratios to deflect by different angles. An exit slit is placed such that it will only transmit the ions with the desired mass/charge ratio. As a result, any other species in the original ion beam can be mostly eliminated in the final beam. The ion beam is then collimated, focused, and accelerated by a large voltage (typically several hundred kV). Deflectors are used to fine tune the beam position before it exits the gun and bombards the front face of the substrate.

The penetration depth of the ions primarily depends on the stopping power of the substrate, which in turn depends on the ion energy and the size of the substrate atoms. In Chapter 3, we calculated the sputter yields of several different target materials and noted that the yield reaches

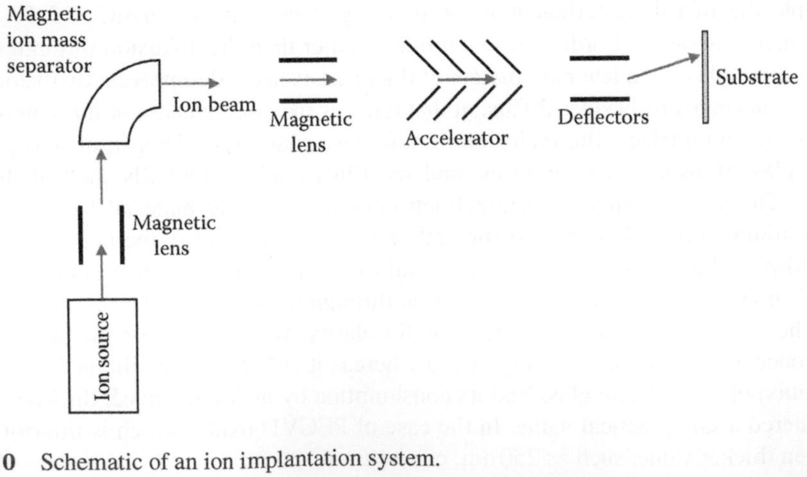

FIGURE 8.10 Schematic of an ion implantation system.

a peak and then declines as the ion energy increases. This decline is due to implantation, where the primary ion gets buried too deep below the surface to cause sputtering. The depth distribution of the ions can be predicted using numerical simulations. A popular free software used by nearly everyone in the field is SRIM (Stopping Range of Ions in Matter) by James F. Ziegler [17], which uses the Monte-Carlo method to trace every ion through the substrate to predict an average statistical distribution. It is fairly easy to use even for a complete novice, but contains enough details and experimental validation for scientific and industrial use.

Figure 8.11 shows the predicted stopping range of a few elements in silicon as a function of ion energy. For example, at 100 keV acceleration, boron and phosphorous have a range of 318 and 134 nm respectively. These are the peak locations of the dopant atoms. The spatial profile of the ions will be determined by the statistical nature of the collisions between the ions and the substrate atoms. Some of the ions will be deflected laterally and will stop at a shallow depth. Some may travel further than the average range. The Monte-Carlo simulation can take many effects into account and predict the longitudinal and lateral statistical distributions (also known as the straggle, or standard deviation). Figure 8.12 shows the longitudinal distribution of boron ions in silicon for an acceleration energy of 100 keV, calculated by tracing the paths of 100,000 ions. The average peak location is at 318 nm and the width of the distribution (straggle) is 777 Å. This becomes the dopant distribution of boron in silicon. The vertical axis is in the units of cm^{-1}. Though that may seem like a strange unit, it becomes meaningful when multiplied by the total ion dose (which is in $ions/cm^2$). The result will be in the usual units for doping concentration of $ions/cm^3$. For example, if the total dose delivered is $1\times10^{15}\ cm^{-2}$, the peak concentration will be $6\times10^{4}\times10^{15}=6\times10^{19}\ cm^{-3}$.

The ions also have a lateral distribution. When all the ions are injected from a single point, the two-dimensional lateral and longitudinal distribution will be as shown in Figure 8.13, which shows a nearly equal longitudinal and lateral straggle.

For compound semiconductors, such as InP and GaAs, the stopping power is calculated by averaging the stopping power of the core atoms, and then adding a correction due to the stopping power of the bonding electrons.

Another effect, known as channeling, occurs when implanting into certain orientations of crystalline substrates. The stopping power of the substrate depends on the ions randomly colliding with the substrate atoms and giving up their energy in a statistically predictable manner. If the ions are

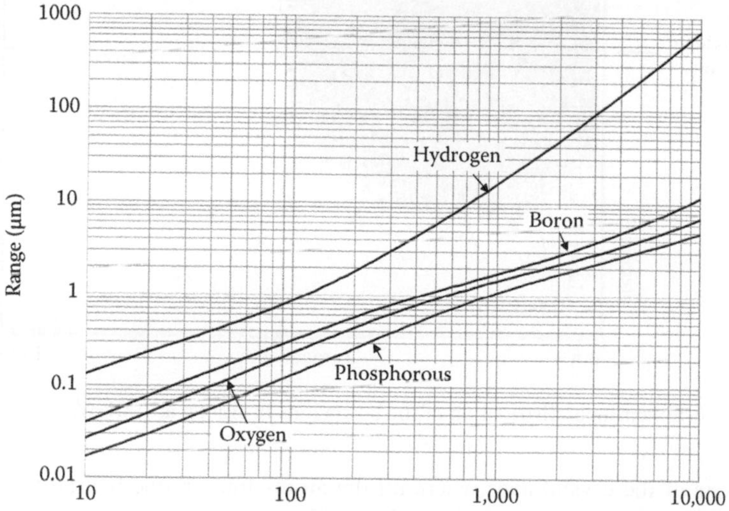

FIGURE 8.11 Stopping range of boron, phosphorous, oxygen, and hydrogen in silicon as a function of ion energy as predicted by SRIM.

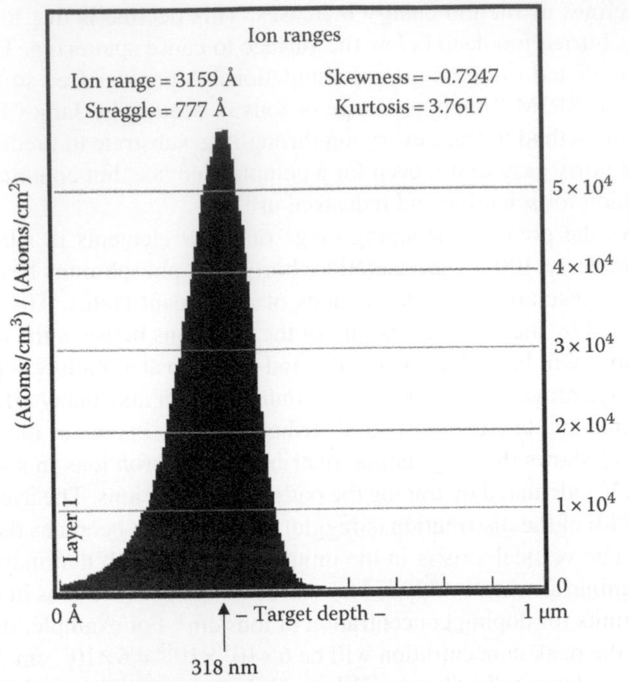

FIGURE 8.12 Plot of the longitudinal distribution of boron atoms in silicon at 100 keV.

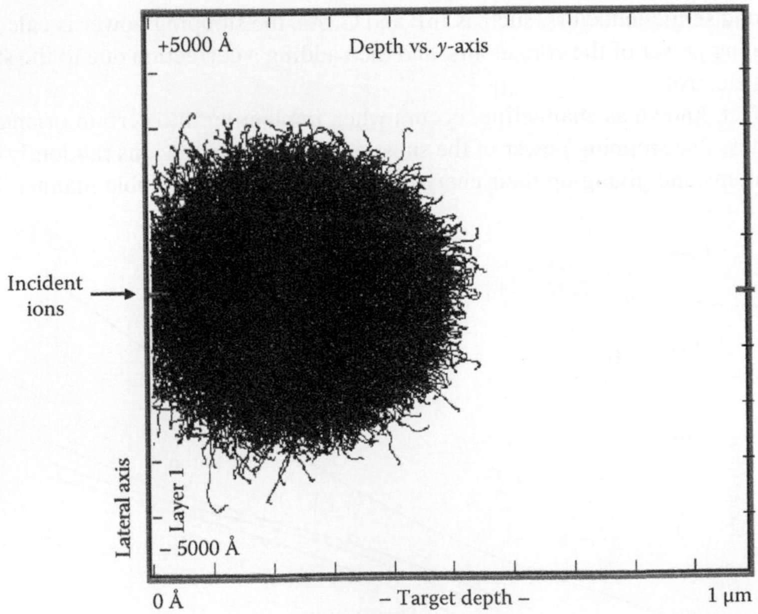

FIGURE 8.13 Two-dimensional distribution of the boron ions in silicon at 100 keV.

incident along one of the crystal axes where all the atoms line up one behind the other, it is possible for the ions to avoid all of these atoms and travel between them through the voids. This makes the stopping range of the ions large and unpredictable. Figure 8.14 illustrates this effect, as well as two well-known techniques, to overcome this problem. The first is to tilt the substrate during the implantation, typically by 2°–5°. This will ensure that none of the crystal voids line up and allow

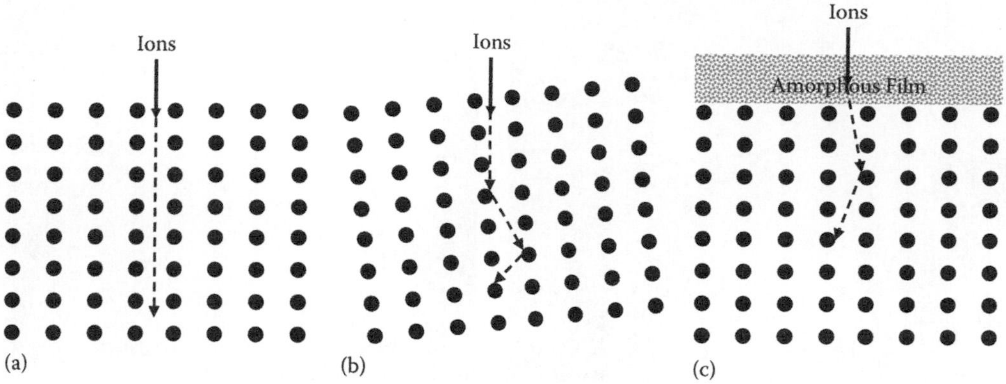

FIGURE 8.14 Ion channeling effect and the techniques used to mitigate it: (a) ion channeling through crystal voids, (b) tilted substrate to avoid channeling, and (c) a thin film used to disperse the incident ions.

channels to appear. This angle is small enough that the longitudinal and lateral distributions will not be significantly distorted in the implanted patterns. The second technique is to use a thin amorphous film to introduce a very small but random angular deviation to the arriving ions. The advantage of the thin film technique is that it also prevents the sputter removal of the substrate species, which is especially important in a compound semiconductor where the differential sputter yield during implant may cause a compositional change in the material.

During transit, each ion will collide with a large number of substrate atoms. It only takes about 20 eV to dislodge a silicon atom from its lattice site. Therefore, at 100 keV, each ion would effectively dislodge over 5000 silicon atoms. At a dose of 1×10^{15} cm^{-2}, this comes to 5×10^{18} dislodged silicon atoms every cm^2 of the surface. All of these silicon atoms will become interstitial, creating a significant disorder in the crystal structure. Furthermore, the implanted ions also come to rest as neutral atoms at random interstitial sites within the crystal, not at substitutional sites. Therefore, before the implanted atoms can perform the electrical functions of doping, they have to become substitutional (which is also referred to as "activating" the dopants). A large number of dislodged silicon atoms at interstitial sites also have to be returned to their lattice positions. These two functions are most often performed by annealing the wafers at a sufficiently high temperature for a short duration. The temperature must be high enough to cause diffusion of the atoms from their interstitial positions to the lattice positions and also for the lattice to relax and accommodate the new implanted atoms. This diffusion distance is normally not more than a few lattice constants, so an extended anneal is not necessary. In highly doped structures, the doped regions could become fully amorphized, so the entire crystal structure has to be rebuilt using the underlying undamaged substrate as a seed. Nevertheless, one of the key benefits of implantation compared to thermal diffusion is the tight spatial distribution of the implanted atoms. It would, therefore, defeat the purpose to introduce a long anneal in a furnace and produce significant thermal diffusion of the implanted profiles. Instead, a rapid thermal anneal (RTA) is often performed by raising the temperature of the wafers to about 1200°C for a brief duration, such as 30 s, and then allowing the wafers to cool quickly. This is sufficient to migrate all of the interstitial atoms to their proper lattice locations [18,19].

Rapid thermal anneal is performed by placing the wafer on a quartz tray designed to have a very low thermal conductivity and thermal mass (typically by placing them on quartz pins, as shown in Figure 8.15), and illuminating with high-intensity halogen lamps. The low thermal mass of the wafer combined with the high incident radiation allows the wafer to rapidly heat up, usually at rates close to 100–200°C/s. A pyrometer or a fast-response thermocouple is used to measure the wafer temperature and adjust the lamp intensity to maintain the desired temperature profile.

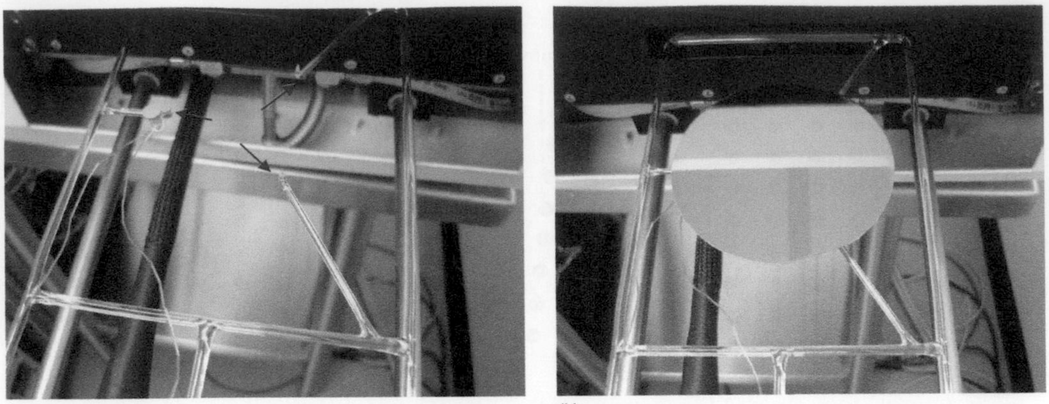

(a) (b)

FIGURE 8.15 Wafer placement on quartz suspension pins in a rapid thermal annealer (RTA): (a) Wafer carrier with the quartz suspension pins and (b) wafer placed on the suspension pins.

After the process, due to the high surface-to-volume ratio of the wafers, ambient cooling is usually sufficient to drop the temperature to below 800°C within a few seconds. Once below 800°C, diffusion becomes negligible and the wafer can cool down normally.

There are tradeoffs between ion implantation and thermal diffusion of dopants. Compared to thermal diffusion, ion implantation has significantly shallower penetration depths. It becomes extremely difficult to achieve doping depths of several microns without MeV levels of acceleration. The ion damage is also a concern. Thermal diffusion, on the other hand, produces almost no crystal damage. As a result, for large geometries and deeper junctions, thermal diffusion still remains the preferred choice.

8.3.2 Masking Materials for Ion Implantation

Another distinct advantage of ion implantation is the flexibility to use a variety of masking materials. Since ion implantation can be performed at room temperature, it becomes possible to use photoresist as a masking material. A photoresist film of several microns thick will easily absorb most of the implanted ions. Nearly all other thin films can also be used, including SiO_2, Si_3N_4 as well as metals. However, unlike in diffusion doping, the masking materials do not generally act as a barrier, but instead they act as a sacrificial absorber. In other words, the selectivity between the masking material and the substrate is not large. In order to act as an effective mask, the film thickness has to be greater than the penetration depth of the dopants in the substrate. In most applications, this is not a serious limitation because the implant depths are usually not more than a few hundred nanometers. A concern, however, is the sputter yield. Even at the high implant energies, some of the masking material can be sputtered, which is a greater concern due to the contamination it creates rather than the loss of the masking material itself.

It is also possible to introduce a thin dielectric film to move the peak location of the doping profile closer toward the surface. This is illustrated in Figure 8.16, where a SiO_2 layer is used. Part of the implant profile falls within the SiO_2 layer, so when it is subsequently stripped, the peak concentration will appear closer to the surface.

8.3.3 Implantation for Silicon-on-Insulator Substrates

In addition to doping, ion implantation is also a critical tool for a number of other applications. One of the largest applications is the manufacture of silicon-on-insulator (SOI) wafers. These are ultra-thin silicon wafers (50 nm to 10 μm thickness) that are attached to another silicon substrate separated

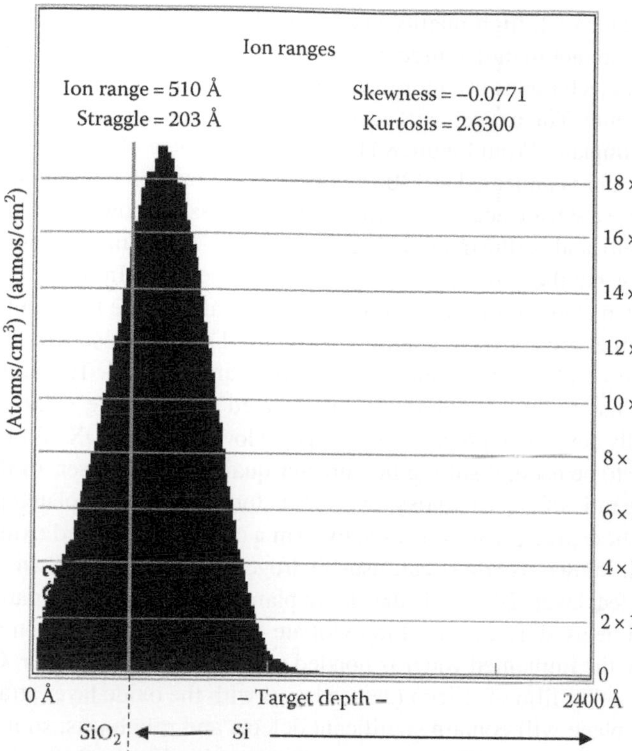

FIGURE 8.16 Placement of a thin film on the substrate to relocate the peak implant concentration closer toward the surface.

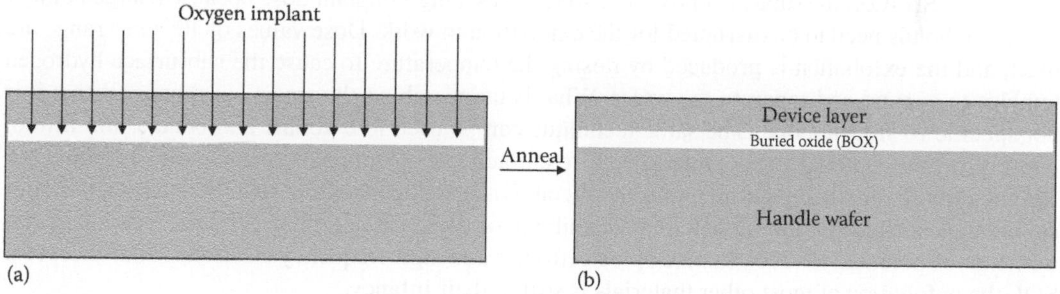

FIGURE 8.17 (a) In the SIMOX process, oxygen ions are implanted to create the buried oxide layer; (b) after a high temperature anneal, the three-layer structure is produced.

by an insulating SiO_2 layer. The thin silicon is referred to as the device layer, the oxide as BOX (for buried oxide), and the bottom silicon substrate as the handle wafer, as shown in Figure 8.17b. SOI was developed to allow silicon CMOS devices to have lower parasitic capacitances, faster switching speeds, and lower operating voltages. The transistors are built on the top device layer and remain electrically isolated from the substrate, and this is the primary reason for the improvement in performance as compared to conventional integrated circuits [20,21]. Though they are mostly manufactured for microelectronics, SOI wafers have also found innovative applications in MEMS and optoelectronics [22].

What is unique about this structure is that the device layer is a high-quality single-crystal silicon, which is a critical aspect for its use in microelectronics. Conventional PVD or CVD thin film growth methods will not produce such high-quality crystalline films on top of amorphous layers such as SiO_2. Therefore, these are not manufactured by growing thin films on top of each other.

One of the dominant techniques used for making SOI is by implanting a large amount of oxygen into a regular silicon wafer. The oxygen will accumulate below the surface at a depth determined by the ion energy during implant. From Figure 8.11, the stopping range of oxygen at 200 keV is 440 nm in silicon. Then, the wafer is annealed at 1300°C to cause the oxygen to combine with neighboring silicon atoms to produce a subsurface layer of SiO_2. This process is known as SIMOX (Separation by Implantation of Oxygen), and is illustrated in Figure 8.17 [23]. The main challenge in this process is not the high ion energy, but the extremely high dose values required. In order to reach the stoichiometric ratio (two oxygen atoms for every silicon), the dose values have to be in the $10^{18}\,cm^{-2}$ range, which is about 1000 times greater than that used in doping. The high dose will also dislodge a large number of silicon atoms and cause the surface to become amorphized. To maintain single-crystal quality, the substrate has to be maintained at an elevated temperature of 500°C–600°C during implant. A subsequently developed process, known as the low-dose SIMOX, allows oxygen implants in the $10^{17}\,cm^{-3}$ range to be used, resulting in a greater quality device layer. In this process, due to the sub-stoichiometric ratio of silicon to oxygen, SiO_2 is formed only as isolated precipitates. When annealed at 1300°C, these precipitates coalesce to form a continuous buried oxide layer.

Another technique, known as smart cut, uses hydrogen implantation on an oxidized silicon to create a subsurface defect layer [24–26]. If the cleave plane of the substrate is parallel to the surface, this implant layer can be used as a catalyst to exfoliate the thin layer of silicon above the implant. Before the exfoliation, the implanted wafer is bonded to a second handle wafer. Controlled exfoliation can then produce a thin film of silicon (device layer) with the oxide layer attached to the handle wafer. The exfoliated plane will contain significant defects and roughness, so it has to be polished and annealed to repair the damage. This process is illustrated in Figure 8.18.

An alternative process, which does not involve ion implantation, is direct bonding and substrate removal [27,28]. In this process, a silicon wafer is oxidized and bonded to a second silicon wafer, and the substrate of one wafer is removed by thinning to create the device layer.

As with SIMOX, the smart cut process also requires a high implant dose because a large number of silicon bonds need to be disrupted for the exfoliation to work. Dose values of $10^{17}\,cm^{-2}$ range are used, and the exfoliation is produced by raising the temperature to cause the subsurface hydrogen bubbles to expand and separate the wafer. What is unique about the smart cut process is that it is not specific to silicon alone. The same technique can be used to exfoliate and bond a thin film of any crystalline material to a completely different substrate. Examples include silicon-on-quartz, silicon carbide-on-silicon, germanium-on-silicon, III–V semiconductors-on-silicon as well as optical crystals such as lithium niobate films on silicon or fused silica [24,26,29]. This technology has tremendous potential to revolutionize optoelectronics and high-frequency electronics, but except for SOI, the exfoliation of most other materials is still in their infancy.

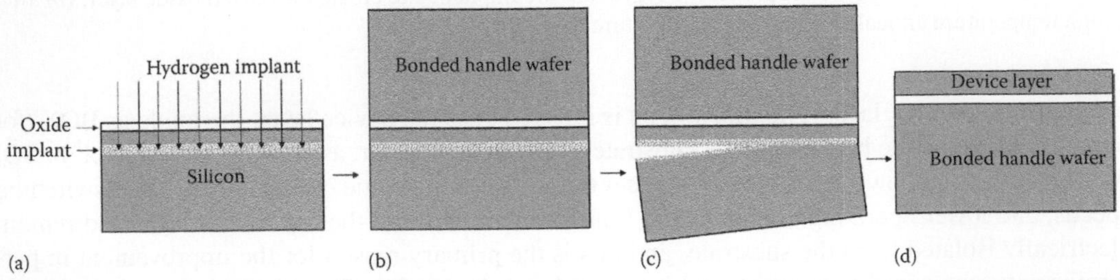

FIGURE 8.18 In the smart cut process, (a) hydrogen is implanted in an oxidized silicon wafer, (b) a handle wafer is bonded, (c) the implanted plane is exfoliated, and (d) the resulting thin film on oxide is annealed and polished.

8.4 THERMAL OXIDATION OF SILICON

Pure silicon can be thermally oxidized to produce high-quality SiO_2. In fact, it is this ability that made silicon a better material for electronics than germanium (though the first diodes and transistors were made from germanium). In ambient air at room temperature, a silicon wafer that has been cleaned in a buffered oxide etch (BOE) solution starts to form a native oxide within minutes, reaching about 5–10 Å in about 30 min [30]. Beyond 1 nm thickness, the oxide growth significantly slows down and comes to a virtual halt. The principal mechanism is believed to be due to the oxygen species diffusing through the oxide layer and reacting with the underlying silicon surface. Therefore, the growth rate will be inversely proportional to the oxide thickness, and directly proportional to the diffusion coefficient of the oxidizing species through SiO_2. When the temperature is raised, the diffusion coefficient becomes significantly larger, so the oxide growth can proceed to greater thicknesses. This is the main idea behind thermal oxidation.

In the simplest implementation, silicon wafers are placed in a quartz furnace (similar to the doping diffusion furnace) and the temperature is raised to about 1300°C in an oxygen ambient. Just like in thermal diffusion, the resulting oxide layer thickness will be strongly related to temperature and only weakly related to the duration of the process. The growth of oxide with pure oxygen gas is referred to as dry thermal oxidation, and when used with water vapor instead of oxygen, the process is referred to as wet thermal oxidation. They involve the following reactions:

$$\mathrm{Si} + \mathrm{O_2} = \mathrm{SiO_2} \tag{8.39}$$

$$\mathrm{Si} + 2\mathrm{H_2O} = \mathrm{SiO_2} + 2\mathrm{H_2}. \tag{8.40}$$

The oxidation rate with water vapor proceeds significantly faster than with oxygen, which is mainly due to the higher solubility of water vapor in the oxide layer. A model for the reaction kinetics was originally described by Deal and Grove in 1965, and it still remains the accepted model for this process [31].

The Deal–Grove model is based on three rate mechanisms for the overall growth rate of SiO_2. Referring to Figure 8.19, the first mechanism is the transport of the oxidation species (oxygen or water) from the gas to the surface of the oxide layer. This is written as

$$J_1 = h_1(C_G - C_s) \tag{8.41}$$

where

J_1 is the flux of the oxidant
C_G is the equilibrium concentration of the oxidant in the free-flowing gas
C_s is the concentration of the gas oxidant on the oxide surface

The concentration of dissolved oxidants in the solid oxide film, C_o, will be a function of the partial pressure P_s at the oxide surface. This is expressed by Henry's law, which states

$$C_o = HP_s \tag{8.42}$$

where H is Henry's constant. For an ideal gas, we have

$$P_s = C_s kT. \tag{8.43}$$

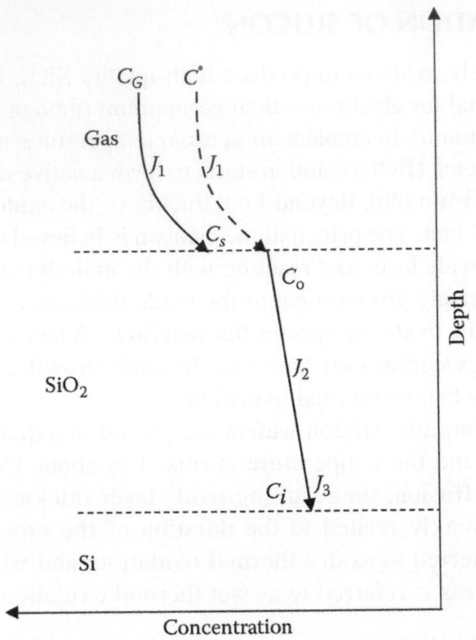

FIGURE 8.19 Concentration of the oxidation species as a function of depth assumed in the Deal–Grove model.

Using Equations 8.42 and 8.43, we can get a relationship between C_s and C_o:

$$C_s = \frac{C_o}{HkT} \tag{8.44}$$

which allows us to express Equation 8.41 as

$$J_1 = h_1\left(C_G - \frac{C_o}{HkT}\right) \tag{8.45}$$

This can be rewritten as

$$J_1 = h(C^* - C_o) \tag{8.46}$$

where

$$h = \frac{h_1}{HkT} \tag{8.47}$$

and

$$C^* = C_G HkT. \tag{8.48}$$

In terms of partial pressures, C^* can be written as

$$C^* = HP_G \tag{8.49}$$

where P_G is the partial pressure of the oxidizing species in the free flowing gas. C^* can be interpreted as the equivalent concentration of the oxidizing species in the free-flowing gas corresponding to the dissolved concentration in the oxide, as depicted in Figure 8.19.

The second rate mechanism is the diffusive transport through the oxide layer. This is the most important factor that limits the overall reaction rate, especially at larger oxide thicknesses, and is written as

$$J_2 = D\left(\frac{C_o - C_i}{x_0}\right) \tag{8.50}$$

where

D is the diffusion coefficient
C_i is the concentration of the oxidant at the silicon surface
x_0 is the oxide layer thickness

The last rate mechanism is the surface reaction rate, which is

$$J_3 = k_i C_i \tag{8.51}$$

where k_i is the rate constant.

At steady state, $J_1 = J_2 = J_3 = J$. Therefore, the overall reaction rate becomes

$$J = \frac{C^*}{\frac{1}{h} + \frac{x_0}{D} + \frac{1}{k_i}}. \tag{8.52}$$

This can be viewed as three resistors in series, with $R_1 = 1/h$, $R_2 = x_0/D$, and $R_3 = 1/k_i$ across a voltage source represented by C^*, as shown in Figure 8.20. The largest of the three will obviously

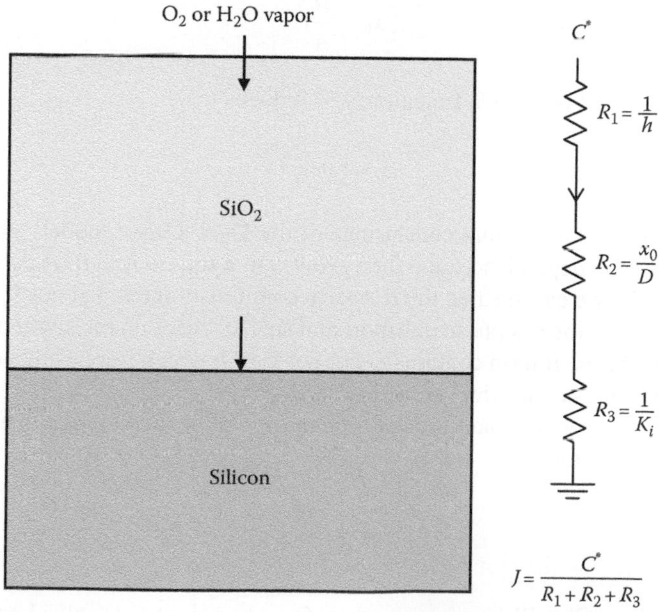

FIGURE 8.20 Representation of the Deal–Grove thermal oxidation model with electrical resistors.

limit the overall reaction rate, which in this case is R_2. In addition, the value of R_2 will increase with increasing oxide thickness, which can be viewed as a nonlinear resistor whose value grows with the total charge delivered through that resistor.

The growth rate of the oxide film will be the oxidation reaction rate divided by the number of oxidation species contained in a unit volume of the oxide film. Representing the latter as N_1, we can write

$$R = \frac{dx_0}{dt} = \frac{J}{N_1} = \frac{C^*/N_1}{\frac{1}{h} + \frac{x_0}{D} + \frac{1}{k_i}}. \tag{8.53}$$

The solution of this differential equation is quite straightforward since it contains only one x_0 term on the right-hand side. Rearranging and integrating results in

$$x_0 = \frac{A}{2}\left[\left(1 + \frac{4Bt}{A^2}\right)^{1/2} - 1\right] \tag{8.54}$$

where the constants in the equation are

$$A = 2D\left(\frac{1}{k_i} + \frac{1}{h}\right) \tag{8.55}$$

$$B = 2D\left(C^*/N_1\right). \tag{8.56}$$

Equation 8.54 has two limiting cases. For very short reaction durations where $t \ll A^2/4B$, the oxide growth is linear with time:

$$x_0 \approx \frac{B}{A}t. \tag{8.57}$$

For long oxidation times, $t \gg A^2/4B$, Equation 8.54 reduces to

$$x_0 \approx (Bt)^{1/2}. \tag{8.58}$$

These are two of the most important conclusions of the Deal–Grove model—the oxide thickness initially has a linear time dependence, and then evolves to a square root time dependence.

From Equation 8.57, we can see that the *B/A* term contains k_i and h, but not D. Therefore, during the linear growth regime, the gas phase transport and surface reaction rates will be the rate-limiting steps. In Equation 8.58, the B term contains D but not k_i or h which implies that diffusion becomes the primary rate-limiting step for thicker oxides.

Furthermore, since h is a gas phase transport factor, it will be much larger than k_i. This allows us to neglect h in the expression for A:

$$A \approx \frac{2D}{k_i}. \tag{8.59}$$

Under this assumption, C^* will become equal to C_o, which is the dissolved concentration of the oxidants in the film.

The parameters k_i and D will have the same exponential dependence on temperature as in Equation 8.4 with different activation energies:

$$D = D_0 e^{-\frac{E_a^D}{kT}} \tag{8.60}$$

$$k_i = k_{i0} e^{-\frac{E_a^{ki}}{kT}}. \tag{8.61}$$

In Table 8.3, the parameters for wet and dry thermal oxidation at 760 Torr partial pressure (or 100% concentration at atmospheric pressure) are shown for a (111) silicon substrate. Furthermore, the Henry's constant H is also a function of temperature, which results in a C^* that will be a function of temperature. The calculated values of C^* are listed in Table 8.4.

We can see that the diffusion of oxygen has a higher activation energy E_a^D than for water vapor. At 1200°C, diffusion coefficient will be 2.8×10^{-8} cm^2/s for oxygen, and 1.9×10^{-9} cm^2/s for water vapor. Therefore, the diffusion rate of water vapor is slower than for oxygen by about one order of magnitude. However, the value C^* is about 400 times larger. This is the main reason for the enhanced oxidation rate with water vapor compared to oxygen.

The rate constant k_i also has a dependence on the silicon substrate's crystal orientation. At low thicknesses, the growth rates are $R_{110} > R_{111} > R_{100}$. At larger thicknesses, the rates switch to $R_{111} > R_{110} > R_{100}$ [32]. Compared to (100), the rate constant for (111) is found to be 45%–77% larger [33,34]. The most commonly used value in the literature is

$$k_{i0}^{(111)} = 1.68 \times k_{i0}^{(100)}. \tag{8.62}$$

TABLE 8.3
Dry and Wet Thermal Oxidation Parameters for (100) Surface Silicon with 100% Concentration of O_2 or H_2O at Atmospheric Pressure

	Dry	Wet
E_a^D (eV)	1.23	0.70
D_0 (cm^2/s)	4.5×10^{-4}	5.0×10^{-7}
$E_a^{k_i}$ (eV)	2.0	1.96
k_{i0} (cm/s)	8.2×10^{4}	3.8×10^{3}
N_1 (cm^{-3})	2.2×10^{22}	4.4×10^{22}

TABLE 8.4
Values of C^* for 100% Concentration of O_2 or H_2O at Atmospheric Pressure

T (°C)	C^* Dry (cm^{-3})	C^* Wet (cm^{-3})
1200	5×10^{16}	2.3×10^{19}
1100	6×10^{16}	2.4×10^{19}
1000	5.9×10^{16}	2.14×10^{19}

This difference between crystal orientations is relevant only in the linear growth regime (Equation 8.57). For thicker oxides (Equation 8.58), the growth rate becomes independent of crystal orientation. Substrate doping also has an effect on k_{i0} [35]. The partial pressure of the oxidation species is also contained in B because it contains C^* (Equation 8.56) which depends on the partial pressure P_G (Equation 8.49). However, for dry oxidation, C^* has been found to depend on the pressure as P_G^n where n is between 0.5 and 1.0. It is presumed that this is due to oxygen decomposing and creating atomic oxygen as well as molecular oxygen species.

Using the values shown in Tables 8.3 and 8.4, the oxidation rate can be calculated from Equation 8.54. This is shown in Figure 8.21. It should be evident that the wet oxidation rate is significantly higher than the dry oxidation rate. The effect of temperature is also dramatic due to the exponential factors in Equations 8.60 and 8.61.

It should be noted that while the Deal–Grove model agrees very well with experimental results for relatively thick oxide films, it underestimates the growth rate at very small thicknesses, especially for dry thermal oxidation. At 100 Å thickness, the actual oxide grows nearly twice as fast as predicted by the model. Even at 500 Å, the measured rates are about 10% higher than the model. This difference is most often handled by including a time offset $t + \tau$ in Equation 8.54 to fit the experimental data. This inconsistency is most likely due to a different oxidation mechanism dominating the initial growth phase. Despite the fact that a large number of device applications today critically rely on ultrathin oxide films, this issue still remains largely unresolved. Some of the proposed theories include increased solubility of oxygen in the oxide film, stress in the oxide film contributing to differences in the reaction parameters, out-diffusion of silicon atoms, and space charge effects [36].

Thermal oxide produced by dry oxidation is easily the highest quality oxide one could obtain on silicon. It is free of pinholes, fully conformal to the silicon structures, highly uniform, acts as an excellent passivation and free of any contaminants. As a result, dry thermal oxidation became the preferred method for producing the gate oxides of CMOS devices. Where thicker oxides were required, such as for isolation between adjacent CMOS devices (also known as field oxide), wet oxidation is used. Wet oxides are slightly inferior to dry oxides, especially with regard to dielectric

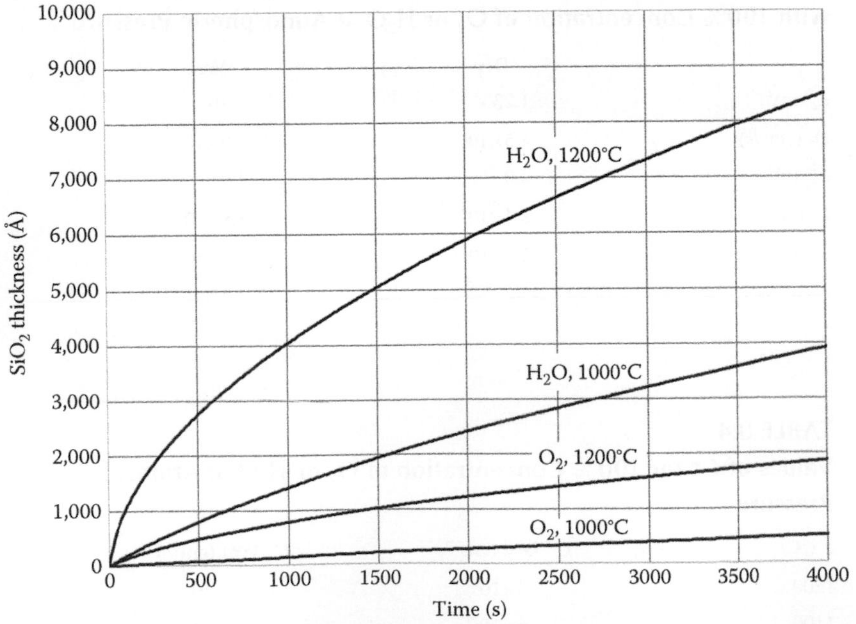

FIGURE 8.21 Wet and dry thermal oxidation film thickness vs. time for 1000°C and 1200°C on a (100) silicon surface.

breakdown and etch resistance. This has to do with the excess hydrogen produced in the reaction which causes vacancy-induced defects. Nevertheless, wet oxide is still a significantly high-quality film compared to all other forms of CVD and PVD oxides. However, thermal oxidation is a high-temperature process, so it can only be used in the front end of line (FEOL) process. Oxides produced in the back end of line (BEOL) have to be low-temperature processes, such as PECVD. LPCVD oxides are moderate quality, but require somewhat higher temperatures. The choice greatly depends on the intended application and its placement in the process sequence. Evaporated oxides have the poorest quality, but are adequate as an optical film. In terms of quality, the ranking of oxides would roughly fall in the order of (1) dry thermal oxide, (2) wet thermal oxide, (3) ALD oxide, (4) LPCVD oxide, (5) PECVD oxide, (6) sputtered oxide, and (7) evaporated oxide.

Thermal oxidation consumes silicon to create SiO_2. Therefore, this is also an etching process, since some silicon is lost during oxidation. The reaction, however, produces a net increase in thickness because each SiO_2 molecule occupies a larger volume than the original silicon atom. The molar volume of silicon (the volume occupied by one mole of silicon atoms) can be calculated by dividing its molar mass by density:

$$V_m = \frac{M}{\rho}. \tag{8.63}$$

For silicon, $M = 28.08$ g/mol and $\rho = 2.33\ \text{g/cm}^3$, which gives $V_m = 12.05\ \text{cm}^3/\text{mol}$. For SiO_2, $M = 60.08$ g/mol and $\rho = 2.2\ \text{g/cm}^3$, which gives $V_m = 27.3\ \text{cm}^3/\text{mol}$. Therefore, the molar volume of silicon is 44% of SiO_2. Since the film grows on a fixed surface area, the change in molar volume has to come entirely from the differences in thickness. Therefore, we can conclude that 44% of the thermal oxide thickness is due to the removal of silicon and 56% of the oxide will be projecting above the original silicon surface. This is illustrated in Figure 8.22.

The equipment required for performing thermal oxidation is relatively straightforward, and the same setup as for thermal diffusion of dopants could be used. The wafers are placed in a quartz tube and are purged with dry nitrogen. Oxygen (or water vapor) is then flowed for a specified duration, and then switched back to dry nitrogen. This allows one to control the process duration accurately without having to account for the warm-up and cool-down phases. Dry thermal oxidation is the simplest since it requires almost no additional setup beyond switching gases. For wet thermal oxidation, nitrogen gas saturated with water vapor can be used. This can be performed, for example, by bubbling nitrogen through a heated water flask using a gas washing bottle, diffusing frit stone and a thermocouple dipped in the liquid to control the temperature. The temperature of the water is important as it will determine its vapor pressure, and hence the growth rate (through the parameter C^* in Equation 8.49). 95°C is a typical temperature used in this process. The tubes leading from the flask to the furnace also have to be heated to a temperature above the water temperature to prevent condensation.

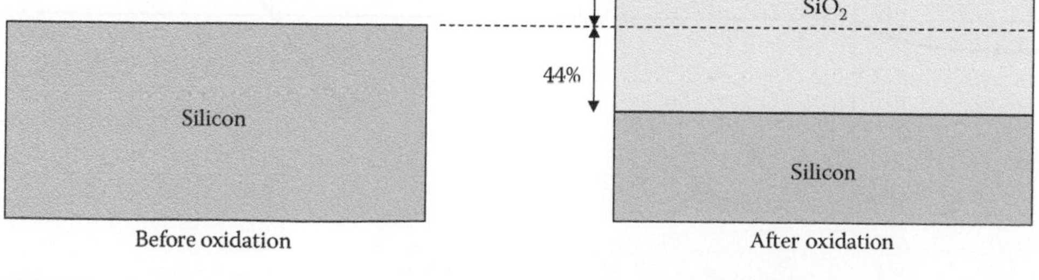

FIGURE 8.22 Subtraction from silicon is 44% of the total thermal oxide thickness.

Alternatively, wet oxidation can also be done by flowing hydrogen and oxygen simultaneously into the reaction tube to produce what is known as pyrogenic steam [37]. This has better reproducibility because the oxidant's concentration can be controlled more precisely than with a water bubbler. However, there are obvious hazards with mixing oxygen and hydrogen.

It is interesting to note that silicon nitride (Si_3N_4) can also be formed by a process known as nitridation, by flowing nitrogen gas instead of oxygen. However, the activation energies are much higher and the growth rates are much smaller than thermal oxidation [38,39]. Any trace amounts of oxygen will quickly form an oxide layer that will prevent the nitrogen from diffusing through the oxide film. Film thicknesses are reported to be less than 500 Å for 1 h at 1200°C. Other gases such as NH_3 and NO have also been used to grow thermal nitridation of silicon [40].

The high quality of thermal oxidation is fairly unique to silicon, and other materials, including the close relatives of silicon such as germanium, and III–V semiconductors do not form high-quality oxides. This is one of the main reasons why silicon has remained the preferred platform for electronic circuits despite the fact that many of the III–V semiconductors have better electronic properties. However, thermal oxidation, anodic processes, and atomic layer deposition on other semiconductors are rapidly being developed, and may some day advance to a stage to compete with silicon.

8.5 METAL CONTACTS TO SEMICONDUCTORS

A metal film deposited on a semiconductor surface can produce different results depending on the type of metal, type of semiconductor, surface conditions, and annealing. When the conduction band is the primary current carrying band (n-type semiconductor), a rectifying contact will result if the Fermi level of the semiconductor is higher than the Fermi level of the metal. A rectifying contact exhibits a highly asymmetrical current–voltage behavior similar to a PN junction diode, and is known as the Schottky diode. An ohmic contact—one with symmetrical current–voltage characteristic—will result if the Fermi level of the semiconductor is lower than the Fermi level of the metal. When the valence band is the primary current carrying band (p-type semiconductor), the opposite happens—a Schottky barrier is produced when the metal Fermi level is higher than the semiconductor, and vice versa. Figure 8.23 shows a qualitative illustration of the current–voltage characteristics of rectifying and ohmic contacts.

Figure 8.24 shows the relevant band diagrams for n-type semiconductor-to-metal junction. The top row (i) is for when the semiconductor Fermi level (E_{fs}) is higher than the metal Fermi level (E_{fm}). The bottom row (ii) is for when (E_{fs}) is lower than (E_{fm}). Column (a) shows the band structure

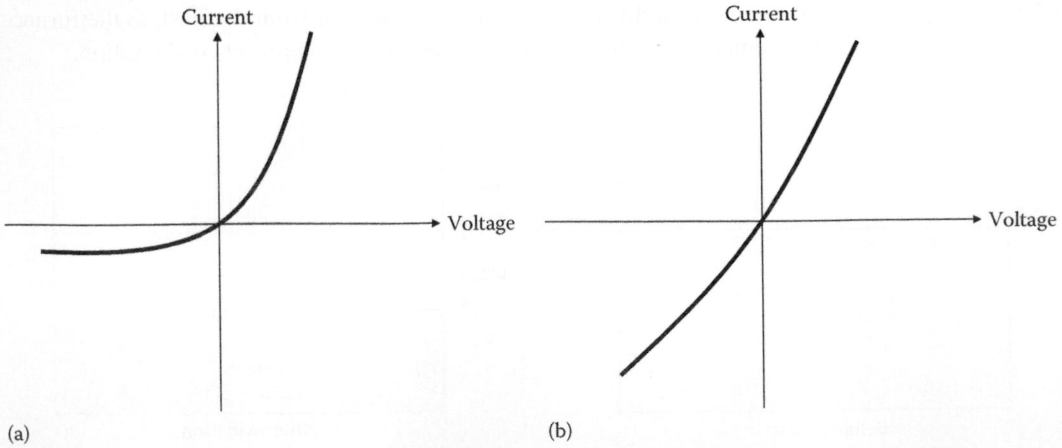

FIGURE 8.23 Current–voltage characteristic of a rectifying contact (a) and an ohmic contact (b).

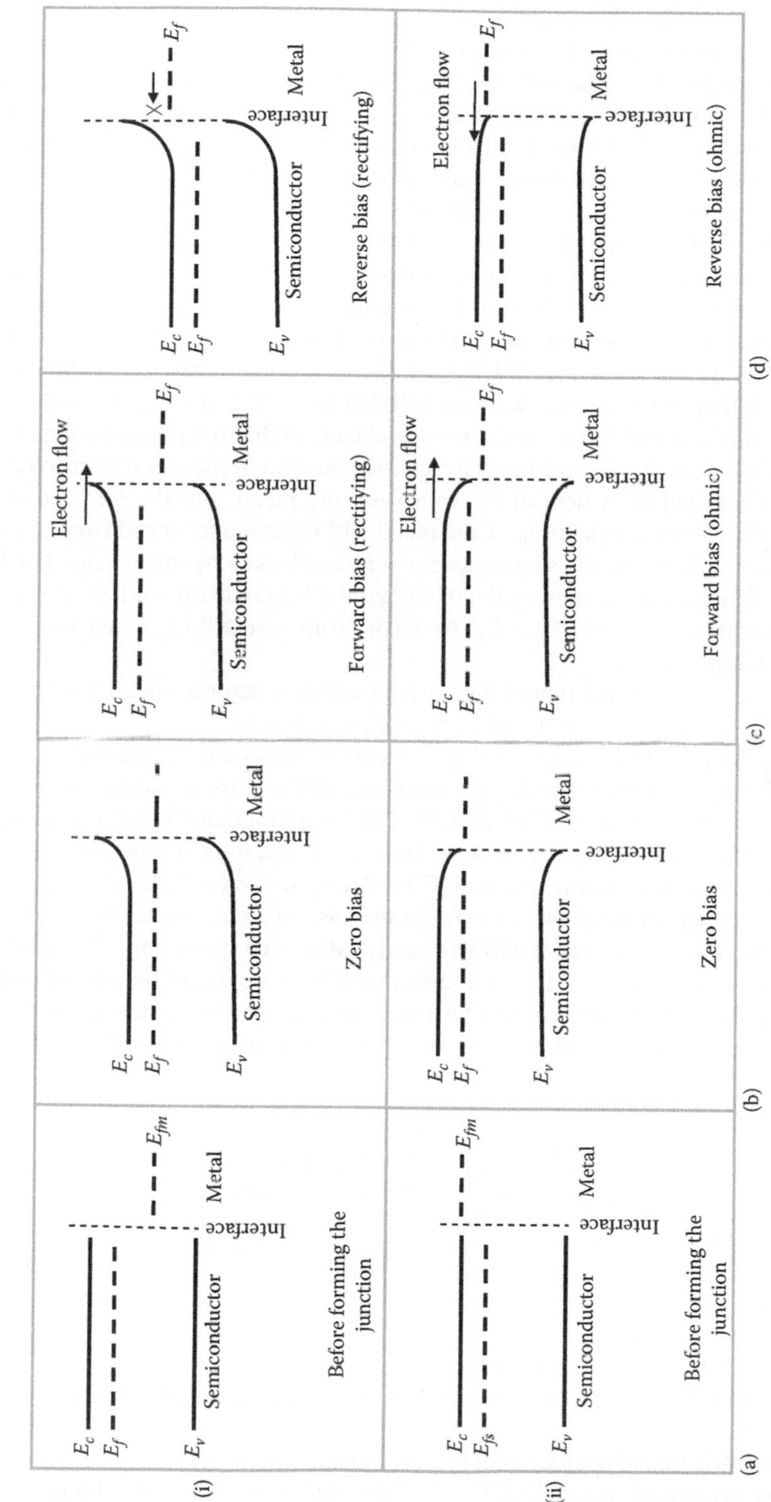

FIGURE 8.24 Metal–semiconductor band diagrams for an n-type semiconductor. The top row (i) is for $E_{fs} < E_{fm}$, and the bottom row (ii) is for $E_{fs} > E_{fm}$. Column (a) is the band diagram before the junction is formed; (b) is after junction is formed under equilibrium; (c) is for a forward applied bias; (d) is for a reverse applied bias.

before the junction is created. When the semiconductor and metal are joined, electrons flow from the material with the higher Fermi level toward the one with the lower Fermi level. This creates a charge imbalance, and a corresponding potential difference. Once the Fermi levels align, net current flow will stop. This condition is shown in column (b). This is the equilibrium condition with no applied voltage. For $E_{fs} > E_{fm}$ (row (i)), electrons on both sides of the junction have a barrier to surmount before any current can flow. The barrier from the metal to the semiconductor will be $\varphi_{bn} = q(\varphi_m - \chi)$, where φ_m is the work function of the metal (energy needed to raise an electron from the Fermi level to the vacuum level) and χ is the electron affinity of the semiconductor (energy needed to raise an electron from the lowest point in the conduction band to the vacuum level) (see Figure 8.26). The barrier from the semiconductor to the metal will be $E_{fs} - E_{fm}$. For $E_{fs} < E_{fm}$ (row (ii)), electrons from the semiconductor do not have a barrier, but the electrons from the metal will encounter a very small barrier. Column (c) shows the case when a bias is applied such that the potential in the semiconductor is raised compared to the metal to induce an electron flow from the semiconductor to the metal. For $E_{fs} > E_{fm}$ (row (i)), the barrier for the electrons in the semiconductor will be lowered, and electrons will flow into the metal. For $E_{fs} < E_{fm}$ (row (ii)), since there was no barrier for electrons in the semiconductor in the first place, current will easily flow into the metal. Column (d) shows when the electron potential in the metal is raised compared to the semiconductor to induce a flow of electron from the metal into the semiconductor. For $E_{fs} > E_{fm}$ (row (i)), the barrier for electrons in the metal will remain unchanged from the zero bias case. Therefore, electron flow into the semiconductor will be blocked by this barrier. For $E_{fs} < E_{fm}$ (row (ii)), the small barrier for electrons in the metal will be lowered still, and electrons will flow into the semiconductor. As a result, $E_{fs} > E_{fm}$ (row (i)) forms a rectifying contact and $E_{fs} < E_{fm}$ (row (ii)) forms an ohmic contact.

Figure 8.25 shows a similar set of figures for a p-type semiconductor-to-metal junction. The barrier from the metal to the semiconductor will be $\varphi_{bp} = E_g - q(\varphi_m - \chi)$.

Table 8.5 shows the work functions φ_s and φ_m of several semiconductors and commonly used metals. The work function here is defined as the energy needed to raise an electron from the Fermi level to the vacuum level, as shown in Figure 8.26. The term "vacuum" is used here to mean an electron that is free from its host material system. In other words, the work function is the energy required to free the electron from the material. Considering the vacuum level as the zero reference, the Fermi level then becomes $-\varphi_s$ or $-\varphi_m$. In the case of semiconductors, the Fermi level will often fall inside the bandgap, and it will be heavily influenced by doping. The electron affinity (χ) is the energy required to raise an electron from the lowest point in the conduction band to the vacuum level and is slightly more useful because it is independent of doping (Table 8.6). Of course, the work function and electron affinity are related, which can be derived to give the following relationship [41]:

$$\varphi_s = \chi + \frac{E_g}{2} - \frac{3}{4} kT \ln\left(\frac{m_{dv}^*}{m_{dv}^*}\right) - kT \ln\left(\frac{n}{n_i}\right) \tag{8.64}$$

where

E_g is the bandgap
n is the electron concentration
n_i is the intrinsic (undoped) electron concentration
m_{dv}^* and m_{dc}^* are the valence and conduction band density-of-states effective masses respectively

In the case of undoped semiconductors, n will be equal to n_i, so the last factor in Equation 8.64 will be zero. For silicon, $m_{dv}^* = 0.55$, $m_{dc}^* = 0.33$, $E_g = 1.12$ eV, and $n_i = 1 \times 10^{10}$ cm^{-3}. Using $\chi = 4.01$ eV, we can get $\varphi_s = 4.56$ eV for undoped silicon. If it is doped at $n = 1 \times 10^{18}$ cm^{-3}, then φ_s will drop by 0.48 eV to become 4.08 eV.

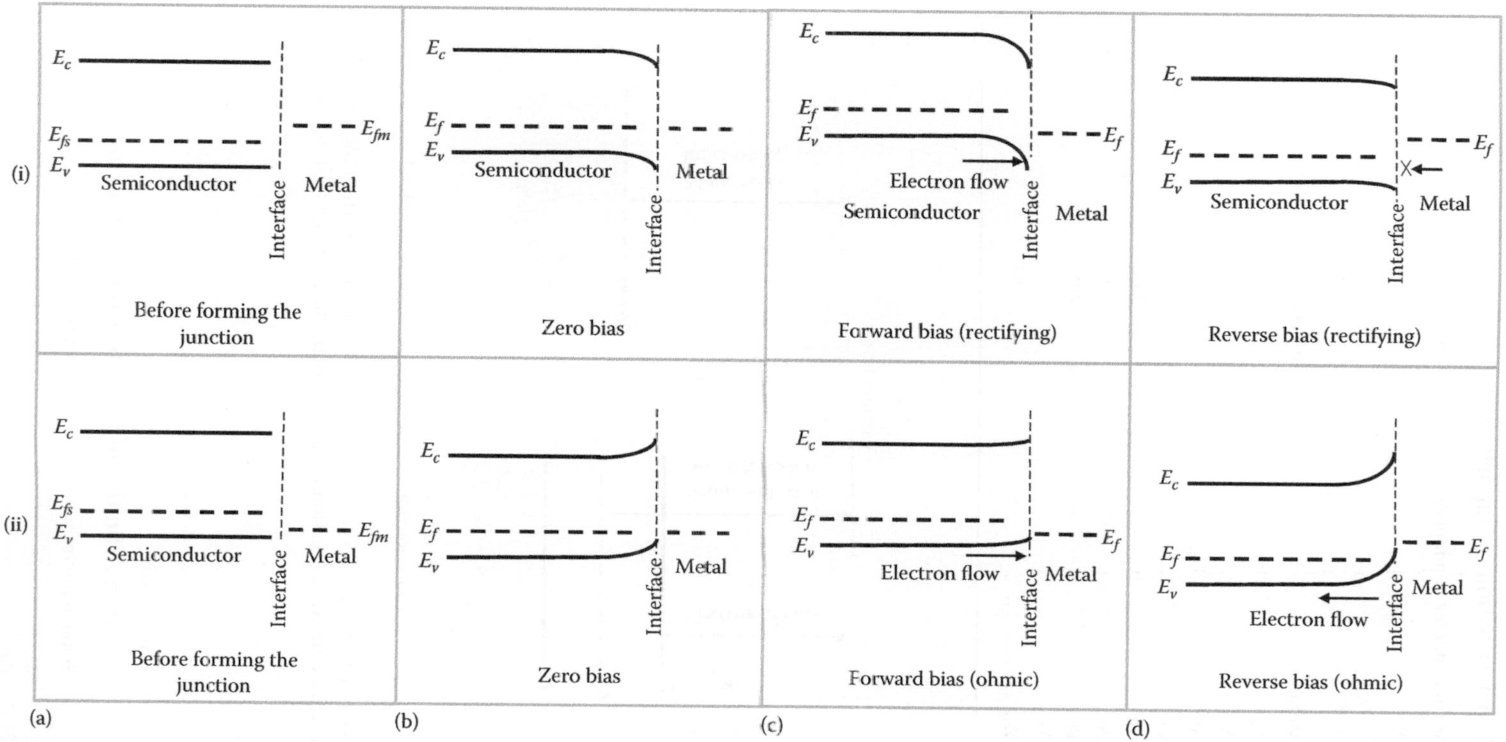

FIGURE 8.25 Metal–semiconductor band diagrams for a p-type semiconductor. The top row (i) is for $E_{fs} < E_{fm}$ and the bottom row (ii) is for $E_{fs} > E_{fm}$. Column (a) is the band diagram before the junction is formed; (b) is after junction is formed under equilibrium; (c) is for a forward applied bias; (d) is for a reverse applied bias.

TABLE 8.5
Work Function of Important Metals and Semiconductors

Semiconductor (Undoped)	Work Function φ_s (eV)
Si	4.56
Ge	5.43
GaAs	4.74
Metal	**Work Function φ_m (eV)**
Ag	4.26
Al	4.28
Au	5.1
Cr	4.5
Cu	4.65
Ni	5.15
Pt	5.65
Ti	4.33
W	4.55

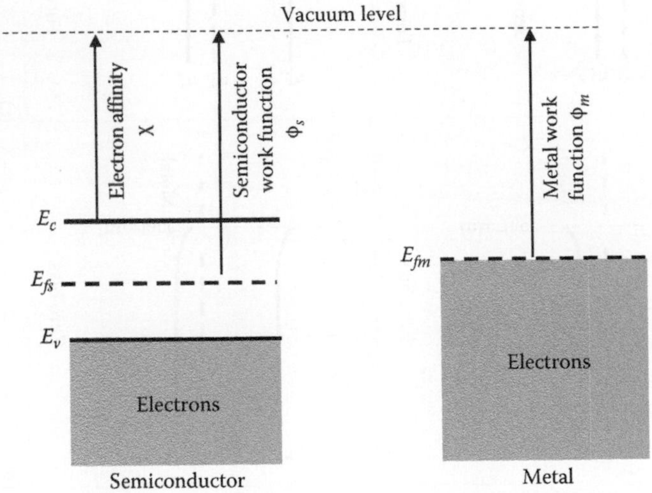

FIGURE 8.26 Work function is the energy required to raise an electron from the Fermi level to the vacuum level. The left figure shows the work function for a semiconductor, and the right figure shows the work function for a metal.

TABLE 8.6
Electron Affinities of Common Semiconductors

Semiconductor	Electron Affinity χ (eV)
Si	4.01
Ge	4.13
GaAs	4.07

For n-type silicon, φ_s will become smaller by as much as 0.3–0.5 to 4.0–4.3 eV. In order to produce an ohmic contact, we need metals with work functions smaller than 4.0–4.3 eV (to make their Fermi levels higher than the semiconductor). Silver, aluminum and titanium are suitable metals with fairly low work functions. The other metals, such as gold, nickel and platinum, would form rectifying contacts. When silicon is doped p-type, the work function will increase by 0.3–0.5 to 4.8–5.1 eV. To make ohmic contacts, the work function of metals has to be larger than 4.8–5.1 eV (to make their Fermi levels lower than the semiconductor). Suitable metals would be gold, nickel and platinum.

This treatment of semiconductor–metal contacts is an idealized scenario. In practice, even before any metal is deposited on a semiconductor, the Fermi level at the surface of the semiconductor will not be the same as it is in the bulk. This is to be expected since there will be many unterminated bonds at the surface that would render their electronic properties significantly different from their interior counterparts. The surface introduces additional electronic states known as interface traps. As a result, the Fermi level at the surface could be lower or higher than the bulk depending on the termination condition of these bonds. For n-type silicon, the metal-to-semiconductor barrier is experimentally found to be $\varphi_{bn} = (0.27q\varphi_m - 0.52)$ eV instead of $\varphi_{bn} = q(\varphi_m - \chi)$. Similarly, for GaAs, it has been found to be $\varphi_{bn} = (0.07q\varphi_m + 0.51)$ eV. Some of the data are shown in Table 8.7. A more complete list of experimentally measured barrier heights for various semiconductor–metal contacts can be found in [41]. These are for clean semiconductor–metal interfaces. Even trace amounts of oxides, nitrides, or other contaminations will further alter these characteristics. An additional effect, known as image force lowering, will also reduce the effective work function of the metal in the presence of an electric field.

The important point to note is that the metal work function φ_m has a lesser influence on the metal-to semiconductor barrier φ_{bn} than the idealized models might predict. In the case of n-type silicon, its influence is only 27%, and for GaAs it is only 7%. As a result, the barrier height of GaAs-to-metal can be considered to be nearly independent of the metal. This effect is known as Fermi level pinning, and it arises due to the extremely high interface trap density in GaAs. In silicon, the interface trap density is smaller, so the metal has a somewhat greater influence. As a result, experimentally measured barrier potential (with silicon or some other semiconductor) is a more relevant quantity than the work function, and this is what is often quoted in the literature.

The interface trap density has a profound effect on semiconductor–metal junctions, and this makes most of the idealized calculations less useful for practical work. Furthermore, the barriers are also influenced by the cleanliness of the surface prior to metal deposition, process conditions during metal deposition, and deposition temperatures [42,43]. They also exhibit changes in characteristics

TABLE 8.7
Measured Barrier Heights φ_{bn} of n-Type Si and GaAs with Various Metals

Metal	n-Si/Metal Barrier (eV)	n-GaAs/Metal Barrier (eV)
Ag	0.83	1.03
Al	0.81	0.93
Au	0.83	1.05
Cu	0.8	1.08
Ni	0.74	0.91
Pt	0.9	0.98
Ti	0.6	0.84
W	0.66	0.8

Source: Data from Sze, S.M. and Ng, K.K., *Physics of Semiconductor Devices*, 3rd edn., John Wiley & Sons, New York, November 3, 2006.

due to aging [42]. As a result, ohmic contacts are most commonly made not only by selecting a metal to produce the smallest barrier height, but also by reducing the barrier width to allow electrons to quantum mechanically tunnel through the barrier. This is known as a tunneling contact or field emission contact, and can be realized by doping the semiconductor very high [44]. The width of the barrier is inversely related to the square root of the doping density:

$$W = \sqrt{\frac{2\varepsilon_s}{qN}\left(V_s - V_a - \frac{kT}{q}\right)} \tag{8.65}$$

where
V_s is the potential difference at the junction due to the alignment of the Fermi levels
V_a is the applied forward bias voltage
N is the doping density
ε_s is the dielectric permittivity of the semiconductor

Assuming $V_s - V_a - (kT/q)$ to be approximately 0.2 V, for a doping level of 10^{19} cm^{-3} the junction width will be approximately 5 nm, which is narrow enough to produce a quantum mechanical tunneling current (Figure 8.27).

Metal–semiconductor contacts are often annealed at elevated temperatures to improve their properties. This causes the atoms from both sides of the interface to diffuse and fill the voids and dangling bonds at the interface. One problem with this technique is that metal atoms can diffuse deep into silicon through interstitial sites. Generally, some techniques are necessary to mitigate this diffusion during the anneal step. The metal can also form a eutectic mixture with the silicon with very low melting temperatures, which will limit the maximum temperature that can be used during anneal.

Aluminum is widely used for making ohmic contacts on silicon, and it can be annealed to improve the contact resistance. However, aluminum also forms a eutectic alloy with silicon with a melting temperature of 577°C. Even at lower temperatures, silicon atoms can diffuse into the aluminum causing voids and spike defects in the silicon crystal [45]. At 550°C, the solubility limit of silicon in aluminum is 1.5% weight. Therefore, one way to mitigate the spike defects is to deposit a film of silicon-doped aluminum. This can be done by co-sputtering or by using an alloy sputter target. Another technique is to deposit a very thin layer of aluminum, perform the high-temperature anneal, and then deposit a thicker aluminum layer on top. The thin layer will limit the maximum amount of silicon that is drawn into the film, and as long as no further annealing is performed after the second deposition, spike defects will not continue.

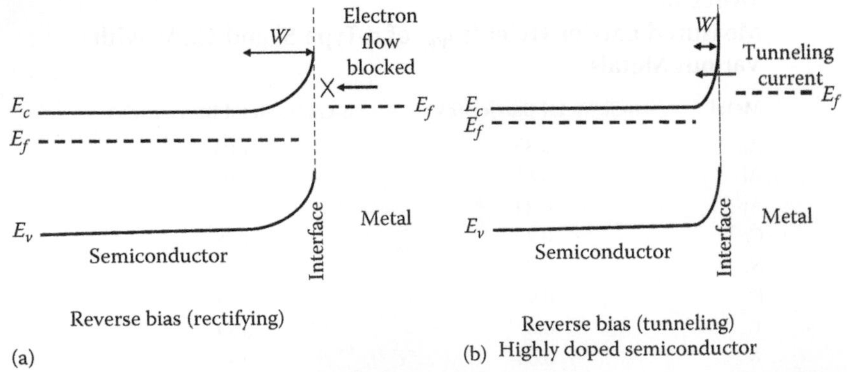

FIGURE 8.27 (a) A moderately doped n-type semiconductor–metal junction under reverse bias where very little current flows and (b) a highly doped n-type semiconductor–metal junction under reverse bias where tunneling current dominates.

TABLE 8.8
Commonly Used Silicides and Their Properties

Silicide	Forming Temperature (°C)	Melting Temperature (°C)	n-Si/Silicide Barrier (eV)
PtSi	300–600	1229	0.87
NiSi	400–600	992	0.67
$TiSi_2$	600–700	1500	0.60
WSi_2	650	2160	0.86
$MoSi_2$	1000	2020	0.69

Sources: Sze, S.M. and Ng, K.K., *Physics of Semiconductor Devices*, 3rd edn., John Wiley & Sons, New York, November 3, 2006; Gambino, J.P. and Colgan, E.G., *Mater. Chem. Phys.*, 52, 99, 1998.

A more attractive solution is the use of silicides as an intermediate film. These are binary compounds that exhibit the beneficial features of both metals and silicon [46,47]. They can sustain high temperatures and have low resistivities and have been used as resistive heating elements and as high-temperature coatings. They are also chemically stable and have low diffusivities in silicon. As a result, they can act as a diffusion barrier between a pure metal film and the underlying silicon substrate. Just like metals, the semiconductor–silicide interface can form rectifying contacts or ohmic contacts depending on the substrate doping and work functions.

Silicides can be deposited by CVD, or they can also be produced by simply depositing the metal on a clean oxide-free silicon surface and then performing a high-temperature anneal to induce the required chemical interaction. Of particular interest are metals such as molybdenum, titanium, tungsten, cobalt, nickel, platinum and palladium. These can form PtSi, NiSi, WSi_2, and $TiSi_2$. They can be formed at relatively low temperatures, but have very high melting temperatures. Annealing must be performed in oxygen-free environments to prevent oxygen molecules competing with the metal to react with silicon, and in many cases have a lower activation energy than the metal–silicon reaction. In Table 8.8, several commonly used silicides and their properties are listed.

REFERENCES

1. M. L. Yu, D. J. Vitkavage, and B. S. Meyerson, Doping reaction of PH_3 and B_2H_6 with Si(100), *J. Appl. Phys.*, 59 (1986) 4032.
2. P. Negrini, A. Ravaglia, and S. Solmi, Boron predeposition in silicon using BBr_3, *J. Electrochem. Soc.*, 125(4) (1978) 609–613.
3. P. Negrini, D. Nobili, and S. Solmi, Kinetics of phosphorus predeposition in silicon using $POCl_3$, *J. Electrochem. Soc.*, 122(9) (1975) 1254–1260.
4. K.-T. Kim and C.-K. Kim, Formation of shallow p^+-n junctions using boron-nitride solid diffusion source, *IEEE Electron Device Lett.*, 8(12) (1987) 569–571.
5. Planar Diffusion Sources, Saint-Gobain Ceramic Materials, Amherst, NY. http://www.bn.saint-gobain.com/planar-diffusion-sources.aspx.
6. BoronPlus and PhosPlus Planar Diffusion Sources, TechneGlas, Perrysburg, OH. http://www.techneglas.com/.
7. F. A. Trumbore, Solid solubilities of impurity elements in germanium and silicon, *Bell Syst. Tech. J.*, 39(1) (January 1960) 205–233.
8. V. E. Boeisenko and S. G. Yudin, Steady-state solubility of substitutional impurities in silicon, *Phys. Status Solidi*, 101(1) (May 16, 1987) 123–127.
9. G. L. Vick and K. M. Whittle, Solid solubility and diffusion coefficients of boron in silicon, *J. Electrochem. Soc.*, 116(8) (1969) 1142–1144.
10. M. L. Barry and P. Olofsen, Doped oxides as diffusion sources I. Boron into silicon, *J. Electrochem. Soc.*, 116(6) (1969) 854–860.
11. S. W. Jones, *Diffusion in Silicon*, IC Knowledge LLC, Georgetown, MA.

12. R. W. Olesinski and G. J. Abbaschian, The B–Si (Boron–Silicon) system, *Bull. Alloy Phase Diag.*, 5(5) (October 1984) 478–484.
13. R. W. Olesinski, N. Kanani, and G. J. Abbaschian, The P–Si (Phosphorus–Silicon) system, *Bull. Alloy Phase Diag.*, 6(2) (April 1985) 130–133.
14. D. M. Brown and P. R. Kennicott, Glass source diffusion in Si and SiO_2, *J. Electrochem. Soc. Solid State Sci.*, 118(2) (February 1971) 313–317.
15. W. H. Press, S. A. Teukolsky, W. T. Vetterling, and B. P. Flannery, *Numerical Recipes in C—The Art of Scientific Computing*, 2nd edn., Cambridge University Press, Cambridge, U.K., 1992.
16. K. A. Ellis and R. A. Buhrman, Boron diffusion in silicon oxides and oxynitrides, *J. Electrochem. Soc.*, 145(6) (June 1998) 2068–2069.
17. J. F. Ziegler, M. D. Ziegler, J. P. Biersack, SRIM—The stopping and range of ions in matter, *Nucl. Instrum. Methods Phys. Res. B*, 268(11–12) (2010) 1818–1823. http://www.srim.org/.
18. P. Vandenabeele and K. Maex, Modelling of rapid thermal processing, *Microelectron. Eng.*, 10 (1991) 207–216.
19. V. Borisenko and P. J. Hesketh, *Rapid Thermal Processing of Semiconductors*, Springer, New York, 1997.
20. J.-P. Colinge, *Silicon-on-Insulator Technology: Materials to VLSI*, Springer Science, Berlin, Germany, 2004.
21. G. K. Celler and S. Cristoloveanu, Frontiers of silicon-on-insulator, *J. Appl. Phys.*, 93 (2003) 4955.
22. W. Noell, P.-A. Clerc, L. Dellmann, B. Guldimann, H.-P. Herzig, O. Manzardo, C. Roman Marxer, K. J. Weible, R. Dändliker, and N. de Rooij, Applications of SOI-based optical MEMS, *IEEE J. Sel. Top. Quant. Electron.*, 8(1) (January/February 2002) 148–154.
23. M. J. Anc, *SIMOX*, Published by the Institution of Electrical Engineers, UK, 2004.
24. C. Maleville and C. Mazuré, Smart-cut technology: From 300 mm ultrathin SOI production to advanced engineered substrates, *Solid-State Electron.*, 48(6) (June 2004) 1055–1063.
25. H. Moriceau, F. Mazen, C. Braley, F. Rieutord, A. Tauzin, and C. Deguet, Smart Cut^{TM}: Review on an attractive process for innovative substrate elaboration, *Nucl. Instrum. Methods Phys. Res. B*, 277 (2012) 84–92.
26. A. Plobl and G. Krauter, Silicon-on-insulator: Materials aspects and applications, *Solid-State Electron.*, 44 (2000) 775–782.
27. J. B. Lasky, Wafer bonding for silicon-on-insulator technologies, *Appl. Phys. Lett.*, 48 (1986) 78.
28. M. Alexe and U. Gösele, *Wafer Bonding: Applications and Technology*, Springer, Berlin, Germany, 2004.
29. C. Wang, M. J. Burek, Z. Lin, H. A. Atikian, V. Venkataraman, I.-C. Huang, P. Stark, and M. Lončar, Integrated high quality factor lithium niobate microdisk resonators, *Opt. Express*, 22(25) (2014) 30924–30933.
30. M. Morita, T. Ohmi, E. Hasegawa, M. Kawakami, and M. Ohwada, Growth of native oxide on a silicon surface, *J. Appl. Phys.*, 68 (1990) 1272.
31. B. E. Deal and A. S. Grove, General relationship for the thermal oxidation of silicon, *J. Appl. Phys.*, 36(12) (December 1965) 3770–3778.
32. E. A. Irene, H. Z. Massoud, and E. Tierney, Silicon oxidation studies: Silicon orientation effects on thermal oxidation, *J. Electrochem. Soc.*, 133(6) (1986) 1253–1256.
33. E. A. Lewis and E. A. Irene, The effect of surface orientation on silicon oxidation kinetics, *J. Electrochem. Soc.*, 134(9) (1987) 2332–2339.
34. J. R. Ligenza, Effect of crystal orientation on oxidation rates of silicon in high pressure steam, *J. Phys. Chem.*, 65(11) (1961) 2011–2014.
35. B. E. Deal and M. Sklar, Thermal oxidation of heavily doped silicon, *J. Electrochem. Soc.*, 112(4) (1965) 430–435.
36. J. Dabrowski and H.-J. Müssig, *Silicon Surfaces and Formation of Interfaces: Basic Science in the Industrial World*, World Scientific, Singapore, 2000.
37. R. R. Razouk, L. N. Lie, and B. E. Deal, Kinetics of high pressure oxidation of silicon in pyrogenic steam, *J. Electrochem. Soc.*, 128(10) (1981) 2214–2220.
38. C.-Y. Wu, C.-W. King, M.-K. Lee, and C.-T. Chen, Growth kinetics of silicon thermal nitridation, *J. Electrochem. Soc.*, 129(7) (1982) 1559–1563.
39. H. Zhu, D. Yang, L. Wang, and D. Due, Thermal nitridation kinetics of silicon wafers in nitrogen atmosphere during annealing, *Thin Solid Films*, 474(1–2) (March 1, 2005) 326–329.
40. S. P. Murarka, C. C. Chang, and A. C. Adams, Thermal nitridation of silicon in ammonia gas: Composition and oxidation resistance of the resulting films, *J. Electrochem. Soc.*, 126(6) (1979) 996–1003.

41. S. M. Sze and K. K. Ng, *Physics of Semiconductor Devices*, 3rd edn., John Wiley & Sons, New York, November 3, 2006.
42. H. C. Card, Aluminum-silicon schottky barriers and ohmic contacts in integrated circuits, *IEEE Trans. Electron Dev.*, 23(6) (June 1976) 538–544.
43. D. C. Northrop and D. C. Puddy, Ohmic contacts between evaporated aluminium and n-type silicon, *Nucl. Instrum. Methods*, 94 (1971) 557–559.
44. A. Y. C. Yu, Electron tunneling and contact resistance of metal-silicon contact barriers, *Solid-State Electron.*, 13(2) (February 1970) 239–247.
45. L. A. Berthoud, Aluminium alloying in silicon integrated circuits, *Thin Solid Films*, 43 (1977) 319–329.
46. J. P. Gambino and E. G. Colgan, Silicides and ohmic contacts, *Mater. Chem. Phys.*, 52 (1998) 99–146.
47. T. P. Chow and A. J. Steckl, Refractory metal silicides: Thin-film properties and processing technology, *IEEE Trans. Electron Dev.*, 30(11) (November 1983) 1480–1497.

9 Metrology for Device Fabrication

Semiconductor metrology is the science of measurement, such as dimensions, film thickness, dielectric constant, resistivity, chemical composition, etc. It is a vast area of science separate from device fabrication; yet, it is intimately related to every nanofabrication process step. After each fabrication process, it is necessary to evaluate and confirm the outcome of that process. If not, the errors from a very early step may not become apparent until much later, after significant time and resources have been expended in processing that wafer. Metrology steps are inserted at several critical points in a fab sequence to determine if a wafer is suitable to proceed ahead or if it should be repaired or scrapped. Metrology is more heavily utilized in the early stages of a nanofabrication process development and is reduced as the process matures.

Semiconductor metrology can be widely categorized into device fabrication metrology and interconnect metrology. The purpose of this section is to give a top level description of the commonly used metrology tools. Specific details of each method, as well as alternative methods, can be found elsewhere in the literature [1].

9.1 SEMICONDUCTOR DEVICE FABRICATION METROLOGY

Metal oxide semiconductor (MOS) transistors are at the core of all device fabrication processes. Therefore, when we say semiconductor devices it is generally intended to mean silicon circuitry designed around MOS transistors. To a lesser extent, bipolar transistors and diodes are also part of this same technology.

9.1.1 Substrate Defect Metrology

Particles adhered to the substrate surface are the largest factor responsible for loss of device yield. Even on pristine unprocessed wafers, there will be particles and surface defects.

A commonly used method to measure defect density is with light scattering [1,2]. A laser beam is normally illuminated on the wafer surface and light is collected from all angles except the specular reflection angle, as shown in Figure 9.1. The scattered light will be an indication of the defect density on the substrate. In order to detect the smallest possible defects on the substrate, a short wavelength ultraviolet laser is ideal. For particles smaller than 0.1λ, Rayleigh scattering theory applies. For larger particles, Mie theory applies. The same technique can also be used for patterned wafers but the scattering from the patterns has to be distinguished from the defects. If the pattern is periodic, the scatter will contain specific diffraction angles. The signal between the diffraction peaks can be used to estimate the defect density.

9.1.2 Lithography Metrology

Lithography remains the most critical and expensive phase of a fab process. In a research laboratory, the most common exposure wavelength is still 365 nm (mercury vapor lamp). In manufacturing, it is currently at 193 nm. Even from 193 nm, achievable dimensions have been continuously shrinking to below 20 nm. Since these structures were created with photons, one might assume that it should be possible to inspect and characterize these structures using photons. However, that is not possible because the resolution needed for inspection is often far greater than the resolution needed for

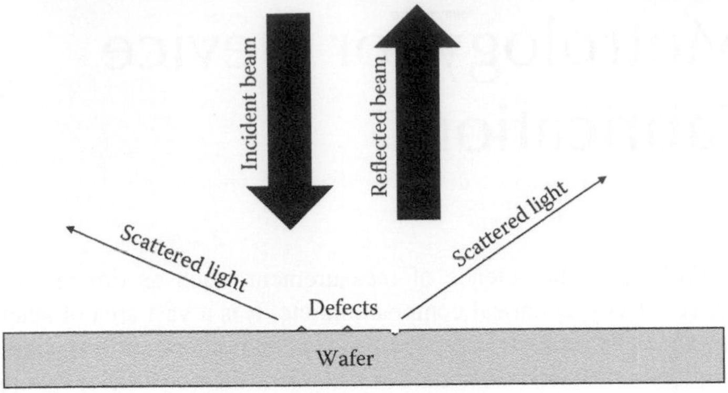

FIGURE 9.1 Defect measurement using a laser beam and scattered light.

fabrication. For example, a 15 nm line will contain line edge roughness and line width variations of the order of 1 nm. This is well beyond what can be accessed with optical microscopy. Scanning electron microscopy (SEM) is by far the most-used tool for dimensional metrology. In addition to nanofabrication, SEM is also widely used in all areas of science [3].

SEM is a relatively old technique but has emerged as one of the most useful and easily accessible tools for nanotechnology. It is used not only in nanofabrication but also in many other areas of sciences where there is a need to acquire high-resolution images beyond what is possible with optical microscopes.

In a conventional optical microscope, the sample is magnified and projected onto an image plane using an illumination source and an optical imaging system. The imaging system projects each point on the sample to a corresponding point on the image plane. The resolution is limited by the wavelength and by the numerical aperture of the system. If for some reason it is not possible to have an imaging system, an alternate method is to construct the image by scanning a finely focused beam on the sample and capture the scattered light. This would not need any imaging optics or image sensors; just a single detector that is synchronized with the scanning system. The resolution in this case will be determined by the spot size of the focused beam. One example of this in optics is the scanning laser radar. However, this is rarely done in conventional optical microscopy because it is much easier to construct an optical image with lenses than to construct a beam scanner and the associated electronics to synchronize it with the detector. Near-field optical scanning microscope is an example of a scanning optical microscopy technique.

The main motivation behind electron microscopy is that the electron wavelength is much smaller than photons, so a significantly higher resolution should be theoretically possible, though in practice the resolution is many orders worse due to the poor focusing optics for electrons as well as scattering mechanisms in the material. However, an electron image sensor analogous to a CCD or CMOS also does not exist. Therefore, instead of attempting to capture the entire image of the sample, a scanning method is used. A finely focused electron beam is scanned across the sample, and the scattered electrons are collected from an electron detector. This is a single-element detector, and not an array. By synchronizing the detector signal with the electron beam scanner, an image of the sample can be constructed.

The primary electrons are generated from a field-emission or thermionic-emission source and accelerated in vacuum to about 1–30 kV. Electromagnetic lenses are used to focus this beam into a small spot, and electromagnetic deflection coils are used to steer and scan the beam by modulating the current through the magnetic coils. The system is nearly identical to the electron beam lithography system that was discussed in Chapter 6, except the acceleration voltages in SEM are much lower because the electrons are not intended to penetrate too deep into the substrate. When the high-energy electrons impinge the sample, they get absorbed and low-energy secondary electrons

(SEs) are emitted. The number of SEs emitted for each primary electron will be a function of the material. Furthermore, these electrons are emitted in random directions. The SE detector is placed at a large angle to the primary beam direction to ensure only the SEs are detected, as shown in Figure 9.2. Therefore, the electrons detected by the SE detector will be a function of the material as well as the topographic features on the sample. Features more within the line of sight of the SE detector will receive more electrons, making them appear brighter. The 2D image will therefore show some material distinctions as well as topographic features. The backscattered electrons can also be detected, which provides more material contrast to the images. Figure 9.3 shows an example of photoresist metrology inspection using an SEM.

The electron detector consists of an electron accelerator, scintillator, and a photomultiplier tube. The received electrons are accelerated and bombarded on a scintillator to generate photons. These photons are then detected by a highly sensitive photomultiplier tube to create an electrical signal.

The spot size of the electron can be smaller than 1 nm, so very-high-resolution images can be obtained. However, as discussed in electron beam lithography, SEM does not operate anywhere

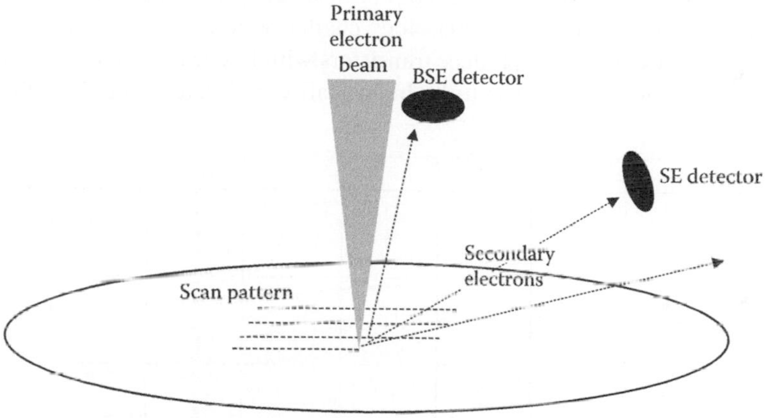

FIGURE 9.2 Scanning electron microscopy showing the secondary electron detector and the backscattered electron detector.

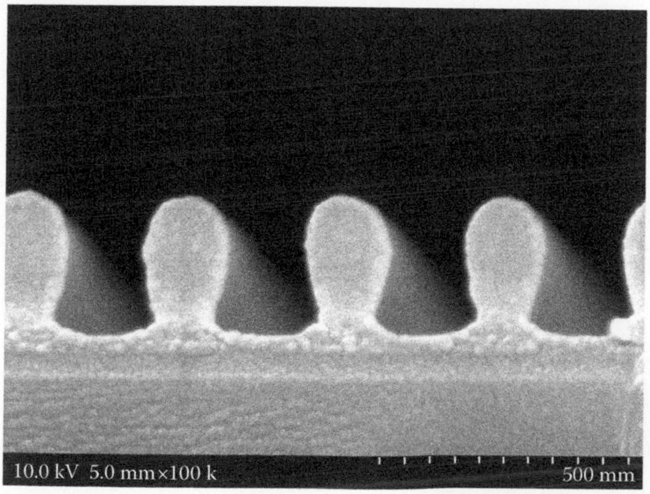

FIGURE 9.3 Example of a photoresist metrology inspection using a scanning electron microscopy image.

near the diffraction limit of the electrons. A 10kV electron has a wavelength of 12pm, so a 1nm spot size is significantly larger than the diffraction limit. This is because electromagnetic lenses suffer from significantly greater chromatic and spherical aberrations than optical lenses.

Samples for SEM have to be conductive with a path to ground to allow the incident electron beam discharge. If not, the sample will charge-up and deflect the beam, causing blurring. Thin photoresists on a silicon substrate can still be imaged without much difficulty because the conductivity of silicon is sufficient to discharge the photoresist. However, photoresist on insulating films such as SiO_2 and Si_3N_4 will require a thin metal coating to prevent charging.

In some SEM systems, x-ray detectors are also used. When electrons bombard the sample surface, x-rays are also emitted. These x-rays will have energies that are characteristic of the material it is being emitted from. This technique is known as energy-dispersive x-ray spectroscopy (EDS) and is used for identifying the material composition. An example of a scan from a metal film consisting primarily of nickel with trace amounts of copper and palladium is shown in Figure 9.4.

9.1.3 Gate Dielectrics

The thickness of the MOS gate dielectric film has been steadily decreasing and is now in the range of 20Å. These are ultrathin films, so the metrology challenge here is to measure the thickness and dielectric constant of these films. To produce transistors with low leakage currents and fast switching speeds, the gate dielectric film has to be high quality and defect-free. For this reason, they

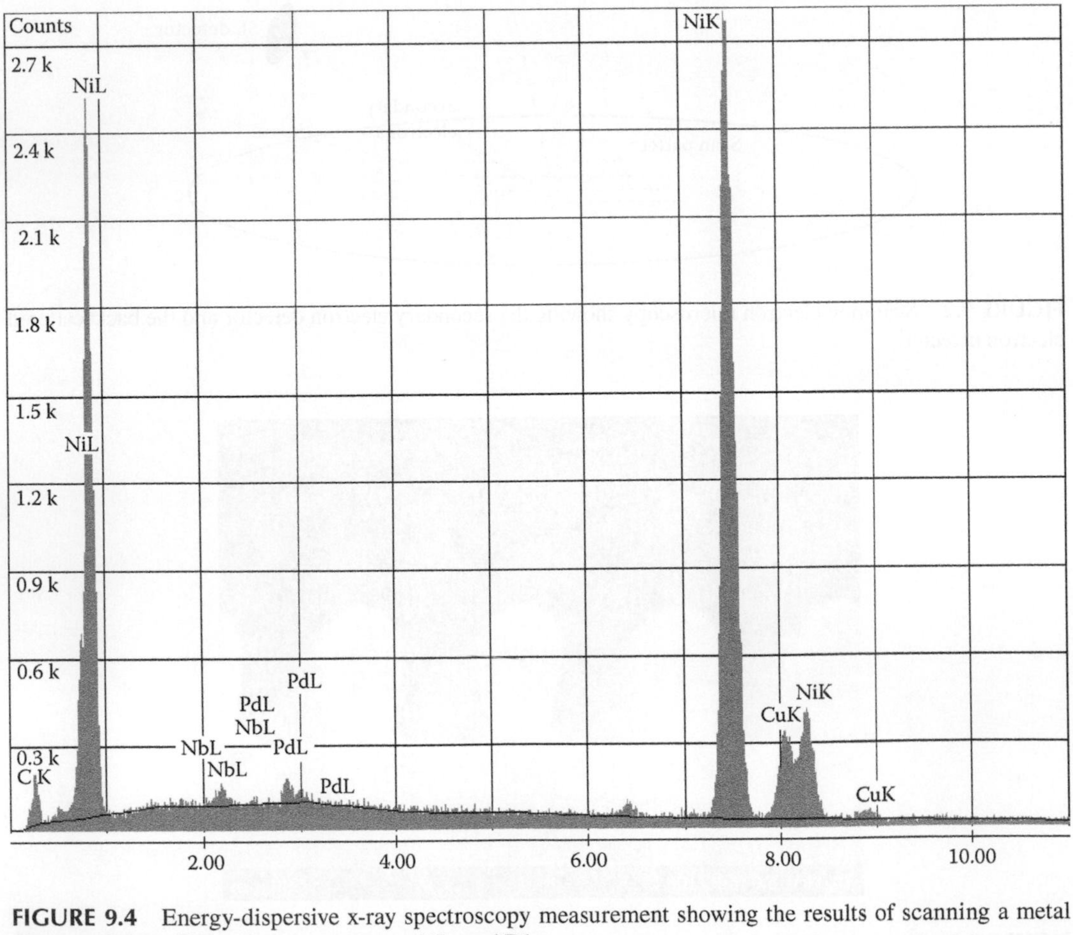

FIGURE 9.4 Energy-dispersive x-ray spectroscopy measurement showing the results of scanning a metal film consisting of Ni and trace amounts of Cu and Pd.

cannot be deposited using PVD techniques because there will be far too many pin holes in the film. Instead, a SiO_2 film is created by thermally oxidizing the substrate. The oxide created this way has the highest density and the fewest defects. For high switching speeds and lower operating voltages, it is desirable to have a dielectric with a higher ε than SiO_2. Recently, alternative dielectrics such as HfO_2 films with very high ε have been developed [4]. These are deposited using atomic layer deposition (ALD) as discussed in Chapter 3, which produces films similar in quality to thermal oxidation.

Spectroscopic ellipsometry is commonly used to characterize the gate oxide films. This technique involves measuring the s and p polarized reflection in the form

$$\frac{R_p}{R_s} = \tan\psi e^{j\Delta}. \tag{9.1}$$

Two spectra $\tan\psi_m(\lambda)$ and $\cos\Delta_m(\lambda)$ are obtained and then fitted to standard dispersion models of dielectrics. Although the principle of ellipsometry is simple, in practice, the fitting routines are complicated and require intensive numerical regression algorithms. Furthermore, the interfacial region between Si and SiO_2 presents a particular problem because its properties differ from the bulk properties on either side of the interface. For thicker films, this interfacial region will only constitute a small fraction, so it can be safely ignored, but for ultrathin films the interface may constitute as high as 20%–50% of the film thickness. Therefore, the analysis of the Si/SiO_2 really becomes a continuously varying multilayer problem, which significantly complicates the ellipsometric measurements.

9.1.4 Metrology for Ion Implantation

Transistors are built by selectively doping the substrate with donors and acceptors to create n- and p-type regions. Doping can be done by thermally diffusing the impurities or by ionizing and accelerating the dopants and implanting them into the substrate, which were discussed in Chapter 8.

After the substrate is implanted and annealed, its resistivity gives the best indication of whether the dopants have been activated. One technique is to use a hot point probe to measure the impurity concentration [5]. The doping density and type of a semiconductor is determined by the polarity and magnitude of the voltage generated by a heated tip in contact with the semiconductor, compared to a cold tip in contact with the same semiconductor. A more common technique for measuring resistance is by driving a current through the semiconductor and measuring the induced voltage. However, this cannot be done by simply placing two measurement probes on the semiconductor. The contact resistance at the metal/semiconductor interface can be significant, and the contact pressure and type of metal will most likely produce unreliable results. Instead, this measurement is most commonly carried out using four probes, usually in a linear configuration. The two outer probes are used for driving the current and the two inner probes used for measuring the voltage, thereby eliminating the contact resistances of the current probes. This is shown in Figure 9.5 [6]. Assuming the lateral dimensions of the film are much greater than the film thickness, and the probe tip distances are all equal, the sheet resistivity can be obtained from

$$\rho_s = \left(\frac{V}{I}\right)\left(\frac{\pi}{\ln 2}\right) = 4.532\left(\frac{V}{I}\right). \tag{9.2}$$

The sheet resistivity ρ_s is specified in Ω/sq, which is the resistance across the opposite edges of any square shape on the surface of the semiconductor. The average bulk resistivity ρ is related to the sheet resistivity as

$$\rho = \rho_s t \tag{9.3}$$

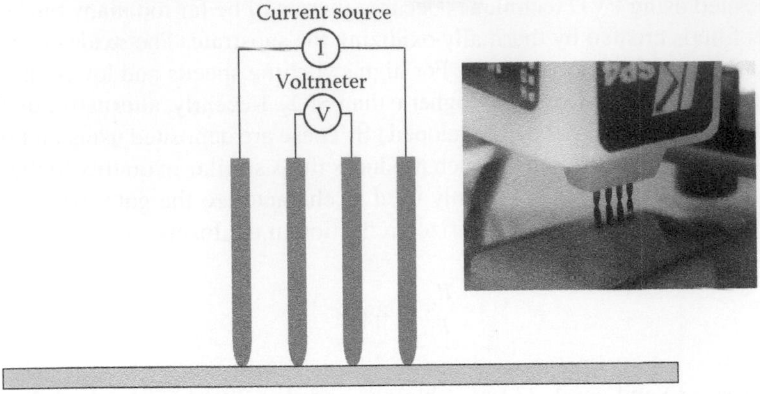

FIGURE 9.5 Four-point probe method for measuring sheet resistances of thin films. The inset shows a photograph of a linear four-point probe head.

where t is the film thickness. The dopant concentration can be extracted from this measurement because the resistivity is related to the doping density as

$$\rho = \frac{1}{\left[q\left(n\mu_n + p\mu_p\right)\right]} \tag{9.4}$$

where

n and p are the electron and hole carrier concentrations,
μ_n and μ_p are the electron and hole mobilities, respectively.

For an n-type doped semiconductor,

$$\rho \approx \frac{1}{[qN_D\mu_n]} \tag{9.5}$$

where N_D is the donor dopant concentration, and for the p-type

$$\rho \approx \frac{1}{[qN_A\mu_p]} \tag{9.6}$$

where N_A is the acceptor dopant concentration. However, the resistivity measurement from the four-point probe gives an average dopant concentration and does not give a direct indication of the depth profile.

Secondary ion mass spectroscopy (SIMS) and spreading resistance profiling (SRP) are two commonly used methods for profiling the depth of the dopant concentration in the substrate. In SIMS, the substrate is sputtered by a focused ion beam (such as argon or oxygen), and the ejected species are ionized, accelerated, and separated using a mass spectrometer for analysis, as shown in Figure 9.6. As the substrate continues to be sputtered, the ejected species will come from deeper within the substrate, allowing a depth profile to be created. This technique has the sensitivity to measure dopant concentrations of 1 part in 10^8.

In SRP, the substrate is beveled by mechanically polishing it at a small angle to reveal the interior of doping profile. Then a sheet resistance measurement is performed by dragging a four-point probe along the taper, as shown in Figure 9.7. The resistance will change with distance due to the changing dopant concentrations. This can then be mapped as a depth profile of the dopant concentration.

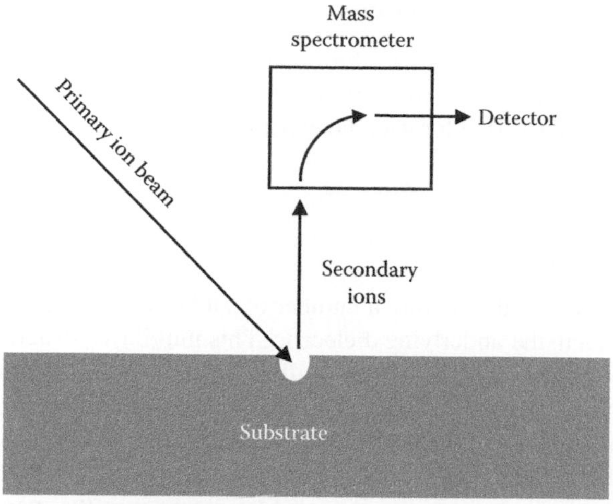

FIGURE 9.6 Secondary ion mass spectroscopy.

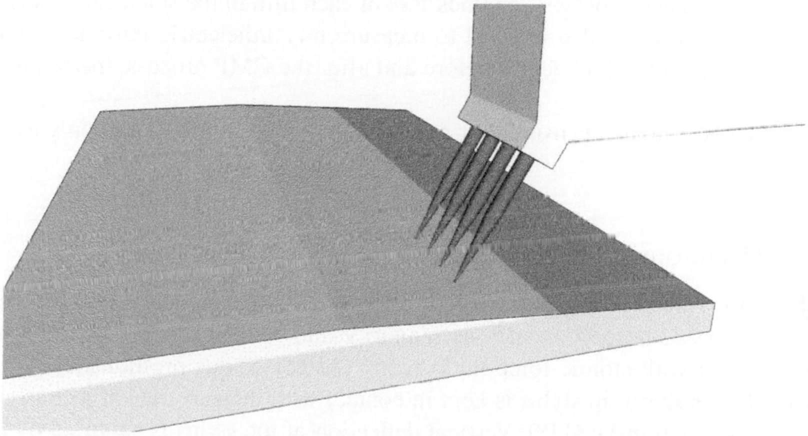

FIGURE 9.7 Spreading resistance measurement with a beveled substrate.

An important difference between SIMS and SRP is that SIMS measures the raw atomic concentration of the dopants, while SRP measures the activated dopant concentration. During implantation, the dopant atoms come to rest at random interstitial sites within the silicon lattice. In order to produce n- or p-doped substrates, these interstitial atoms have to move to a proper lattice site and substitute for a silicon atom, which happens only after thermal annealing. SIMS does not make a distinction between interstitial and substitutional dopants, while SRP only measures the effect of the substitutional atoms. In many applications, one would need to do both measurements to determine the as-implanted dopant concentration and the number of activated dopant atoms.

9.2 INTERCONNECT METROLOGY

On-chip interconnects today are fabricated using the Damascene process, which was described in Chapter 6. A dielectric is deposited, patterned, and etched, and then a metal layer (usually copper) is deposited. Finally, the structure is planarized by chemical–mechanical polishing (CMP) leaving the metal only inside the etched grooves of the dielectric. Unlike the MOS gate dielectric films where high ε is desired, in this case, low-ε dielectrics are required to keep the trace capacitances low.

9.2.1 Low-ε Dielectric Film Metrology

The dielectric film is characterized using ellipsometry and reflectometry. The latter is a simpler technique than ellipsometry because it uses normal incidence beams. However, it is not accurate for ultrathin films such as the ones used for gate dielectrics. Interconnect dielectrics are thicker; hence, both techniques are suitable.

9.2.2 Metal Layer Metrology

Interconnect metal layers typically require a number of sublayers, or silicides, to improve adhesion and to reduce diffusion into the underlying dielectric. This multilayer structure is opaque to optical wavelengths, so ellipsometry and reflectometry will be unsuitable. The technique that has recently become successful is picosecond ultrasonic sonar [7]. A sub-picosecond laser pulse is used to generate an acoustic wave on the top metal surface by thermal expansion. This acoustic wave propagates primarily downward due to the isotropic thermal stress caused by the laser pulse. As the wave propagates downward, the acoustic impedance of each film determines the magnitude of the acoustic reflections generated at each metal interface. When the reflected acoustic pulses reach the top metal surface, they will change the optical reflectivity by a small amount. By measuring the time delay between the reflected acoustic pulses, the thickness of each film in the stack can be determined.

Picosecond ultrasonics can also be used to measure metal/dielectric interfaces. This is useful for measuring the metal film thicknesses before and after the CMP process, inside and outside the etched trenches.

Sheet resistance measurements using four-point probe method is also commonly used to characterize the metal films.

9.2.3 CMP Metrology

An advantage of the Damascene process is that it leaves a flat surface for subsequent photolithography steps. The surface properties such as planarity and roughness can be characterized with stylus profilometry or with atomic force microscopy (AFM). Stylus profilometry is a simple and fast technique where a sharp tip stylus is kept in contact with the substrate at a constant force and the substrate is laterally translated [8]. Vertical deflection of the stylus is captured by illuminating a laser beam at the end of the stylus arm and measuring the deflection of the reflected beam as the stylus moves up and down. Height variations as small as 10 Å can be detected, and is limited mainly by background noise and vibrations. The scan can be linear or two-dimensional, and the results can be plotted as a histogram to indicate the roughness amplitudes. This technique is independent of the surface material. It works equally well on metal surfaces, dielectric surfaces, and soft film surfaces. The major disadvantage of stylus profilometry is the poor lateral resolution. The radius of the stylus tip places a limit on the smallest feature that can be tracked. The tip radiuses are typically on the order of 1–2 μm. The roughness histogram will become unreliable for features approaching the radius of the stylus. The maximum aspect ratio of the structures that can be measured is also a function of the stylus radius. We can show that the lateral features must be greater than a certain value in order for the stylus tip to reach the bottom of the feature (Figure 9.8). This can be shown to be

$$W > 2\sqrt{H(2r - H)} \tag{9.7}$$

where

H is the depth of the feature

r is the stylus radius

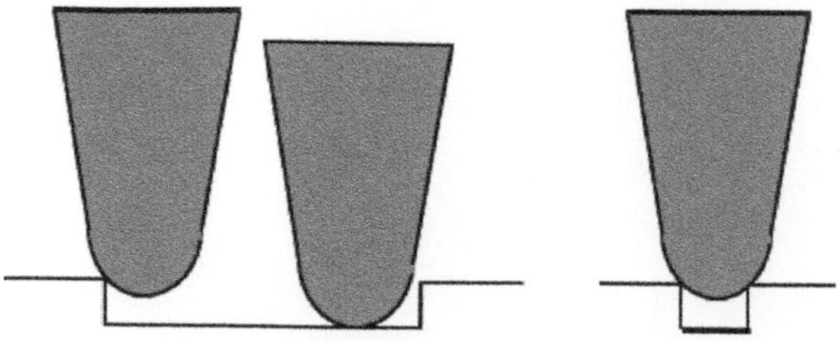

FIGURE 9.8 Illustration of a stylus profiler showing the effect of the tip radius on the feature sizes being measured.

For example, using a 2 μm radius stylus, a 1.5 μm deep feature has to be larger than 3.9 μm wide.

In AFM, a cantilever with an extremely small tip radius of the order of a few nanometers is used. As a result, the lateral resolution is much higher than a stylus profiler. The tip detects weak short range forces of the atoms on the surface. It can be used in contact mode or noncontact mode. In contact mode, it works very similar to the stylus profilomter. In the noncontact mode, an oscillation is induced on the cantilever. At very close distances to the substrate (1–10 nm), the oscillation frequency will shift due to the van der Waals forces. The tip is then moved toward or away from the substrate surface to maintain a constant oscillation frequency. The topography of the surface can then be constructed from the vertical displacement profile of the tip.

REFERENCES

1. A. C. Diebold, *Handbook of Silicon Semiconductor Metrology*, Marcel Dekker, New York, 2005.
2. K. Takami, Defect inspection of wafers by laser scattering, *Mater. Sci. Eng. B*, 44(1–3) (February 1997) 181–187.
3. I. M. Watt, *The Principles and Practice of Electron Microscopy*, Cambridge University Press, Cambridge, U.K., 1996.
4. R. Chau, S. Datta, M. Doczy, B. Doyle, J. Kavalieros, and M. Metz, High-κ/metal-gate stack and its MOSFET characteristics, *IEEE Electron Device Lett.*, 25(6) (June 2004) 408–410.
5. G. Golan, A. Axelevitch, B. Gorenstein, and V. Manevych, Hot-probe method for evaluation of impurities concentration in semiconductors, *Microelectron. J.*, 37 (2006) 910–915.
6. F. M. Smits, Measurement of sheet resistivities with the four-point probe, *Bell Syst. Tech. J.*, 37(3) (May 1958) 711–718.
7. J. Bryner, D. M. Profunser, J. Vollmann, E. Mueller, and J. Dual, Characterization of Ta and TaN diffusion barriers beneath Cu layers using picosecond ultrasonics, *Ultrasonics*, 44(Suppl. 22) (December 2006) e1269–e1275.
8. P. M. Lonardo, D. A. Lucca, and L. De Chiffre, Emerging trends in surface metrology, *CIRP Ann. Manuf. Technol.*, 51(2) (2002) 701–723.

Index

M

Q

T

V

W

Printed and bound by CPI Group (UK) Ltd, Croydon, CR0 4YY
01/11/2024
01782601-0002